Nonlinear Optics
and Optical Computing

ETTORE MAJORANA
INTERNATIONAL SCIENCE SERIES
Series Editor:
Antonino Zichichi
European Physical Society
Geneva, Switzerland

(PHYSICAL SCIENCES)

Recent volumes in the series:

Volume 40   DATA ANALYSIS IN ASTRONOMY III
Edited by V. Di Gesù, L. Scarsi, P. Crane,
J. H. Friedman, S. Levialdi, and M. C. Maccarone

Volume 41   PROGRESS IN MICROEMULSIONS
Edited by S. Martellucci and A. N. Chester

Volume 42   DIGITAL SEISMOLOGY AND FINE
MODELING OF THE LITHOSPHERE
Edited by R. Cassinis, G. Nolet, and G. F. Panza

Volume 43   NONSMOOTH OPTIMIZATION AND RELATED TOPICS
Edited by F. H. Clarke, V. F. Dem'yanov,
and F. Giannessi

Volume 44   HEAVY FLAVOURS AND HIGH-ENERGY COLLISIONS
IN THE 1–100 TeV RANGE
Edited by A. Ali and L. Cifarelli

Volume 45   FRACTALS' PHYSICAL ORIGIN AND PROPERTIES
Edited by Luciano Pietronero

Volume 46   DISORDERED SOLIDS: Structures and Processes
Edited by Baldassare Di Bartolo

Volume 47   ANTIPROTON–NUCLEON AND ANTIPROTON–
NUCLEUS INTERACTIONS
Edited by F. Bradamante, J.-M. Richard, and R. Klapisch

Volume 48   SAFETY, ENVIRONMENTAL IMPACT, AND ECONOMIC
PROSPECTS OF NUCLEAR FUSION
Edited by Bruno Brunelli and Heinz Knoepfel

Volume 49   NONLINEAR OPTICS AND OPTICAL COMPUTING
Edited by S. Martellucci and A. N. Chester

Volume 50   HIGGS PARTICLE(S): Physics Issues and Experimental
Searches in High-Energy Collisions
Edited by A. Ali

A Continuation Order Plan is available for this series. A continuation order will bring delivery of
each new volume immediately upon publication. Volumes are billed only upon actual shipment.
For further information please contact the publisher.

Nonlinear Optics
and Optical Computing

Edited by
S. Martellucci
The Second University of Rome
Rome, Italy
and
A. N. Chester
Hughes Research Center
Malibu, California

Plenum Press • New York and London

Library of Congress Cataloging-in-Publication Data

Course of the International School of Quantum Electronics on Nonlinear
  Optics and Optical Computing (13th : 1988 : Erice, Italy)
    Nonlinear optics and optical computing / edited by Martellucci and
  A.N. Chester.
       p.   cm. -- (Ettore Majorana international science series.
  Physical sciences ; v. 49)
    "Proceedings of the Thirteenth Course of the International School
  of Quantum Electronics on Nonlinear Optics Computing, held May
  11-19, 1988, in Erice, Italy"--T.p. verso.
    Includes bibliographical references and index.

     ISBN-13: 978-1-4612-7900-6   e-ISBN-13: 978-1-4613-0629-0
     DOI: 10.1007/978-1-4613-0629-0

     1. Nonlinear optics--Congresses. 2. Optical data processing-
  -Congresses.   I. Martellucci, S. II. Chester, A. N. III. Title.
  IV. Series.
  QC446.15.C68 1988
  621.39'1--dc20
                                                          90-7306
                                                          CIP

Proceedings of the Thirteenth Course of the International School of
Quantum Electronics on Nonlinear Optics and Optical Computing,
held May 11–19, 1988, in Erice, Italy

© 1990 Plenum Press, New York
Softcover reprint of the hardcover 1st edition 1990

A Division of Plenum Publishing Corporation
233 Spring Street, New York, N.Y. 10013

PREFACE

The conference "Nonlinear Optics and Optical Computing" was held May 11-19, 1988 in Erice, Sicily. This was the 13th conference organized by the International School of Quantum Electronics, under the auspices of the "Ettore Majorana" Center for Scientific Culture. This volume contains both the invited and contributed papers presented at the conference, providing tutorial background, the latest research results, and future directions for the devices, structures and architectures of optical computing.

The invention of the transistor and the integrated circuit were followed by an explosion of application as ever faster and more complex microelectronics chips became available. The information revolution occasioned by digital computers and optical communications is now reaching the limits of silicon semiconductor technology, but the demand for faster computation is still accelerating. The fundamental limitations of information processing today derive from the performance and cost of three technical factors: speed, density, and software. Optical computation offers the potential for improvements in all three of these critical areas:

Speed is provided by the transmission of impulses at optical velocities, without the delays caused by parasitic capacitance in the case of conventional electrical interconnects. Speed can also be achieved through the massive parallelism characteristic of many optical computing architectures;

Density can be provided in optical computers in two ways: by high spatial resolution, on the order of wavelengths of light, and by computation or interconnection in three dimensions. In general, optical computer architectures avoid the yield and rework difficulties often associated with densely packaged electronic devices; and,

Software is facilitated in optical computers, as in some electronic computers, through the use of highly parallel architectures, and through the use of adaptive, self-programming configurations analogous to networks.

Before optical computer benefits can be realized in practice, considerable development is needed of the devices, structures, and architectures which only exist in research laboratories today. Fortunately, a strong foundation already exists in these areas, and this book treats both the fundamental devices and the computing architectures which will make possible the advanced computers of the future. Due to the peculiar characteristics of this rapidly developing field, we did not interfere with the original manuscripts in editing this material and wanted only to arrange it without reference to the chronology of the conference into five categories:

1) "Optical Nonlinearities and Bistability", a group of five papers emphasizing nonlinear Fabry-Perot resonators and other bistable structures, which could serve as basic logic and memory elements;

2) "Quantum Wells and Fast Nonlinearities", four papers describing quantum well structures and the fast nonlinear effects they exhibit;

3) "Optical Computing, Neural Networks, and Interconnects", a set of four papers covering optical computer architectures, optical interconnects, and their practical implementation;

4) "Materials and Devices", three papers treating lasers, nonlinear fibers, and nonlinear effects at surfaces, as possible elements for optical computation; and

5) "Suggestion for Further Reading", two papers containing an extensive annotated bibliography on nonlinear optical activity and nonlinear eigen-polarizations, and additional selected references on optical computing using phase conjugation.

These papers, and the further references therein, should form a useful guide to today's research results, and the basis for future advances in optical computers.

Before concluding, the Editors acknowledge Miss R. Colussi, who voluntereed to retype and revise the entire volume, as well as the continuous assistance of the Plenum Editor (J. Curtis); they wish to express sincere appreciation to Prof. A. M. Scheggi, the Scientific Secretary of the conference, and to Mrs. V. Cammelli for the very specialized assistance in the successfull organization of the conference. Thanks are also due to the organizations who sponsored the conference; among them, the Ettore Majorana Centre for Scientific Culture, whose support made the conference possible.

The Directors of the Int'l School of Quantum Electronics:

A. N. Chester
Vice-President and Director
Hughes Research Center
Malibu, California (USA)

S. Martellucci
Professor of Physics
The Second University of Rome
Rome (Italy)

November 15, 1989

CONTENTS

OPTICAL NONLINEARITIES AND BISTABILITY

Theory of optical bistability and optical memory 3
W. J. Firth

Dynamic operation of nonlinear waveguide devices for
      fast optical switching 21
S. Laval and N. Paraire

Optical bistability in coupled-cavity
      semiconductor lasers 37
H. D. Liu

Semiconductor bistable etalons for digital
      optical computing 51
A. Miller, I. T. Muirhead, K. L. Lewis, J. Staromlynska
      and K. R. Welford

Three photon ionization of Na atoms and
      related plasma phenomena 63
F. Giammanco

QUANTUM WELLS AND FAST NONLINEARITIES

Ultrafast dynamical nonlinearities in
      III-V semiconductors 83
J. L. Oudar

Nonlinear optical materials and devices 99
N. Peyghambarian, S. W. Koch, H. M. Gibbs, and H. Haug

Excitonic optical nonlinearities, four wave mixing and
      optical bistability in multiple quantum well structures . . 119
A. Miller, P. K. Milsom and R. J. Manning

Nonlinear optics of a single slightly-relativistic
      cyclotron electron 131
A. E. Kaplan and Y. J. Ding

OPTICAL COMPUTING, NEURAL NETWORKS, AND INTERCONNECTS

Principles of optical computing 151
A. W. Lohmann

Optical associative memory 159
K. Hsu and D. Psaltis

Optical computing using phase conjugation 173
G. J. Dunning and C. R. Giuliano

Nonlinear photorefractive effects and their application
    in dynamic optical interconnects and image processing . . . 197
L. Hesselink, J. Wilde and B. McRuer

MATERIALS AND DEVICES

Ultrafast all-optical switching in optical fibers 217
S. Trillo, S. Wabnitz, and B. Daino

Nonlinear coupling to ZnS, ZnO and SDG planar waveguides:
    theory and experimental study 229
G. Assanto

Nonlinearity at an interface 247
F. Bloisi, L. Vicari, S. De Nicola, A. Finizio, P. Mormile
    G. Pierattini, A. E. Kaplan, S. Martellucci, J. Quartieri

SUGGESTION FOR FURTHER READING

Polarization instability in crystals with nonlinear
    anisotropy and nonlinear gyrotropy 253
N. I. Zheludev

Selected references on optical computing using
    phase conjugation 265
G. J. Dunning and C. R. Giuliano

Index 269

OPTICAL NONLINEARITIES AND BISTABILITY

THEORY OF OPTICAL BISTABILITY AND OPTICAL MEMORY

W.J. Firth

Dept. of Physics and Appl. Physics, Univ. of Strathclyde
Glasgow G4 0NG, Scotland, U.K.

I. INRODUCTION

Optical bistability (OB) characterises an optical system with two possible state outputs for a single input[1-3]. This phenomenon, and its possible applications to optical information processing, was proposed nearly two decades ago[4]. It has since been demonstrated in a wide variety of systems and media, including vacuum and malt whisky!

Two objective of this presentation is to review the basic physics of OB with emphasis on those features and systems most relevant to information processing. There will thus be a bias towards small systems exploiting electronic nonlinearities, especially semiconductors. The seminal theoretical and experimental work in two level systems such as Na vapour[1-3], and the important field of instabilities in OB systems[5], will be largely neglected: see Lugiato for a detailed review[5].

This short review begins with perhaps the simplest model of optical bistability[6] which displays rather general features[7] in a direct manner. Most OB systems involve a resonant cavity, and the basic theory of absorptive and dispersive OB in cavities is presented, leading on to questions of optimisation[8] and the mapping of OB output states on to logic functions[9]. I then analyse switching dynamics, which leads to slowing-down phenomena[10] and gain-bandwidth considerations[11]. The possibility of competing nonlinearities (e.g. electronic and thermal) is considered, which can lead to self-oscillation[12] but also has important implications for design of optical processors[13].

The minimum size of OB devices is determined by the transverse coupling mechanisms: both diffractive and diffusive mechanisms are analysed, and hysteresis loops and beam profiles obtained numerically are discussed[3,14]. Diffusion-dominated OB presents an interesting and practical limit in which key features of optical memory devices can be analysed. In particular switching-wave phenomena imply that large-area memory and image processing

Nonlinear Optics and Optical Computing
Edited by S. Martellucci and A. N. Chester
Plenum Press, New York, 1990

devices must be "pixellated" into arrays of OB elements[15,16], while minimum pixel spacings in such arrays can be inferred by analogy with nonlinear dynamical systems[17]. Similar considerations arise for diffractive coupling, with the extra feature that a self-focussing type of nonlinearity leads to the spontaneous formation of soliton-like structures[18-20]. Both as "intrinsic pixels" and in their own right as nonlinear wave phenomena, these are among the most significant current developments in the theory of optical bistability.

II. SIMPLE MODEL

As a simple and instructive model for OB, consider a slice of material, thickness L, on which a plane wave of intensity I_0 is incident. If the absorption coefficient α depends on temperature ϕ, then

$$\frac{dI}{dz} = -\alpha(\phi)I \tag{1}$$

Clearly ϕ will in turn depend on the absorbed energy, and may be crudely modelled thus:

$$\frac{d\phi}{dt} + \Gamma(\phi-\phi_0) = I_0 f(\phi) \tag{2}$$

where ϕ_0 is the ambient temperature, while $f(\phi)$ is proportional to the bulk absorption rate $\sim(1-e^{-\alpha L})$ and Γ^{-1} is the thermal time constant. In steady state, (2) leads to

$$(\Gamma/I_0) (\phi-\phi_0) = f(\phi) \tag{3}$$

Similar equations to (3) govern most OB systems (ϕ, not T, is used so that these equations can be used below in other contexts), and graphical solution of such equations is instructive (Fig. 1). Each possible steady state is represented by the intersection of a line through (ϕ_0,0) of slope $\sim I_0^{-1}$, with the response function $f(\phi)$. There are evidently multiple intersections for I_0 large enough provided $f(\phi)$ is sufficiently steeply increasing over a suitable range of ϕ. For bounded $f(\phi)$, there are an odd number of intersections: linearisation of (2) around these roots enables their stability to be determined. Suppose that ϕ_s solves (2), and let

$$\phi = \phi_s + \varepsilon e^{\lambda t}$$

Substituting into (2):

$$(\lambda+\Gamma)\varepsilon e^{\lambda t} = I_0 f'(\phi_s)\varepsilon e^{\lambda t} + 0(\varepsilon^2)$$

and hence

$$\lambda = I_0(f'(\phi_s) - \Gamma/I_0).$$

If $\lambda<0$, the perturbation damps out, and ϕ_s is stable, and vice versa. By

4

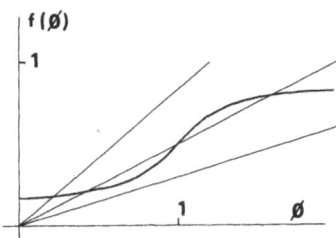

Fig. 1. Graphical solution of Eq. 3. The straight line's intersections with the response function give one, three or
more solutions.

comparison with (3) and Fig. 1 it can be seen that when the line cuts $f(\phi)$
"from below" the corresponding state is dynamically stable, and vice
versa[2]. Thus the middle intersection in Fig. 1 belongs to an unstable
state, while the other two are stable, and we indeed have optical bistability.

While not necessarily very practical, this simple model illustrates
key features of OB:

(i) nonlinearity - of the light-matter coupling function $f(\phi)$

(ii) feedback   - symbolycally (1) and (2) represent a feedback loop:

$$\left(\begin{array}{c} \text{light} \\ \text{matter} \end{array} \right)$$

Both features seem necessary for OB: in fact the feedback is commonly
explicit, as when the nonlinear medium is enclosed in a cavity structure.
In such a structure the response is sensitive to the phase, as well as the
intensity, of the optical field. OB can then arise from nonlinear refraction (Kerr effect), with the cavity Airy function as the appropriate response function, and also from saturable absorption[4]. These schemes are
known as dispersive and absorptive OB respectively[5].

III. CAVITY OPTICAL BISTABILITY

Cavity OB has been analysed many times[1-5] so a complete treatment is
not necessary here, but it is worthwhile to develop the usual model.

Consider a two-level atomic system enclosed in a ring cavity. On the
assumption that the polarisation may be adiabatically eliminated (certainly valid in condensed media) the Maxwell-Bloch equations reduce to a pair
of coupled delay-differential equations[2].

$$E(t+t_r) = A + Re^{i\theta}E(t)\exp\left[-\frac{\alpha L}{2}(1-i\Delta)D(t)\right] \tag{4}$$

5

$$\tau \dot{D}(t) = 1 - D(t) - |E(t)|^2 [1 - \exp(-\alpha L D(t)]/\alpha L \qquad (5)$$

$E(t)$ is the scaled field amplitude on entrance to the nonlinear medium, which has length L and small-signal absorption coefficient α. E is scaled to the saturation intensity I_s. $D(t)$ is the spatially-averaged population inversion which is seen from (5) to be normalised to unity for $E \to 0$, and τ is its decay time. The cavity has round trip time t_R, mistuning parameter θ and $R = (R_1 R_2)^{\frac{1}{2}}$ where R_1, R_2 are the input and output mirror reflectivities. A is the scaled input field entering the cavity, and thus equal to $(1 - R_1)^{\frac{1}{2}} E_{in}$. Finally Δ is the scaled detuning from the atomic resonance: $\Delta = 0$ gives a purely absorptive effect, while $|\Delta| \gg 1$ makes dispersive effects dominant.

Other nonlinear media give qualitatively similar equations, perhaps with α and Δ empirically determined, as in semi-conductors[21].

For condensed media in short cavities, $t_R \ll \tau$ is normal, so that the left side of (4) may be approximated by $E(t)$. $E(t)$ may then be eliminated from (5) to yield, in terms of the convenient new variable $\phi = (1 - D(t))$

$$\dot{\phi} + \tau^{-1}\phi = I_0 f(X) \qquad (6)$$

where $I_0 = |E_{in}|^2 I_s$, and
$$f(\phi) = \frac{(1 - R_1)(1 - e^{-\alpha L} e^{\alpha L \phi})/\alpha L}{\tau I_s |1 - b\exp[\alpha L(1 - i\Delta)\phi/2]|^2} \qquad (7)$$

where $b = Re^{i\theta} \exp[-\alpha L(1 - i\Delta)/2]$.

Equation (6) is evidently formally identical to (2), and thus OB will exist provided $f(\phi)$ has a suitable form. Since the numerator of (7) decreases monotonically with ϕ, and is zero at $\phi = 1$ (complete saturation), OB relies on the denominator more than compensating this decrease. The denominator is, physically, the ratio of the input to the internal intensities, and exhibits a cavity resonance structure. Fig. 2 shows $f(\phi)$ for two important special cases, $\Delta = 0$ (purely absorptive OB) and $\Delta = 15$ (largely dispersive OB). In the latter case $f(\phi)$ is oscillatory, leading to multiple intersections for I_0 large enough. In fact, in the dispersive or Kerr - limit: $|\alpha L| \to 0$, $|\Delta| \to \infty$, $|\alpha L \Delta| \sim 1$, $f(\phi)$ becomes simply the Airy function governing the spectral response of an optical cavity, while the nonlinearity can be described by an intensity-dependent refractive index: $n = n_0 + n_2 I$. This is a favourable and practically-important limit.

The above analysis assumes a ring cavity configuration, whereas OB experiments, especially those in solids, generally use a Fabry-Perot or folded cavity. To a large extent, this leads to the same OB phenomena as the ring cavity, but with rescaled parameters: in particular the inversion is driven by both forward and backward propagating fields, while the fields accumulate nonlinear phase and amplitude changes over 2L instead of L: both favourable effects. Standing-wave effects in principle lead to a spatial modulation of the inversion with period $\lambda/2$: this seriously com-

6

plicates the analysis, but fortunately in many cases the inversion population is sufficiently mobile that diffusion "washes out" the population
grating, leading back to quasi-ring-cavity behaviour. A more subtle change
is that the space-time averaging implicit in D^5 is no longer valid, and
both field and population have to be considered as function of z, the
longitudinal coordinate, necessitating approximations or numerical solution or both.

Finally, it should be noted that the above analysis works just as
well for αL negative, i.e. an amplifying medium, as for an absorbing
medium, provided that $|b| < 1$; $b = 1$ is just the laser threshold. This is
important for a number of practical reasons, not least of which is that OB
in semiconductor laser amplifiers has extremely low switching powers and
energies[21-23].

Bistability evidently implies a binary memory capability: other logic
functions can be performed in OB or closely-related states. For example,
the reflected and transmitted beams from a Kerr cavity have the basic
response features sketched in Fig. 3, whose shape clearly allows digital
optical logic to be performed[9].

Sequences of input-output curves such as those in Fig. 3 are obtained
experimentally by tilting, heating or otherwise varying the cavity mistuning θ. One finds a maximum loop size at fixed I_0, and a minimun I_0
below which OB vanishes, corresponding to coalescence of the three intersections in Fig. 1. It is then of interest to optimise this minimum with
respect of other parameters. Miller[8] and Wherrett[24] have analysed this
problem for Kerr cavities and find that it is possible to decouple the
microscopic and macroscopic aspects, so that the material nonlinearity and
cavity can be separately optimised. Miller finds that $R\,e^{-\alpha L}$ is optimum
for passive cavities, while Wherret emphasises that the contrast of OB
loops is much larger in reflection than transmission if αL is appreciable.
Adams[22] considers the behaviour of OB amplifiers, which have lower thresholds for small R, in contrast with passive cavities, where large R and
low absorption gives lowest thresholds[24].

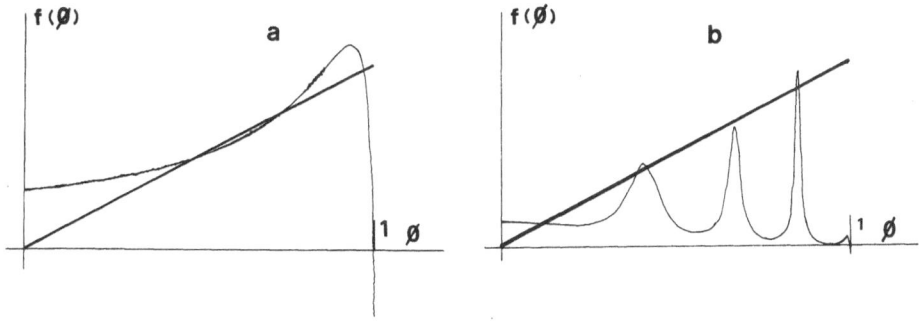

Fig. 2. Response function $f(\phi)$ for a 2-level system in an optical
cavity - equation (7) - for R=0.9, αL=3 and Δ=0 (left);
Δ=15 (right). Intersecting line indicates bistability.

IV. SWITCHING DYNAMICS

The cycle time of the OB system described by (1,2) will be of order Γ^{-1}, i.e. controlled by the thermal time constant. With good design this can be submicrosecond, but faster cycling is generally sought via elec- tronic excitations in, especially, semiconductors. One can infer from (2), however, that $I_0/\Gamma \sim$ constant, i.e. that bandwidth and power are roughly proportional, which is a good rule-of-thumb. This rule works also for (7), since $I_s \sim \tau^{-1}$ (ref. 5).

It follows from our stability analysis of (2) that fluctuations around the steady state ϕ_s damp at a rate

$$\Gamma_e = \Gamma - I_0 f'(\phi_s).$$

Γ_e changes sign at a switch point, where the line is tangent to $f(\phi)$ in Fig. 1, and is thus small in its neighbourhood, which leads to the phenomena of critical slowing-down and critical fluctuations well known in phase transition theory. At the switch point the perturbation analysis has to be taken to second order, leading to a prediction that when the inten- sity is stepped from below the threshold value I_{th} to above that value, the switching times scale as $(I_0-I_{th})^{-\frac{1}{2}}$, which has been confirmed in a number of systems[1-3,25].

In the "transphasor" regime[26], where the characteristic is not quite bistable, the gain spectrum for a small modulation of I_0 at frequency ω can be calculated as

$$|\text{gain}| \sim \Gamma/(\Gamma_e^2 +\omega^2)^{\frac{1}{2}}$$

which shows that high gain is available close to a switch point, but only at the cost of a bandwidth narrowing in proportion. Nardone and Mandel[27] calculate nonlinear corrections to this gain which remove the singularity as $\Gamma_e \to 0$.

Critical slowing down is manifested when the input is stepped to a new, constant value. An interesting alternative is to consider an "address pulse", i.e. a temporary step in I_0 designed to switch the OB device from its lower to its upper state. The first order character of (2) means that whether or not the device switches depends entirely on the value of ϕ at the end of the address pulse. If the three intersections with the line describing the "hold" level I_0 are $\phi_\ell, \phi_m, \phi_u$, then if at the end of the address pulse $\phi > \phi_m$, the device will switch i.e. $\phi \to \phi_u$. Conversely $\phi < \phi_m$ leads to a decay back to ϕ_ℓ. For $\phi_\ell \sim \phi_m$, ϕ is small, and the system "dwells" a long time close to ϕ_m. Mandel has analysed this phenomenon, and terms it "slowing-down". He shows that the dwell-time scales logarithmically, in contrast to critical slowing-down. The effect has been demonstrated exper- imentally[28]. This shows, incidentally, that the unstable branch is phys- ically observable, and can be traced out by analysis of slowing-down for different I_0 and address-pulse area.

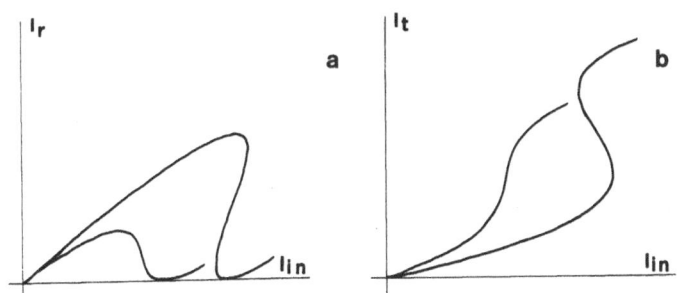

Fig. 3. Reflected (left) and transmitted (right) intensity versus
input intensity for a Kerr cavity, for R-0.5, $\alpha L=0$ and
$\theta=-1.6$ (bistable); $\theta=-1.1$ (monostable).

We have so far assumed that only a single nonlinear effect is active.
In practice, there may be several: in particular in semiconductors electronic OB will inevitably be accompanied by thermal effects, which can
also influence the optical properties of the device. Suppose then that two
mechanisms are operative, one fast (electronic) and the other slow (thermal), and that they have opposing effects. The dynamics can then be approximated by still using (2), but regarding the slow effect as a "bias"
of the response function $f(\phi)$. Since e.g. the thermal load will be dif-
ferent for the states ϕ_ℓ, ϕ_u, these will have different bias, say θ_ℓ, θ_u
corresponding to translation of $f(\phi)$ in Fig. 1.

Consider now a fast "electronic" switch-up. Initially the bias will
still be close to θ_ℓ, but the device will begin to heat up, and θ will
drift towards θ_u. This means that $f(\phi)$ will move to the right in Fig. 1.
If $|\theta_u - \theta_\ell|$ is large enough, OB will be lost, and the device will switch
back down (fast). It will then cool down, $f(\phi)$ will move left, and the
device will eventually switch up once more. This sequence can repeat in-
definitively as a self-sustaining oscillation.

This effect was predicted by McCall[12], and has been observed in a
number of OB systems - see Chapter 6 of Gibbs' book[2] for a review.

The presence of slower, possibly opposed, contributions to nonlinearity is an important factor in the design of OB systems. As we have noted,
the strength of a given nonlinearity increases as its response time is
lengthened, and vice versa. Since the response times of both electronic
and thermal excitations are sensitive to surface effects, the balance between these contributions to OB may be radically altered by system geometry. For example, "pixellation" of material to obtain high packing densities of OB devices (see below) will normally affect the response times of
both electronic and thermal excitations, and careful design will be necessary to avoid instability in such systems.

This last consideration leads naturally to the topic of transverse
effects, which determine the minimum size of OB devices, and together with
the response time control the holding powers and switching energies which
will determine the viability of OB systems.

9

V. TRANSVERSE EFFECTS IN OPTICAL BISTABILITY

Transverse coupling, both within a single bistable element and between adjacent elements, has a major influence on the nature and quality of optical bistability in practical systems. Diffraction is the obvious coupling mechanism, but in many systems the excitation is mobile enough for transverse diffusion to be significant. This is the case for semi-conductors such as InSb[26,29], where the excitation is an electron-hole plasma and, indeed, for thermal devices such as interference filters[9].

Basically, these transverse effects require the addition of a dif-fraction term to (1) and a diffusion term to (2) or their generalisation. For illustration, we concentrate on Kerr nonlinearity and Fabry-Perot cavities.

In this case, the propagation equations for the forward and backward (scaled) field amplitudes $F(\underline{r})$, $B(\underline{r})$ and excitation density $h(\underline{r})$, in an etalon of thickness L, may be expressed as[30]

$$\frac{\partial F}{\partial z} = -\left[\frac{\alpha L}{2} + ih(\underline{r}) + \frac{1}{2} i \mathpalette\overbar{\jmath} \nabla_t^2\right] F \tag{8}$$

$$-\frac{\partial B}{\partial z} = -\left[-\frac{\alpha L}{2} + ih(\underline{r}) + \frac{1}{2} i \mathpalette\overbar{\jmath} \nabla_t^2\right] B \tag{9}$$

$$(\ell_D^2 \nabla_t^2 - 1)h = -4\mathpalette\overbar{\jmath}\,\mathrm{sgn}(n_2)(|F|^2 + |B|^2) \tag{10}$$

Here α is the linear absorption coefficient and $\mathpalette\overbar{\jmath} = L/kw^2$, where we assume a gaussian input beam of width w, and scale transverse distances to w. ∇_t^2 is the transverse Laplacian, and l_D the diffusion length for the medium excitation. These equations, together with Fabry-Perot boundary conditions on F and B, and transverse and surface boundary conditions for h, are the basis of our model for steady-state OB.

A brief discussion of possible approaches to solution of these partial differential equations is appropriate: even in the steady state they require substantial computer effort, especially in two transverse dimensions. For the moment, consider $\ell_D=0$, i.e. the "diffraction-only" case, in which (8) and (9) are third order in $|F|$ and $|B|$.

Perhaps the simplest approach is mode-expansion, thereby reducing the system to a number of ode's equivalent to the number of modes analysed. Because, however, third order linearity couples three modes to a fourth, the computation grows as N^4 (see reference 31), and is thus only useful if one mode is dominant, in which case the technique can work well[32] and may even yield analytic results[33].

Even more straightforward is the finite-difference technique, which is flexible and direct, and has proved useful for problems with cylindrical symmetry[34].

Undoubtely the method of choice, however, is the split-step FFT method, developed for OB problems by Moloney and co-workers (see below). Since propagation is trivial in Fourier space, while the nonlinearity is trivial in real space, the method consists in replacing the nonlinear medium by a sequence of slices of equivalent strength, between which linear propagation is assumed. Successive Fourier transforms shuttle the field beteween $\underline{r}$- and $\underline{k}$-space, and use of the FFT algorithm gives the method an NlogN scaling with the number of transverse points used. This advantage is particularly significant for the two-transverse dimension calculations recently undertaken by Molonely[35], which even so require substantial CPU time, even on a Cray. With some extra overhead, the method can be adapted for cylindrical symmetry as the Fast Hankel Transform or FHT method, results of which we now examine.

Fig. 4 shows results of FHT calculations based on equations (1-3), but neglecting diffusion, for parameters typical of InSb etalons[26]. These results are in essentially complete agreement with the finite-difference results of Weaire et al[34], which lends confidence in the reliability of both codes.

Fig. 4 shows a rather small hysteresis loop which disappears alto- gether on focusing down. The upper-branch near-field profile (inset) is decidedly non-gaussian and far from ideal for an optical processing chain.

Bistable loops observed in InSb are much larger and squarer than those in Fig. 4. It was proposed[36] that diffusion ought to be important in InSb, and measurement of ℓ_D at $\cong 60\mu$m confirmed this[29]. With this value in the full code incorporating both diffraction and diffusion, we obtain results exampled in Fig. 5.

It is clear that diffusion increases the OB threshold for a given beamwidth, but bistability persists to much smaller spot sizes as a com- pensation. Furthermore, the hysteresys loops in Fig. 5 are nearly ideal, even at small spot sizes, with excellent vertical and horizontal contrast. The (inset) profiles at 15μm spot size are nearly gaussian at each corner of the hysteresis loop, so that cascading of such devices ought to be straightforward.

These desirable features are due to the fact that the refractive index change across the beam itself is relatively modest for beams nar- rower than ℓ_D, with consequent reduction in the phase distortion across the beam which causes the structure in the diffraction-only profile. At the same time, the hysteresis loops approach the large, squarish shapes found in plane-wave OB theories. The penalty is that the refractive index change can extend transversely much further than the beam itself, leading to crosstalk in close-packed arrays of OB devices - see below.

VI. DIFFUSION DOMINATED DEVICES

The full code used to generate Fig. 5 demands considerable re- sources. Fortunately, over much of the range, a simple model in which the

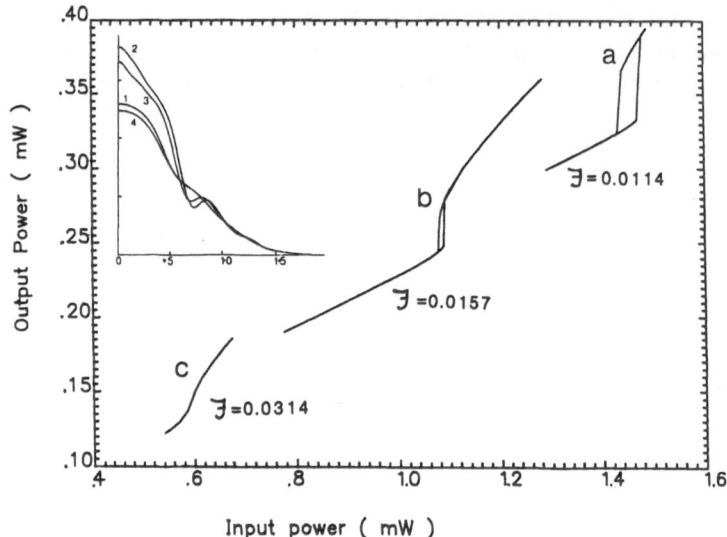

Fig. 4. Diffraction-only calculations for $\alpha L=0.5$, $\phi_0=0$, $R_F=R_B=0.36$ and $n_2=7.0\times10^{-4}$ cm^2/W. Inset profiles for $\mathfrak{J}=0.0157$: (1)-(2) before and after switching up, (3)-(4) before and after switching down. Corresponding spot sizes are $\cong100$ μm for typical InSb etalons.

diffraction terms in (8,9) are dropped, generates essentially identical results. In this model[15] the field equations can be integrated exactly, and (10) converted to an equation for the nonlinear phase shift ϕ:

$$-\ell_D^2 \nabla_t^2 \phi + \tau \frac{\partial\phi}{\partial t} + \phi = \frac{I(\underline{r},t)}{1 + F \sin^2(\phi+\phi_0)} \qquad (11)$$

We have also introduced the response time τ of the nonlinearity: integration of (8,9) demands that it be long compared to the cavity response time, which condition is well satisfied for most solid-state OB devices. For convenience we have scaled the input intensity, and now allow it to be a function of time as well as $\underline{r}$. The parameters F and ϕ_0 on the right of (10) represent the cavity finesse and mistuning respectively: the whole term, which is the only nonlinear one in (10), describes the Airy funtion relation between the input intensity and the internal intensity.

We have applied (10) to a variety of "single-beam" problems, including spotsize dependence of minimum switching power (which is zero in this model, i.e. diffraction governs the ultimate limit); spotsize dependence of recovery time (which can be $\ll\tau$ for beams with $w\ll\ell_D$); and the influence of transverse effects on the cavity optimisation calculations of Wherrett[24] - which is, in fact, slight.

An important general phenomenon that emerges directly from (11) is the existence of switching-waves. If $I(\underline{r},t)$ is constant (plane-wave excitation), then it is easy to show that (11) possesses travelling-wave solutions. Among these solutions are waves asymptotic in one direction to

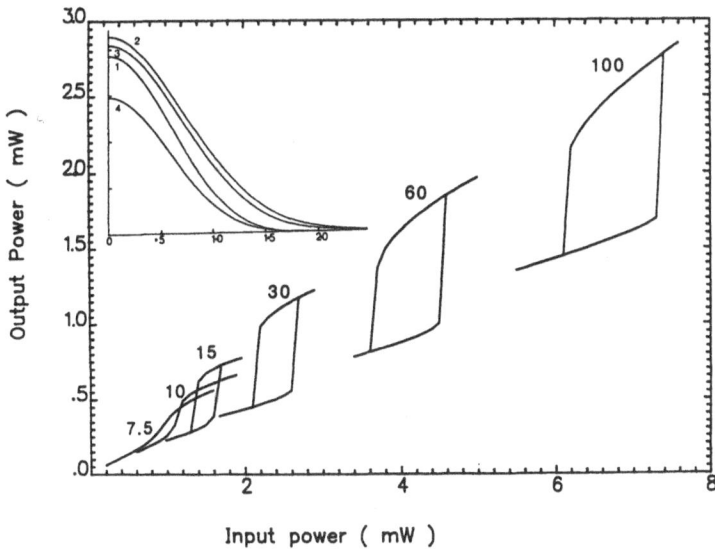

Fig. 5. Diffusion-and-diffraction calculations; same parameter
values as Fig. 4 with ℓ_D=60μm. The number beside each loop
is the corresponding value of w_0 in μm for the assumed parameters. The inset profiles are for w_0=15μm.

the lower fixed point ϕ_ℓ, and in the other to ϕ_u. As it propagates, this
wave progressively switches the whole area to the same state: which one is
determined by the sign of the velocity. It turns out[15] that if the line
and response function in Fig. 1 intercept more area in the lower lobe then
ϕ_ℓ is the "winner" and vice versa. Only in the special case of equal areas
can ϕ_ℓ and ϕ_u co-exist. This phenomenon extends to finite beams: initia-
tion of switch-up in the beam-centre leads to propagation outwards of a
switching wave until essentially the entire OB area is switched up.

It follows that a uniform etalon uniformly illuminated cannot act as
an optical memory - or rather acts as a one-bit memory. Useful optical
information storage requires "pixellation": the material or the pump beam
or both must be spatially-modulated so as to "pin" switching-waves at the
pixel boundaries. Material pixellation leads to very device-dependent
analysis, which is being actively studied[16]. Here, for simplicity, we
consider "optical pixellation", arising from spatial modulation of the
pump beam. Diffusive coupling, as in (11), and restriction to linear
arrays are further simplifications which enable some analytic progress in
the analysis of the global memory states of OB arrays[17].

Fig. 6 shows some typical results where $I(\underline{r},t)$ represents a linear
array of nine gaussian beams. For a spot size of $\ell_D/2$, and spacing $2\ell_D$, a
pulse which kicks the central element to the upper branch causes all the
others to switch up also, in a sort of switching-wave, while if the spacing is increased, the central element can stay "off" when all the others
are kicked "on". This last case illustrates the key requirement for independent pixel action in OB arrays: that a single pixel can be stable in
the opposite state from all the others.

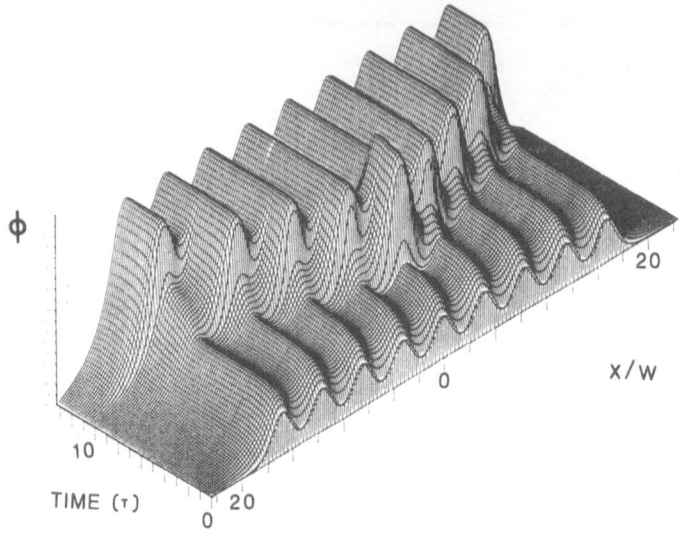

ϕ

20

x/w

0

10

TIME (τ)   20

0

Fig. 6. Effect of address pulses in an array of nine OB devices
with $w_0 = \ell_D/2$ and spacing $2\ell_D$. See ref. (15) for discussion.

The stability of steady-state solutions $\phi_s(\underline{r})$ to (11) can be examined
by setting

$$\phi(\underline{r},t) = \phi_s(\underline{r}) + \Psi(\underline{r})e^{\lambda t}$$

and linearising (11) for small (and thus bounded) Ψ:

$$-\ell_D^2 \nabla_t^2 \Psi + (1+\lambda\tau)\Psi = -V(\underline{r})\Psi \tag{12}$$

where

$$V(\underline{r}) = - \frac{\partial}{\partial\phi} \left(I(\underline{r})/\left[1+F\sin^2(\phi+\phi_0)\right]\right)\Big|_{\phi=\phi_s(\underline{r})}$$

This is an equation of Schrodinger form[37], with $V(\underline{r})$ the "potential
energy", and "energy eigenvalue"

$$E = -(1+\lambda\tau).$$

Since it is Hermitian, E is real, and thus ϕ_s is unstable if Ψ describes a sufficiently deeply bound state, for which $E \leqslant -1$ and thus $\lambda \geqslant 0$.
This quantum analogy is particularly illuminating for the case where $I(\underline{r})$
is spatially periodic, as in OB arrays: the eigenfunctions Ψ can then be
expressed, using the Bloch theorem, as $u(\underline{k},\underline{r})e^{i\underline{k}\cdot\underline{r}}$ where $u(\underline{k},\underline{r})$ has the
lattice period. The energy E is split into bands for the case where $\phi_s(\underline{r})$
is periodic (e.g. all pixels "on"). The crucial case of a single pixel to
all the others corresponds to a "point defect" or "impurity", and will thus

have a localised wavefunction with energy lying below the band: the critical
case is when it lies sufficiently deep to have E<-1, and engender instability. (An array with pixels "on" and "off" at random corresponds to an alloy
in this picture).

Since power requirements are less for beamwidths much less than ℓ_D,
while pixel independence demands spacings d comparable to ℓ_D, it seems
reasonable to examine the case in which $I(\underline{r})$ is an array of delta
functions:

$$I(\underline{r}) = \sum_{\text{sites}} P\delta(\underline{r}-\underline{r}_r)$$

where P is the power per pixel. For a one-dimensionale array, ϕ_s is finite
on every site, and for the uniform solution has the value A where

$$f(A) = \frac{A}{P} 2\tanh(d/2\ell_D) \tag{13}$$

Here f(A) is the cavity response function (Airy function in (11)). This
equation can be solved graphically exactly as in the plane wave case, the
effect of the new geometry being merely to scale the slope of the
intercepting line. One can solve the Schrodinger equation (12) for this
problem, and we find that the stability condition is exaclty analogous to
the plane wave case, i.e. if $\frac{\partial A}{\partial P} > 0$ the uniform solution is unstable and
vice versa. The two-dimensional δ-function array is complicated by the
fact that P→0 as w→0. By careful handling of the limits it is possible,
however, to derive a relationship analogous to (13) but with a much more
complex scaling factor: the stability then follows the normal rule.

Non-uniform solutions must normally be obtained numerically. For a
linear δ-function array the nonlinear phase at the nth pixel is given by

$$A_{n+1} = 2A_n \cosh(d/\ell_D) - Pf(A_n)\sinh(d/\ell_D) - B_n$$
$$B_{n+1} = A_n \tag{14}$$

a two-dimensional mapping which is directly analogous to stroboscopic maps
arising in the analysis of "kicked" systems in Hamiltonian dynamics[38].
Such systems are generically chaotic: in this case chaos means spatial
chaos. At first sight surprising, spatial chaos is actually a natural and
necessary requirement for a satisfactory OB memory array. Such a memory
must have steady states in which it is impossible to infer the state of any
pixel from the state of its neighbours - pixel independence. But chaos in
the dynamical systems described by maps like (14) means that it is
impossible to predict the future motion from the past. This is evidently
the same type of restriction, and the link is even stronger when one
considers that caotic dynamics is frequently analysed by establishing 1:1
correspondence between the motion and symbol sequences, including binary
digit sequences (see reference (38) for a broad treatment of such dynamical
systems). Again, this is precisely analogous to the correspondence between
the spatial states $\phi_s(\underline{r})$ and the binary information they represent.

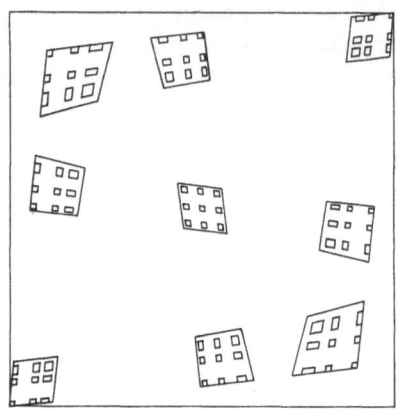

Fig. 7. Illustration of the formation of the set of bounded solutions to the mapping (14). The outer square is $A_{\ell} < A, B < A_u$ in
the (A,B) plane, and the bounded set is the limit of the
infinitely-nested small "squares"[17].

Fig. 7 illustrates these features for the case of the response function of Fig.1. Almost all initial pairs (A_0, B_0) used to seed (14) diverge
as n tends to $\pm\infty$. The "bounded set" of finite sequences turns out to be the
limit set of the infinitely nested "squares" of Fig. 7: a Cantor set. Each
member of this set can be identified by assigning it "0", "1" or "2"
according to whether the left, right or centre square is chosen at each
level (lower, upper or centre for $n \rightarrow -\infty$), and this sequence is the content
of this optical memory.

This argument ignores stability: it turns out that all sequences
containing one or more "2"'s is dynamically unstable. Such states corresponds to the middle branch OB states. In fact a simple general rule is
that if one defines an "instability strip" between the values of ϕ at which
the tangent to $f(\phi)$ is parallel to the "OB line", then any sequence for
which A_n never enters this strip is stable, and thus a possible memory
state. These stable states are all representable by sequences of "0" and
"1", i.e. binary strings.

A key question for arrays is the maximum allowable pixel packing
density. In the light of the above discussion, this is the minimum pixel
separation at which (14) has chaotic solutions. A sufficient condition for
this can be derived geometrically from Fig. 1. Draw tangents from ϕ_{ℓ} and ϕ_u
to $f(\phi)$, and let r_{ℓ}, r_u be the ratios of their slopes to that of the "OB
line". Then the minimum pixel separation for spatial chaos and thus func-
tional optical memory is d_{min}, where

$$\cosh(d_{min}/\ell_D) = r/(r-1)$$

and r is the smaller of r_{ℓ}, r_u. This formula typically gives d_{min} as two
or three diffusion lengths, consistent with Fig. 6: for the case where
$f(\phi)$ is a step function (edge-filter function), d_{min} can be calculated
analytically and is $\ell_D \ell n 3$[17].

While these particular results are specific to the diffusive-coupling model (14), it seems likely that qualitatively-similar features will arise in more realistic models, say with material pixellation, and featuring mixed thermal and electronic nonlinearity and diffusion, and also diffraction. In particular, the link between chaos and memory seems likely to have a very general application.

VII. SPATIAL SOLITONS IN OPTICAL BISTABILITY

To conclude this discussion of transverse effects, and indeed this review, we consider purely diffractive coupling, but now for the case of a self-focussing nonlinearity, $n_2 > 0$ for a Kerr medium. It is well-known that indefinite propagation in a self-focussing Kerr medium leads to catastrophic self-focussing: this singular behaviour is modified in an OB cavity because the cavity losses mean that the cavity field is continually smoothed by its replenishment with the smooth input beam. Even so, Moloney and co-workers have found it useful to consider saturable nonlinear refraction as well as the true Kerr effect in their modelling[19]. An intensity dependent refractive index described by

$$n = n_0 + n_2 I/(1+I/I_s)$$

becomes essentially independent of intensity for very large fields, and thus catastrophic self-focussing is avoided: importantly also smoothing of the field profiles reduces the size of the computational problem.

For a broad Gaussian beam input, and in one cartesian dimension, with unidirectional propagation (ring cavity), Moloney and co-workers find OB switching as expected followed by a switching-wave spreading of the beam much as in the diffusive case[15] except that here the velocity of the switching wave depends on n_2 (if $n_2 < 0$, the index gradients at the switching edge bends light into the lower branch region, and speeds up the switching wave, while for $n_2 < 0$ the reverse is the case, and the switching wave can even be halted).

The significant new feature for $n_2 > 0$ is that the switched-up region is not stable, but spontaneously organises itself into a deeply-modulated pattern. This effect seen in the first computations of the Tucson group[39], was subsequently explained[18] as soliton-like structures directly analogous to the known soliton solutions of the nonlinear Schrodinger equation, with the transverse dimension here playing the role of time. Originally explained as originating in the diffraction from the sharp amplitude gradient in the on-off transition region, it has recently been shown[19] that these "solitary wave" patterns are the end-product of a modulational instability of the plane-wave "on" solution. This follows the independent demonstration by Lugiato and Lefever[20] of spontaneous structure emerging in an approximate model of OB, the so-called mean-field model[5]. Both results evidently involve spontaneous breaking of the transverse translational symmetry.

Moloney and co-workers have extended these calculations to model a full two-transverse-dimension system. This shows a further symmetry-breaking, this time azimuthal, as the initial ring-pattern breaks up into isolated peaks. Recent calculations show that these peaks grow, migrate, and decay in a complex and sustained "dance"[35].

It is interesting to speculate on whether these solitary waves can act as pixels in an OB array with diffractive coupling: it is plausible that a spatial modulation of the input beam could pin the waves into fixed positions. The question then arises as to how much modulation is required to accomplish this and, equally important, to allow these waves to be independently "off" and "on" as in the foregoing discussion of diffusive coupling.

It should be added that there has been no direct demonstration of these solitary waves, because most OB media have $n_2 < 0$, or are diffusive, or both. Nor does the analysis allow for counter-propagation, but there are indications that spatial pattern formation occurs spontaneously with counter-propagation, even for the self-defocussing case $n_2 < 0$.

VIII. CONCLUSION

This review has attempted to summarise what the author considers to be those features of OB theory most relevant for optical computing. The emphasis has been on qualitative and general features, which has largely excluded discussion of particular mechanisms for optical nonlinearity and of particular devices and architectures for applications. Noise effects are another significant omission. These topics are covered in detail elsewhere in these Proceedings and in the cited literature. Within these limitations, I hope to have shown that the marriage between fundamental physics and potential applications has, in the case of OB theory, been a very fruitful one. Furthermore, it seems that the field is still very rich, particularly in the realm of transverse effects.

This work is supported in part under a Twinning Contract of the European Community.

REFERENCES

1. C. M. Bowden, M. Ciftan and H. R. Robl (eds.), "Optical Bistability", Plenum (1981)
    C. M. Bowden, H. M. Gibbs and S. L. McCall (eds.), "Optical Bistability II", Plenum (1984)
    H. M. Gibbs, P. Mandel, N. Peyghambarian and S. D. Smith (eds.), "Optical Bistability III", Springer (1986)
    N. Peyghambarian, W. J. Firth and A. Tallet (eds.), "Optical Bistability IV", Edition de Physique (1988)
2. H. M. Gibbs, "Optical Bistability - Controlling Light with Light", Academic (1985)

3. P. Mandel, S. D. Smith and B. S. Wherrett (eds.), "From Optical Bistability Towards Optical Computing", N. Holland (1987)
4. A. Szoke, V. Daneu, J. Goldhar and N. A. Kurnit, Appl. Phys. Lett., 15, 376 (1969)
    H. Seidel, U. S. Patent 3,610,731 filed May 19, 1969, granted October 5, 1971
5. L. A. Lugiato, "Progress in Optics", XXI, E. Wolf ed., North-Holland, 69 (1984)
    R. W. Boyd, M. G. Raymer and L. M. Narducci (eds.), "Optical Instabilities", Cambridge U.P., (1986)
6. D. A. B. Miller, J. Opt. Soc Am, B1, 857 (1984)
7. J. A. Goldstone and E. Garmire, Phys. Rev. Lett., 53, 910 (1984)
8. D. A. B. Miller, IEEE J. Quant. Elec., QE-17, 306 (1981)
9. S. D. Smith, A. C. Walker, F. A. P. Tooley and B. S. Wherrett, Nature, 325, 27 (1987)
10. R. Bonifacio and L. A. Lugiato, Opt. Commun., 19, 172 (1976)
11. F. A. P. Tooley, W. J. Firth, A. C. Walker, H. A. Mackenzie, J. J. E. Reid and S. D. Smith, IEEE J. Quant. Elec., QE-21, 1356 (1985)
12. S. L. McCall, App. Phys. Lett., 32, 284 (1978)
13. P. W. Smith, in: "Optical Bistability IV", N. Peyghambarian, W. J. Firth and A. Tallet (eds.), Editions de Physique (1988)
14. W. J. Firth, I. Galbraith and E. M. Wright in: "Optical Bistability III", H. M. Gibbs, P. Mandel, N. Peyghambarian and S. D. Smith (eds.), 193, Springer (1986)
15. W. J. Firth and I. Galbraith, IEEE J. Quant. Elec., QE-21, 1399 (1985)
16. D. Frank and B. S. Wherrett, Opt. Eng., 26, 53 (1987)
17. W. J. Firth, Phys. Lett., A125, 375 (1987)
18. D. W. McLaughlin, J. V. Moloney and A. C. Newell, Phys. Rev. Lett., 51, 75 (1983)
19. H. Adachihara, J. V. Moloney, W. D. McLaughlin and A. C. Newell, J. Math. Phys., 29, 63 (1988)
20. L. A. Lugiato and R. Lefever, Phys. Rev. Lett., 58, 220 (1987)
21. M. J. Adams, H. J. Westlake, M. J. O'Mahony and I.D. Henning, IEEE J. Quant. Elect., QE-21, 1498 (1985)
22. M. J. Adams, Int. J. Electron., 60, 123 (1986)
23. W. F. Sharfin and Dagenais, Appl. Phys. Lett., 48, 321 (1986)
24. B. S. Wherrett, IEEE J. Quant. Elec., QE-20, 646 (1984)
25. H. A. Al-Attar, H. A. Mackenzie and W. J. Firth, JOSA B3, 1157 (1986)
26. D. A. B. Miller and S. D. Smith, Opt. Commun., 31, 101 (1979)
27. P. Nardone and P. Mandel, in "Optical Bistability III", H. M. Gibbs, P. Mandel, N. Peyghambarian and S. D. Smith (eds.), 53, Springer (1986)
28. J. Y. Bigot, A. Daunois and P. Mandel, Phys. Lett., A123, 123 (1987)
29. D. J. Agan, H. A. Mackenzie, H. A. Al-Attar and W. J. Firth, Opt. Lett., 10, 187 (1985)
30. W. J. Firth, I. Galbraith and E. M. Wright. JOSA B2, 1005 (1985)
31. W. J. Firth and E. M. Wright, Opt. Comm., 40, 233 (1982)
32. L. A. Orozco, M. G. Raizen, A. T. Rosenberger and H. J. Kimble, in: "Optical Bistability III", H. M. Gibbs, P. Mandel, N. Peyghambarian and S. D. Smith (eds.), 307, Springer (1987)

33. L. A. Lugiato, report to EEC Twinning Meeting, Aussois, France, March 1988

34. D. L. Weaire, J. P. Kermode and V. M. Dwyer, Opt. Commun., 55, 223, (1985)

35. J. V. Moloney, report to EEC Twinning Meeting, Aussois, France, March 1988

36. W. J. Firth, E. Abraham, E. M. Wright and B. S. Wherrett, Phil. Trans. Roy. Soc. Lond., A313, 299 (1984)

37. N. N. Rosanov, Sov. Phys. JETP, 53, 47 (1981)

38. J. Guckenheimer and P. Holmes, "Nonlinear Oscillations, Dynamical Systems and Bifurcations of Vector Fields", Springer (1983)

39. J. V. Moloney, F. A. Hopf and H. M. Gibbs, Phys. Rev., A25, 3442, (1982)

DYNAMIC OPERATION OF NONLINEAR WAVEGUIDE DEVICES

FOR FAST OPTICAL SWITCHING

S. Laval and N. Paraire

Institut d'Electronique Fondamentale - Univ. Paris-Sud
CNRS-UA 2, Batiment 220, 91405 Orsay Cedex, France

INTRODUCTION

All-optical logic devices are key elements for optical computing systems, and current research aims at optimizing these devices by decreasing their size, switching times and operating power. These parameters are strongly coupled together, and according to the application which intended one or another may be emphasized. Although the massive parallelism offered by optics is often referred to as one of the major advantages of optical computing, ultrafast switching devices are also highly desirable in order to further increase the capabilities of optical processing systems. Dynamic properties are therefore essential and must be analysed. Let us also note that faster switching generally requires higher optical power, and one important goal is to improve the sensitivity of the device, in order to avoid using excessive intensities.

The search for short switching times excludes the use of hybrid bistable devices, in which the response time is limited by the electronic feedback loop. So long as intrinsic systems are considered, limitations come both from the nonlinear material properties, and the optical feedback mechanism. The usual structure for optical bistable devices is a Fabry-Perot cavity filled with a nonlinear medium. The switching time is then limited by the cavity build-up time and by the medium response time. The former requires very thin resonant cavities, while the latter calls for fast nonlinearities which are still sufficiently strong to allow operation at low intensities.

During the recent years, increasing interest has developed in the use of nonlinear guided waves to obtain optical bistability[1,2]. An optical waveguide can be considered as a very thin resonant cavity[3] whose thickness is on the order of the light wavelength, and in which the build-up time is therefore very short. Furthermore, as light is totally reflected on the waveguide walls, the waveguide is equivalent to a high Q-factor cavity.

Nonlinear Optics and Optical Computing
Edited by S. Martellucci and A. N. Chester
Plenum Press, New York, 1990

It is well known that light power density can be considerably enhanced in a guiding structure, and this leads to much stronger nonlinear effects.

To take advantage of the fast response time available with such devices, short laser pulses have to be used and stationary operating conditions do not hold. The characteristic times of the structure have to be compared with the light pulse duration, and the observed phenomena strongly depend on their relative values[4,5]. As thermal effects develop in the nanosecond range in most devices, these must also be taken into account in the analysis of the dynamic operation of optical switching devices.

I. OPTICAL BISTABILITY IN GUIDED-WAVE STRUCTURES

I.1. Stationary Linear Properties of the Structures

When nonlinear guiding structures are considered, two kinds of use must be clearly distinguished: either propagation over large distances is used to obtain large changes in optical path to accumulate the effects of nonlinearities, or the waveguide is considered only as a thin highly resonant cavity.

In the first case, many devices have been developed which employ integrated electro-optic modulators or directional couplers[6], most of them being hybrid systems. Some intrinsic distributed feedback systems have also been considered[7,8].

We are mainly interested here in the second point of view[9]. In an optical waveguide, light is totally reflected back and forth by the interfaces between the substrate and the external medium. Since light must enter the waveguide, this is generally achieved using either a high refractive index prism or a diffraction grating coupler. If the waveguide is used as a resonant cavity, the coupler covers its entire surface, unlike the devices where propagation effects are emphasized, and this is responsible for cavity damping. Thus, the device can be represented in a simplified way as in Fig. 1 by a coupling system and a rear mirror, with a thin layer sandwiched between, which may exhibit nonlinear optical properties. Let n be the refractive index of the layer, n_e that of the external medium, and n_s that of the substrate. As an electromagnetic plane wave $\vec{E} = \vec{E_i} \exp j(\omega t - \vec{k} \cdot \vec{r})$ is incident on the coupler, light can be partially coupled into the layer, and partially reflected by the structure. There may also be a transmitted wave if a grating coupler is used.

According to the EM field continuity conditions, the component parallel to the interfaces of the wavevector, $k_{//}$, is conserved through the whole structure. This is written as:

$$k_{//} = (2\pi/\lambda_0)n_p \sin\theta_p$$

for prism coupling and

$$k_{//} = (2\pi/\lambda_0)n_e \sin\theta_e + pk \tag{1}$$

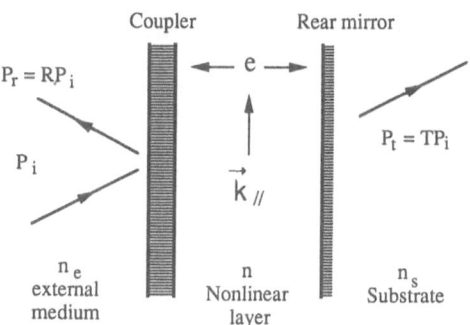

Fig. 1. Schematic representation of a nonlinear waveguide with its associated coupler used as a highly resonant cavity.

for grating coupling (with $k_0 = 2\pi/d$, where d is the diffraction grating period).

When the angle of incidence is varied, minima in the reflected intensity are observed, corresponding to the excitation of guided waves in the film and the coupling of light energy into the waveguide.

If the absorption in the layer is low and if the perturbation induced by the coupler is negligible, these resonance occur when $k_{//}$ is close to the propagation constant β_m of an eigenmode of the waveguide given by:

$$e \sqrt{n^2 k^2 - \beta_m^2} - \phi_s^m - \phi_e^m = m\pi \tag{2}$$

where e is the nonlinear film thickness, $k = 2\pi/\lambda_0$, $\phi_s(m)$ and $\phi_e(m)$ are the differences in phase occuring for plane waves propagating in the film with a longitudinal wave vector component β_m, when reflecting from the substrate or from external medium.

An analogous result is obtained with a Fabry-Perot (FP) cavity, as the resonance condition to obtain constructive interference is similar. Indeed in the case of a FP, β_m is usually taken equal to zero (the incidence angle is kept constant at $\theta = 0$) and resonance is obtained when e or λ is varied.

The main advantage of the waveguide structure compared with the Fabry-Perot interferometer is that, as stated before, it presents natural highly reflective interfaces which allows an easy realization of a resonant cavity whose optical thickness can be as short as $\lambda/2$, i.e., a submicron thickness. This is not only important for the response time of the device, but also for the minimum power required to obtain optical bistability under stationary operation conditions, as the power density inside the cavity is inversely proportional to the thickness.

I.2. Stationary Nonlinear Operation

Let us now consider the nonlinear properties of such structures, first under stationary conditions. As the wavevector component parallel to

the interfaces is conserved, its value in the guiding layer is given
either by:

$$k_{//} = (2\pi/\lambda)n \sin\theta = (2\pi/\lambda)n_p \sin\theta_p$$

for a prism coupler, with $n_p > n > n_s$, n_e where θ and θ_p define the propagation direction in each material with respect to the normal at the inter-
face, or by:

$$k_{//} = (2\pi/\lambda)n \sin\theta = (2\pi/\lambda)n_e \sin\theta_e + p\, 2\pi/d$$

for a grating coupler of period d used in the diffraction order p, θ_e
being the incidence angle, and n_e the refractive index in the external
medium.

In both cases, if the refractive index n in the waveguide core varies
and shifts from its linear value n_0, according to eq. (2) the propagation
constants β_m of the modes change, and the incidence angle θ_p or θ_e for
which resonance occurs is different. This is represented schematically in
Fig. 2. From this figure, it can be seen that for a given incidence angle,
the reflection coefficient of such a structure varies with the refractive
index. The R(n) curve also exhibits a minimum for a given value of n. For
a nonlinear material, the refractive index varies with the light power
density P in the medium, and we may consider that the variations are
given, as in the case of Kerr effect, by:

$$n = n_0 + n_2\, P. \tag{3}$$

Assuming that the light power in the waveguide corresponds to the
fraction of light which is not reflected (i.e., there is no transmitted
wave) its mean value is given by:

$$P = [1 - R(n)]\, P_i \cos\theta_e /e \tag{4}$$

where P_i is the incident light intensity, θ_e the incidence angle and e the
waveguide thickness.

The reflection coefficient can then be expressed as:

$$R(n) = 1 - [\, e(n-n_0)/n_2 P_i \cos\theta_e\,] \tag{5}$$

For a given incident intensity, the value of R may be found graphically as the intersection between the curve R(n) and the straight line
corresponding to eq. (5) (Fig. 3), as is usually done for other kinds of
bistable devices[10]. If the resonance value n_r of n is shifted far enough
from n_0, three intersection points may appear when P_i is increased, two of
them corresponding to stable states of the device, and bistable operation is
predicted.

Since we hope to obtain a device which can be conveniently inserted
in a monolithic system, a grating coupler has significant advantages, and
the following discussion will focus on such structures.

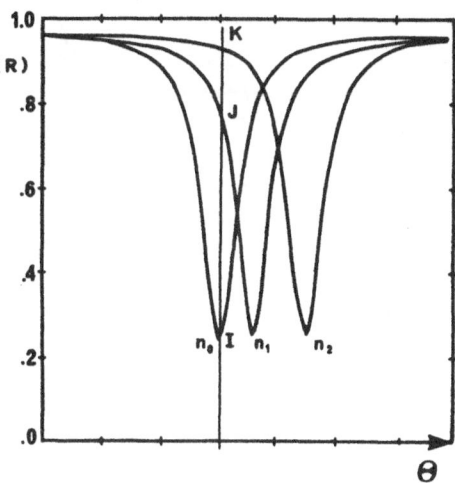

Fig. 2. Reflection coefficient R of a waveguide structure versus the incidence angle θ for different values of the refractive index.

In order to obtain high contrast switching, the reflectivity variations with θ_e or with n must be as large as possible, and configurations in which R can vary form 1 to 0 are most desirable. Thus, it is first necessary to optimize the coupling efficiency, which is mainly related to the grating modulation depth. The optimization consists in maximizing the electromagnetic field intensity in the nonlinear material.

To properly take into account the periodic modulation of one of the interfaces of the waveguide, a rigorous approach by numerical integration of Maxwell's equations must be used[11,12]. This allows us to characterize the linear properties of the structure: determination of the incidence

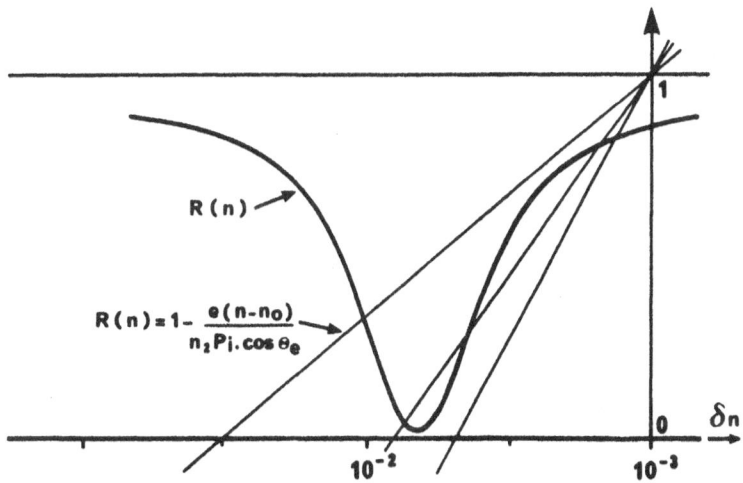

Fig. 3. Operating point of nonlinear waveguide in the bistable regime.

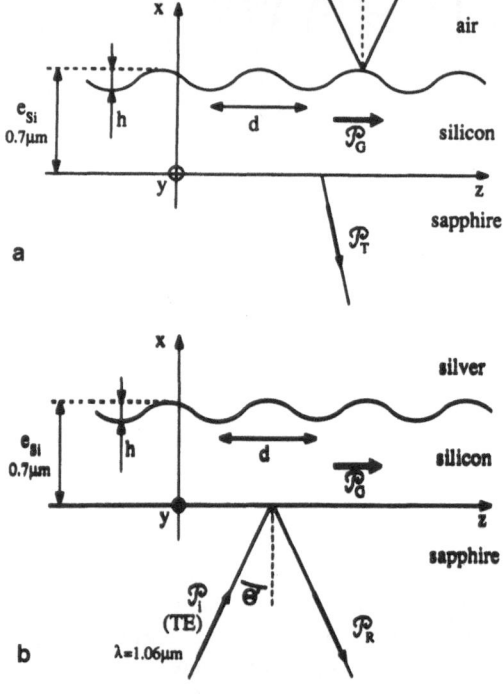

Fig. 4. Experimental devices: (a) structure of device A: Air -
Silicon - Sapphire; and, (b) structure of device B: Silver -
Silicon - Sapphire.

angles $\theta_e(r)$ for resonant excitation of guided waves for a given period d of
the grating grooves, and modulation depth a for a maximum coupling efficiency. It has already been established that a metallic-coated dielectric
grating may lead to a zero reflection coefficient and a strong enhancement
of the EM field in the waveguide[13]. Thus, two kinds of devices need to be
considered: a substrate/nonlinear-material/external-medium for the so-
called "A - configuration" and a substrate/nonlinear-material/metal for the
B one (Fig. 4). In the examples given in this paper, the nonlinear material
is epitaxially-grown silicon, deposited on a sapphire substrate, and option-
ally covered with silver.

Let us assume translation symmetry in the propagation direction and
use plane-wave models. This may introduce some difficulties in describing
the observed phenomena. In particular, if no absorption is accounted for
in structure B, since the grating acts simultaneously as input and output
coupler, under stationary conditions no variations of the reflectivity with
the incidence angle can be predicted. Thus the absorption cannot be neglected in the calculations, and the relevant parameter is then the absorbed
power: $P_{abs} = (1-R)P_i$. For the A configuration, part of the incident wave is
transmitted through the structure, and the energy absorbed in the waveguide
is given $P_{abs} = (1-R-T)P_i$.

An example of the transmission and reflection coefficient variations versus the incidence angle are given in Fig. 5, together with the absorbed power, in the case of the A-configuration. Even for the optimum modulation depth of the grating, the $R(\theta)$ and $T(\theta)$ curves are non-symmetric. The field intensity enhancement in the waveguide[12] is on the order of 50.

The B configuration gives better results. There is no transmitted wave, as the metal layer acts as a mirror. The modulation depth can be accurately adjusted to get a zero value of the reflectivity at resonance (Fig. 6). The field intensity enhancement can reach values[12] on the order of 250.

The lowest critical power for optical bistability is obtained if the field enhancement at resonance is maximum, but also if the angular width of the resonance curve is the smallest. Thus, we can benefit by using only the lowest order propagation modes of the structure, which are narrower r than the higher order ones. In this case, the guide has to support only a few modes, to avoid leakage from the mode under consideration to other case, and this is quite compatible with the small thickness requirement.

From these calculations, the reflectivity can also be represented versus refractive index variation, as in Fig. 3, and the hysteresis loops which characterize the bistable properties can be derived[12].

Up to this point, the phenomena have been described assuming implicitly stationary conditions with respect to time. However, since fast switching is desirable, pulsed excitation is required, and dynamic properties must be investigated.

II. TEMPORAL BEHAVIOR OF NONLINEAR WAVEGUIDE DEVICES

The dynamic characteristics of nonlinear devices depend basically on

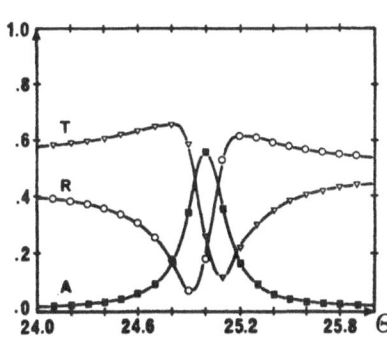

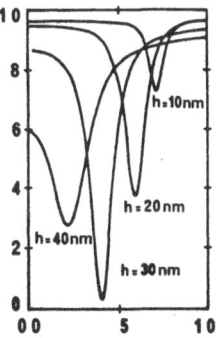

Fig. 5. Reflection (R), transmission (T) and absorption (A=1-R-T) efficiencies versus the incidence angle θ for structure A.

Fig. 6. Reflectivity versus the incidence angle θ for various values of the diffraction grating modulation depth h, in structure B.

the physical processes which are at the origin of the involved nonlinearities. Strong nonlinear effects are generally observed when light-matter
interaction has a resonant character, i.e., when there is energy exchange
between light and matter. These is known as "dynamic nonlinearities", compared to parametric interactions in which only virtual transitions between
eigenstates of the material system occur[14]. Since real transitions are
involved in dynamic nonlinearities, the processes are dissipative.

The most widely studied materials for optical bistable devices in
recent years are semiconductors, as they exhibit strong nonlinearities due
to various physical phenomena such as free carrier generation, exciton
screening, absorption saturation, etc., which lead to important changes in
the optical properties of the material.

Thus, let us consider a semiconductor layer used as a waveguide, as
described in Fig. 4, and analyse the refractive index changes induced by
illumination of the material.

In the following, it is assumed that illumination is uniform on the
device, so that no carrier diffusion or drift parallel to the interfaces
needs to be accounted for. If $\delta N(x,t)$ is the excess free carrier density
in the x-abscissa plane of the nonlinear layer, these carriers may locally
recombine with phonon emission, may absorb incident light, and may diffuse
in the layer thickness due to the carrier concentration gradient. As the
carrier diffusion in the film depth occurs in very short times (a few ten or
hundred picoseconds, according to the carrier mobility), the excess carrier
density can be averaged throughout the layer thickness e:

$$\delta N = (1/e) \int_0^e \delta N(x)dx$$

For time scales larger than the diffusion time, the phenomena associated with the excess free carriers can be described by x-independent
quantities, and the mean value δN will be used hereafter.

If the photon energy in the incident beam is larger than the energy
band gap of the semiconductor, electron-hole pairs are created, and this
induces an electronic variation of the refractive index which is negative[15]:

$$\delta n_e(t) = -K_e \ \delta N(t) \tag{6}$$

where $\delta N(t)$ is the density of excess free carriers present in the medium at
time t, and K_e a coefficient which depends on the light wavelength with
respect to the material energy gap.

For room temperature operation in structures formed with a single semiconductor material, exciton effects are negligible. The absorption saturation is also much more efficient at low temperature, as the width of the
carrier energy distribution function is reduced, and will not be considered
here.

The excess free carrier density at any time t results from generation
and recombination processes:

$$d\delta N/dt = (\alpha_0/\alpha)\,|P_{abs}(t)/eh\nu\,|- R_{eh}\,\delta N(t) \tag{7}$$

where P_{abs} is the total light power absorbed per unit surface of the structure; it depends on the coupling conditions, i.e. on incidence angle and refractive index values. α_0/α is the ratio of the power absorbed in the semiconductor to P_{abs}; it depends on the photocarrier density: in the examples given in the preceding section, this ratio takes into account the fraction K of the light absorbed in the metal layer in the B configuration, and the free carrier absorption with a cross-section σ:

$$\alpha = \alpha_0(1 + K) + \sigma\,\delta N \tag{8}$$

R_{eh} is the recombination coefficient, and it depends on the excess carrier density $\delta N(t)$. If the de-excitation process is assumed to be exponential with a time constant τ, R_{eh} is given by the following equation:

$$R_{eh}\,(\delta N)\,\delta N = \delta N/\tau \tag{9}$$

In any semiconductor, electron-hole recombination is more or less related to phonon emission, and yields lattice heating. This is very important in semiconductors with indirect-band structure, such as silicon, as no radiative transitions can occur, but it also exists in direct-band structure materials such as III-V compounds, in particular if deep levels are present in the energy band gap. The lattice temperature changes induce a variation of the refractive index which is positive in most cases:

$$\delta n_T(t) = (\partial n/\partial T)\,[\delta T(t)] \tag{10}$$

This strongly influences device operation and must be taken into account to properly describe or project the performance of nonlinear structures. As a first approximation, absorption changes with temperature are neglected and only dispersive effects are considered.

As the device temperature varies, some geometrical factors also change, such as the waveguide thickness e or the grating period d. These two factors add to the refractive index variations to alter the propagation mode in semiconductor layer, and we can estimate their relative influence on the incidence angle variation required to maintain resonant excitation of a mode m:

$$\delta\theta_m = \left[(\partial\theta/\partial n)_{e,d}\right]\delta n_T + \left[(\partial\theta/\partial e)_{n,d}\right]\delta e + \left[(\partial\theta/\partial d)_{e,n}\right]\delta d \tag{11}$$

For the TM_0 mode of a silicon waveguide with e=0.7 μm, d=0.33 μm, using $\partial n/\partial t = 2 \times 10^{-4}$°K^{-1} from experimental results, and a linear dilatation coefficient equal to[16] 2.6×10^{-6}°K^{-1}, we find that the angular shift due to thermal refractive index variations is $(\partial\theta/\partial n)_{e,d} \cong 10^{-2}$ deg°K^{-1}, whereas the shift related to thermal expansion are of the order of 5×10^{-4} deg°K^{-1}. Thus the thermally induced geometrical variations can be neglected.

To investigate the dynamic properties of the device, the heat sources must be physically defined to describe their kinetics, and the boundary conditions for heat exchange must be specified.

Assuming that band-to-band recombination is totally non-radiative, the heat produced per unit time by these transitions, per unit surface of the device, is given by:

$$[d(\Delta Q)/dt]_R = R_{eh} \, \delta N \, e \, E_g \qquad (12)$$

where E_g is the band gap energy of the material, and R_{eh}, which is given by eq. (9), characterizes the time evolution of this process, mainly governed by the time constant τ. Some other processes can be considered as instantaneous : metal layer heating in the B configuration, and free carrier thermalization within the conduction band, following either generation with an excess energy $(h\nu - E_g)$ or free carrier absorption. Other processes, as avalanche or Auger effects, are neglected. The total heat created per unit surface of the device can be written:

$$d\Delta Q/dt = R_{eh} \, \delta N \, e \, E_g + (P_{abs}/\alpha) \, |(\alpha - \alpha_0) + \alpha_0 \, (h\nu - E_g)/h\nu \qquad (13)$$

It can be shown that the temperature rise δT in the guiding layer is fairly independent of the transverse spatial coordinate x, i.e., the thermalization within the layer is fast (a few hundred picoseconds), and the device can then be characterized as a whole by an equivalent heat capacity per unit volume $(\rho C_p)_e$, with thermal exchange with a semi-infinite substrate[5]. Then ΔQ can be expressed as:

$$\Delta Q = (\rho C_p)_e \, e \, \delta T \qquad (14)$$

From equations (6), (7), (10), (13) and (14), the refractive index variations from electronic effects and from thermal phenomena can be derived:

$$d(\delta n_e)/dt = K_e \left[R_{eh} \, \delta N - (\alpha_0/\alpha) \, (P_{abs}(t)/eh\nu) \right] \qquad (15)$$

$$d(\delta nT)/dt = \left[(\partial n/\partial T) \, (1/\rho C_p)_e \right] \{ R_{eh} \, \delta N \, e \, E_g +$$

$$+ (P_{abs}(t)/\alpha) \left[(\alpha - \alpha_0 + \alpha_0 \, (h\nu - E_g)/h\nu \right] \} \qquad (16)$$

Both nonlinearities δn_e and δn_T contribute to the feedback required for optical bistability, as they both depend on the absorbed power P_{abs} which, as stated before, is a function of the coupling conditions, and in particular of the refractive index $n = n_0 + \delta n_e + \delta n_T$. The two contributions may have different time evolutions, depending on the carrier recombination time, and on the device parameters which determine heat dissipation and light coupling efficiency. If c.w. light is sent into the device, stationary conditions are reached, and bistability can be expected as previously discussed. On the other hand, if pulsed light excitation is used, the pulse duration needs to be compared with the various characteristic times of the device. The ratio $\delta n_e/\delta n_T$ is a function of time, and so long as neither of the two effects is negligible, a dynamic competition between them occurs and non stable state can be reached; i.e., no optical bistability can be defined, but switching can however lead to quasi-bistable effects.

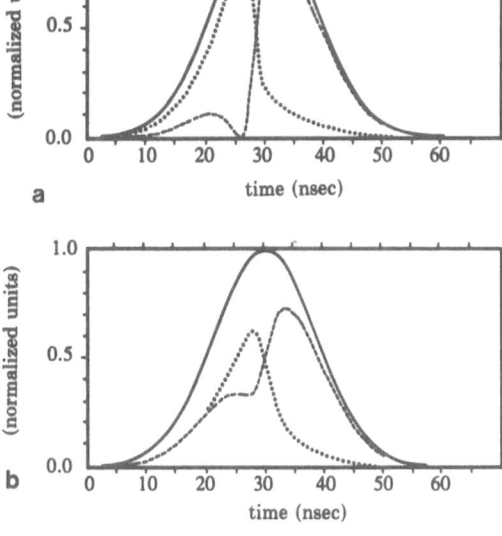

Fig. 7. Temporal variations of the exciting pulse (solid line),
reflected beam intensity (dashed line) and guided light
intensity (dotted line) in structure B, with $\Delta\theta=0.04°$, $P_{ic}=$
$= 250$ kW/cm^2 for two grating modulation depths: (a) h=41 nm;
and (b) h=26 nm.

III. NUMERICAL AND EXPERIMENTAL RESULTS FOR SILICON ON SAPPHIRE DEVICES

Experiments have been performed with silicon on sapphire devices (A or
B) as described in Fig. 4, using a Q-switched Nd:Yag laser which delivers
light pulses at $\lambda=1.06$ µm, whose duration is 20 ns.

Calculations have also been made to analyse temporal transients, using
the model defined in the previous sections. The incident light power is
assumed to vary with time according to a Gaussian function:

$$P_i(t) = P_{ic} \; f(t) = P_{ic} \; \exp\left|-4t^2/(\Delta t)^2\right|$$

The influence of several parameters have been investigated: coupling
efficiency, light pulse duration Δt, carrier lifetime τ, and initial angular
detuning with respect to low light level resonance $\Delta\theta = \theta - \theta_0$.

The coupling efficiency is strongly related to the grating modulation
depth. In the B configuration, the latter can be adjusted to get zero
reflected intensity at resonance. In this case, high contrast switching is
observed (Fig. 7a): for the TE$_0$ mode the reflected efficiency changes from 0
to 0.82 for a peak incident power P_{ic}=250 kW/cm^2, with an initial detuning
$\Delta\theta=0.04°$. If the grating modulation depth is decreased from its optimum
value (41 nm) to 26 nm, then the linear reflectivity minimum is 0.35, and
the contrast in nonlinear operation is strongly reduced (Fig. 7b). Thus it

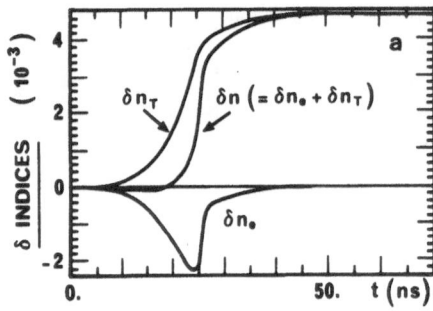 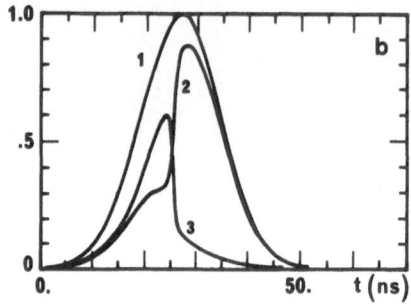

Fig. 8. Theoretical curves drawn for structures B with $\theta-\theta_0=0.04°$, $\tau = 180$ ps, $\Delta t=20$ ns and $P_{ic}=500$ kW/cm^2: (a) temporal variations of the silicon refractive index (solid line), its
electronic (dotted line) and thermal (dashed line) components; and, (b) temporal variations of the exciting pulse
(solid line), reflected (dashed line) and guided (dotted
line) intensities.

is important, from the experimental point of view, to realize devices with
grating parameters as close as possible to the optimum values.

The light pulse duration and the carrier lifetime cannot be considered
independently. Their ratio determines the relative importance of thermal and
electronic effects. Experiments have been performed with various samples, in
which the carrier lifetime was about 180 ps and 800 ps respectively, with a
fixed light pulse duration of 20 ns.

When $\tau\cong180$ ps, lattice heating develops within a few hundred picoseconds, and no pure electronic effect related to the carrier creation can
be observed with a nanosecond light pulse. Numerical calculations show that
electronic and thermal effects first cancel one another (Fig. 8a). Then
thermal effects clearly predominate, yielding, for $\theta>\theta_0$, a switching whose
duration is shorter than that induced by any one of these two effects. This
can be explained by the fact that, since the incidence angle is fixed,
because of refractive index changes the detuning may increase; then, the
light coupled into the waveguide is reduced and fast carrier recombination
shortens the switching time. The computed light intensities, in the waveguide, and for the reflected beam, exhibit subnanosecond switchings (Fig.
8b) which are quite similar to the observed ones[17].

If the carrier lifetime is increased, heat generation is delayed and
two successive switchings can be observed for sufficiently large angular
detuning from resonance and incident power (Fig. 9a). The calculated electronic and thermal contributions to the index variations (Fig. 9b) show that
the first switching is mainly due to electronic effects, i.e., to carrier
creation in the waveguide. The second one is physically explained as in the
previous case by combined thermal and electronic variation of the refractive
index. The switching times are both subnanosecond. The larger the incident
power, the earlier the first switching happens, and the longer the low-level
reflectivity lasts.

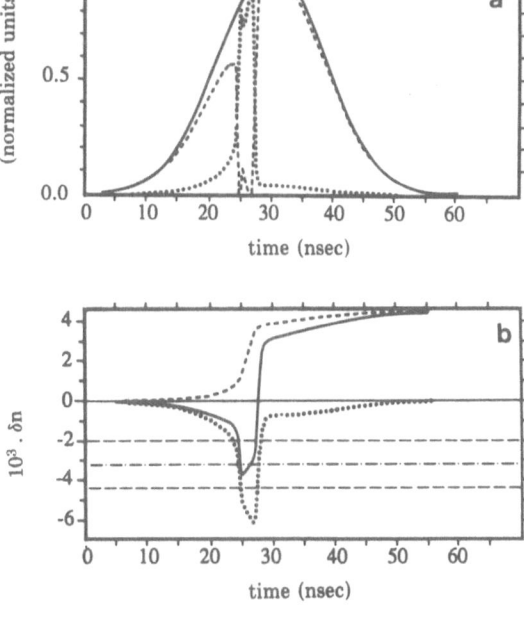

Fig. 9. Theoretical curves drawn for structure B with $\theta-\theta_0=-0,20°$, $\tau=500$ ps, $\Delta t=20$ ns and $P_{ic}=750$ kW/cm^2 : (a) temporal variations of the exciting pulse (solid line), reflected (dashed line) and guided (dotted line) intensities; and, (b) temporal variations of the silicon refractive index (solid line), its electronic (dotted line) and thermal (dashed line) components.

For smaller initial incidence angle detuning, some switchings can still be observed but the switching times are longer[5].

Experiments have also been performed with devices in which $\tau\cong800$ ps, in the A-configuration. Because the incident beam is not a plane wave, light issuing from a small spot on the device is observed to avoid spatial integrating effects. Both TE_0 and TM_0 modes have been considered. Either the reflected, the transmitted, or the guided intensity variations have been measured. For large values of $\Delta\theta$, two fast switchings are observed; the larger the initial detuning, the earlier the emergence of the first switching, and the longer the delay between them. Examples are given in Fig. 10, where reflection and transmission coefficients are reported versus time within a 20 ns light pulse.

It is worth noting that the two switching appear for different values of the instantaneous incident power, and can both occur before the light pulse reaches its peak intensity. If the output intensity of one beam is plotted versus the input power a hysteresis loop is obtained, but its shape is strongly dependent on the ratio $\Delta t/\tau$, because dynamical effects are responsible for these phenomena.

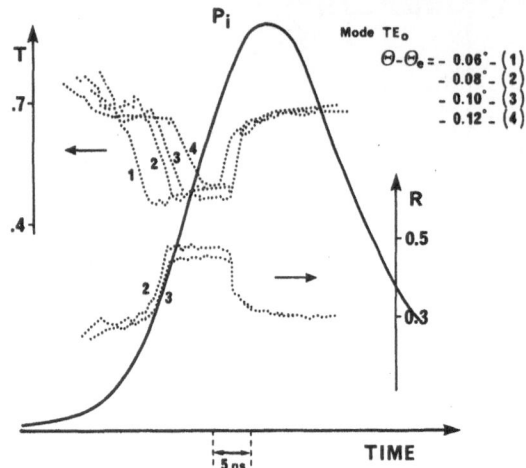

Fig. 10.   Experimental results obtained with structure A on the TE_0
mode: temporal variations of the transmission (T) and
reflection (R) coefficients for various angular detunings
with $\tau=800$ ps, $\Delta t=20$ ns and $P_{ic}=1,7$ MW/cm^2 (solid line: the
exciting pulse).

CONCLUSION

The analysis of dynamic operation of optical switching devices which
has been carried out shows that physical phenomena are quite different in
the c.w. and pulsed regimes. In most devices, thermal effects make an important contribution, and compete with pure electronic index variations which
are much faster. Determination of optimal parameters and operating conditions must take into account both electronically and thermally induced variations of the optical properties of the structure, and these variations depend
not only on the instantaneous light intensity, but also its temporal behavior. Under these conditions, a real optical bistability cannot be reached.

ACKNOWLEDGEMENTS

The authors are grateful to A. Koster, H. Chelli, A. Niepceron, F.
Pardo, H. Sauer and D. Berard for their contribution to the experimental
and computed results presented here.

REFERENCES

1. H. A. Haus, Nonlinear Optical Waveguide Devices in: "Nonlinear
        Optics: Materials and Devices", C. Flytzanis and J. L. Oudar,
        eds., Springer Verlag (1986)
2. G. I. Stegeman, C. T. Seaton, W. M. Hetherington III, A. D. Bordman
        and P. Egan, Nonlinear Guided Waves in: "Nonlinear Optics: Materials
        and Devices", C. Flytzanis and J. L. Oudar, eds., Springer Verlag
        (1986)

3. F. Pardo, "Etude et réalisation d'un dispositif bistable optique basé sur l'excitation d'un mode guidé couplé par réseau dan une couche mince de silicium sur saphir", Thèse de Doctorat, Orsay: 99 (1986)

4. F. Fidorra, M. Wegener, J. Y. Bigot, B. Honerlage, and C. Klinghirn, Journ. of Luminescence, 35:43 (1986)

5. H. Sauer, N. Paraire, A. Koster and S. Laval, JOSA B5:443 (1988)

6. H. M. Gibbs, "Optical Bistability: Controlling light with light", Academic Press, New York (1985)

7. M. Okuda and K. Onaka, Jpn J. Appl. Phys., 17:1105 (1978)

8. H. G. Wingful, J. M. Marburger and E. Garmire, JOSA, 69:1421 (1979)

9. H. Kogelnik, Theory of Dielectric Waveguides in: "Integrated Optics", T. Tamir. ed., Springer Verlag (1975)

10. P. W. Smith and D. A. B. Miller, Laser Focus, 18:77 (1982)

11. M. Neviere, The homogeneous problem in: "The Electromagnetic Theory of Gratings", R. Petit, ed., Springer Verlag, Berlin (1980)

12. P. Vincent, N. Paraire, M. Neviere, A. Koster and R. Reinisch, JOSA, B2:1106 (1985)

13. M. Neviere and R. Reinisch, Phys. Rev., B26:5403 (1982)

14. J. L. Oudar, Transient Nonlinear Optical Erffects in Semiconductors in: "Nonlinear Optics: Materials and Devices", C. Flytzanis, and J. L. Oudar, ed., Springer Verlag (1986)

15. J. Tauc, Progress in Semiconductors, 9:87 (1965)

16. S. M. Sze, "Physics of Semiconductor Devices", Wiley, New York (1981)

17. F. Pardo, H. Chelli, A. Koster, N. Paraire and S. Laval, IEEE Journ. Quant. Elect., 23:545 (1987)

OPTICAL BISTABILITY IN COUPLED-CAVITY SEMICONDUCTOR LASERS

Hong-Du Liu

Department of Physics, Peking University
Beijing, China

I. INTRODUCTION

Optical bistability is a rapidly expanding field of current research, and recent progress in this field[1-3] has stimulated renewed interest in bistable semiconductors lasers[4-22]. Optical bistability in semiconductor lasers is an interesting subject because of their many advantages: they have optical gain, they can be controlled both electronically and optically, and they can be operated faster than nano- to sub-nanoseconds. In a commonly used scheme, inhomogeneous current injection along the laser cavity (using a segmented contact) is used, so that part of the cavity acts as a saturable absorber. This scheme is usually found to be limited in the current and temperature range for which bistability is observed. Moreover, since the "OFF" state in this case is below threshold, the turn-on delay will be limited by the spontaneous carrier lifetime leading to long delay times of several tens of nanoseconds; and since the "ON" state is by necessity close to threshold, the available output power is limited. An extensive review, including absorptive and dispersive bistability in semiconductor injection lasers, has been presented by Kawaguchi[20].

On the other hand, optical bistability in the circulating optical radiation in coupled-cavity semiconductor lasers has many interesting features. In contrast to the case of absorptive bistability, both sections of coupled-cavity laser operate above threshold, thus leading to bistability with very short switching times and high output power, and observable at any temperature. The mechanism responsible for the bistable behaviour in this case is attributable to nonlinearities associated with gain saturation above threshold, which are strongly affected by the mutual coupling of the two Fabry-Perot cavities of the lasers. It is also interesting to note that the cleaved (or etched) coupled-cavity (c^3) laser has been studied[23-29] as a stable single-longitudinal mode light source under high-speed direct modulation. However, in that case one section is operated near or below threshold and gain-saturation related bistability does not generally occur.

Nonlinear Optics and Optical Computing
Edited by S. Martellucci and A. N. Chester
Plenum Press, New York, 1990

In our discussion, optical bistability in cleaved-coupled-cavity (c^3) semiconductor heterostructure lasers is taken as a specific example to discuss and analyse. We first treat the rate equations governing the steady and dynamic behaviour of coupled-cavity semiconductor lasers, then theoretical results on the bistable characteristics of the c^3 laser are given, and the experimental observations are described, followed by a summary and conclusions. An appendix is attached which derives the rate equation phenomenologically for the complex optical field amplitude.

II. RATE EQUATIONS

II.1. Rate Equation for Optical Field[30,31]

A coupled-cavity semiconductor laser, shown schematically in Fig. 1, has to Fabry-Perot sections, which are optically coupled through their mutual feedback, but simultaneously and independently driven. The coupling element in this case is simply an air gap of width d (usually a few micrometers).

For obtaining the rate equations governing the static and dynamic characteristics of a c^3 laser, several assumptions are made to simplify the analysis. We first assune single longitudinal mode operation. This is justified by the fact that the side mode suppression ratio, defined as the intensity ratio of the main mode to the most prominent side mode, can reach 20-35 dB under pulsed conditions[23,24]. We also neglect lateral variations of the optical field and the carrier density in the two cavities. In the plane-wave approximation, the optical field in each cavity is

$$E(z,t) = \frac{1}{2} (E_j(t) e^{-i\omega t} + C.C.) \sin(k_j z) \qquad (1)$$

where $E_j(t)$ is the complex field amplitude, ω is the laser frequency, $k_j = \bar{\eta}_j \omega/c$, c is the vacuum velocity of light, and $\bar{\eta}_j$ is the effective refractive index experienced by the laser mode. However, the refractive index $\bar{\eta}_j$ varies with the injectied carriers and so k_j is not constant. As a further simplification, the facet losses are assumed to be distributed in the cavity, and then we can obtain the phenomenological rate equation for an uncoupled laser, plus a term responsible for the intercavity coupling. This approach is simple and physically appealing (see Appendix), and a more rigorous derivation of the rate equation for optical field can be carried out using the analysis given in ref. 32. The complex field amplitude $E_j(t)$ satisfies the rate equation[30]

$$\frac{dE_j}{dt} = \frac{i\eta_j}{\eta_{gj}} (\omega - \Omega_j) E_j + \frac{1}{2} (G_j - \nu_j)(1-iR)E_j + \nu_j \kappa_j E_{3-j} \qquad (2)$$

where j=1,2 for cavity 1 and 2, respectively, $\Omega_j = m_j \pi c/\eta_j L_j$ is the cavity resonance frequency (m_j is an integer), $G_j = \Gamma_j v_{gj} g_j$ is the stimulated emission rate, and $\nu_j = v_{gj} \alpha_j$ is the photon decay rate, Γ_j is the confinement factor, $v_{gj} = c/\eta_{gj}$ is the group velocity, η_{gj} is the group index, g_j is local

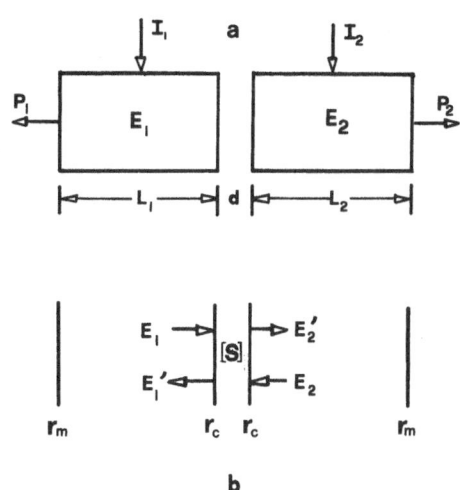

Fig. 1. (a) Schematic illustration of a c³ semiconductor laser.
The coupling is provided by an air gap of width d; (b)
schematic illustration of coupling via the air gap
described by a scattering matrix (S). r_m and r_c are the
amplitude reflection coefficients at the facets indicated.

electronic gain, and $\alpha_j = \alpha_j^m + \alpha_j^{int}$ includes the mirror loss α_j^m and the
internal loss α_j^{int} due to scattering and free carrier absorption. R is
the antiguiding parameter (or denoted as α, the linewidth broadening factor),
$R = -2(\omega/c)(\partial\eta_j/\partial g_j) = \Delta\eta'/\Delta\eta''$, where $\Delta\eta = \Delta\eta' + i\Delta\eta''$ is the refractive index
change due to injected carriers.

The coupling coefficient κ_j in Eq. (2) depends on the characteristics
of the coupling element. In a c³ laser it depends o the air-gap width and
the diffraction losses in the gap. An expression for κ_j can be deduced
phenomenologically by considering the coupling process via the air-gap. The
effects of the coupling gap can be described by a scattering matrix[25], and
so we have (see Fig. 1b)

$$E' = s_{11}E_1 + s_{12}E_2 = s_{11}(E_1 + s_{12}E_2/s_{11}) \tag{3}$$

Now the field entering cavity 1 is $s_{11}(E_1 + s_{12}E_2/s_{11})$, instead of $r_c E_1$
in the absence of coupling (in this case s_{11} reduces to r_c) where r_c is
the reflection coefficient of the facet at the gap. We see that the coupling
process results in an additional term $(s_{12}E_2/s_{11})$, which represents the
contribution from cavity 2. Since this happens at a rate of $1/\Delta t$, where $\Delta t = $
$= 2L_1/v_{g1} = 2\eta_{g1}L_1/c$, the last term in Eq. (2) should be $(cs_{12}/2\eta_{gj}L_j s_{11})E_{3-j}$.
A direct comparison gives

$$\kappa_j = (2\alpha_j L_j)^{-1}(s_{12}/s_{11}) \tag{4}$$

$$s_{12}/s_{11} = T\exp(i\theta). \tag{5}$$

Both the amplitude T and phase θ vary significantly with the coupling
gap width d[25,26]. The phase θ plays an important role in mode discrimina-

tion in a c^3 laser. To discuss optical bistability we choose $\theta = \pi/2$, which is the case for narrow gaps ($d \lesssim \lambda/4$) or when d is an odd multiple of $\lambda/4$ (in the latter case, κ_j is imaginary). Note that the latter case is the least interesting from the standpoint of mode selectivity[24-26]. In a c^3 laser with a few-micrometer-wide air-gap, the coupling parameter $T \lesssim 1$.

II.2. Rate Equation for Carrier Density[30,33]

The gain $G_j = \Gamma_j v_{gj} g_j$ in each cavity is related to the injected carrier density n_j. In the usual linear approximation

$$g_j = a(n_j - n_0) \tag{6}$$

where a is the gain coefficient and n_0 is the carrier density required to achieve transparency. They are assumed to be the same for both cavity sections. This is justified since both sections are usually identical in their structure and material composition. As the two cavities of the laser are electronically isolated, the carrier density satisfies the usual rate equation[33] if diffusion is neglected

$$\frac{dn_j}{dt} = \frac{I_j}{qV_j} - \left(\frac{n_j}{\tau_j} + \frac{G_j |E_j|^2}{\hbar\omega}\right) \tag{7}$$

where I_j is the device current, q is the electronic charge, V_j is the active volume, τ_j is the total carrier lifetime from various radiative and non-radiative (Auger) recombination processes. In the above equation, the first term represents the contribution from injection, the second term describes the consumption of free carriers caused by the recombination, and the last term accounts for stimulated recombination.

Eqs. (2) and (7) form a set of rate equations to model a coupled-cavity laser; however, they are not suitable for the treatment of multimode phenomena because in our case a single longitudinal mode operation has been assumed.

II.3. Dimensionless Form of Rate Equations[13,30]

A c^3 laser is usually formed by cleaving the laser chip approximately in the middle, and the two sections have identical structure. The length difference $|L_1 - L_2| \simeq 10$ to 20 μm, and the laser oscillates at a frequency which is nearly coincident with the cavity mode frequency such that $\Omega_1 \simeq \Omega_2$. In other words, L_j, v_j, V_j, τ_j, Γ_j, Ω_j and κ_j are assumed to be the same for both sections in our further discussion and so the subscripts j on these parameters are ignored in the following discussions.

For obtaining results in a simple form, we follow Agrawal[13,30] to define

$$\bar{t} = vt$$
$$E_j = \sqrt{Ps} \; A_j \; e^{-i\phi_j} \tag{8}$$
$$N_j = n_j/n^{th}$$

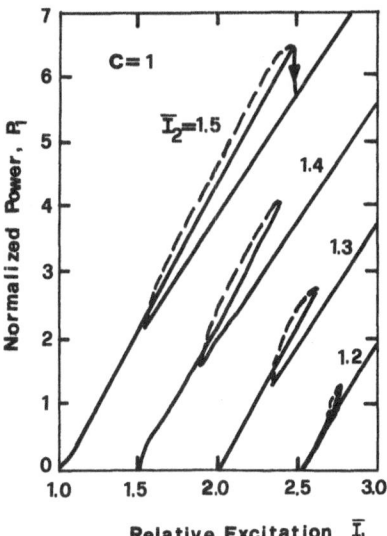

Fig. 2. Calculated output power P_1 vs $\bar{I}_1$ for several values of $\bar{I}_2$.
For clarity, successive curves have been displaced horizon-
tally by an amount of $I_1^{th}/2$. The dashed portion of each
curve indicates an unstable branch which is not accessible
experimentally.

where P_s is the saturation intensity (see below) and n^{th} is the threshold
carrier density in each cavity (assumed to be the same) in the absence of
coupling. By using Eq. (8), the rate equations (2) and (7) can be written
in the following dimensionless form

$$\dot{A}_j = ((N_j - 1)/2\beta) A_j \pm \kappa A_{3-j} \sin(\phi_2 - \phi_1) \tag{9}$$

$$\dot{\phi}_j = (R/2\beta)(N_j - 1) - (\omega - \Omega)/\nu - \kappa \frac{A_{3-j}}{A_j} \cos(\phi_2 - \phi_1) \tag{10}$$

$$\nu\tau\dot{N}_j = \bar{I}_j - (N_j + (N_j - 1 +\beta) A_j^2) \tag{11}$$

where the dot represents the time derivative, and other parameters are
$\bar{I}_j = I_j/I^{th}$, $I^{th} = qVn^{th}/\tau$, $\beta = 1-(n_0/n^{th})$, $P_s = \hbar\omega/(a\tau\Gamma)$.

In the absence of coupling ($\kappa=0$), Eqs. (9) - (11) may be solved for
the case of steady-state to yield

$$N_j = 1, \qquad \omega = \Omega, \qquad A_j^2 = (\bar{I}_j - 1)/\beta \tag{12}$$

where A_j^2 is the emitted power (in normalized units). For $\kappa \neq 0$, it is
convenient to define

$$P_j = A_j^2, \qquad \varepsilon_j = (N_j - 1)/\beta, \qquad \phi = \phi_2 - \phi_1 \tag{13}$$

and then we obtain from Eqs. (9) - (11)

41

$$\dot{P}_j = \varepsilon_1 P_1 + (CP_1 P_2)^{1/2} \sin\phi \tag{14}$$

$$\dot{P}_2 = \varepsilon_2 P_2 - (CP_1 P_2)^{1/2} \sin\phi \tag{15}$$

$$\dot{\phi} = \frac{R}{2}(\varepsilon_2 - \varepsilon_1) - (CP_1 P_2)^{1/2} \frac{(P_1 - P_2)}{2P_1 P_2} \cos\phi \tag{16}$$

$$\nu\tau\dot{\varepsilon}_1 = (\bar{P}_1 - P_1) - (1 + P_1)\varepsilon_1 \tag{17}$$

$$\nu\tau\dot{\varepsilon}_2 = (\bar{P}_2 - P_2) - (1 + P_2)\varepsilon_2 \tag{18}$$

where the coupling parameter $C = 4|\kappa|^2$, and $\bar{P}_j = (\bar{I}_j - 1)/\beta$. By using Eq. (4), C is given by

$$C = (\alpha L)^{-2} |s_{12}/s_{11}|^2 \tag{19}$$

For $\alpha = 50$ cm^{-1}, $L = 200$ μm, $C \cong |s_{12}/s_{11}|^2$, and C varies significantly with the coupling gap width d.

III. BISTABILITY AND HYSTERESIS

In the steady state, elimination of ϕ in Eqs. (14) - (18) results in

$$\varepsilon_1 P_1 + \varepsilon_2 P_2 = 0 \tag{20}$$

$$CP_1 = P_2 \varepsilon_2^2 (1 + R^2 (\frac{P_1 + P_2}{P_1 - P_2})^2) \tag{21}$$

$$\varepsilon_j = (\bar{P}_j - P_j)/(1 + P_j) \tag{22}$$

Equations (20) - (22) can be used to obtain the emitted power P_j and the carrier density N_j in each cavity for a given value of driving current I_1 and I_2.

Let us now consider the operation of a c^3 laser in which both cavities are pumped above threshold. The calculated power P_1 and P_2 are plotted in Fig. 2. and Fig. 3 as a function of I_1 for several values of the driving current I_2 in cavity 2. In the calculation, C is used as a parameter and R=3, $\beta=0.25$ for the 1.3 μm InGaAsP channeled-substrate buried-heterostructure lasers[34]. From Figs. 2 and 3, we can see the bistable behaviour in the light intensity characteristics, in which the dashed portion of each curve indicates an unstable branch that is not accessible experimentally. For clarity, the successive curves in Fig. 2 have shifted horizontally by an amount of $I_1^{th}/2$. In Fig. 3 as I_2 increases, the hysteresis width increases rapidly until eventually the laser remains on its lower branch (or P_1 on its upper branch in Fig. 2) throughout the operating range of I_1. The important feature is that hysteresis width ΔI_1 and the magnitude of intensity change ΔP_2, can be controlled by adjusting I_2 in the cavity 2. We also note C, the coupling parameter given in Eq. (19), influences the bistable behaviour of the c^3 laser.

42

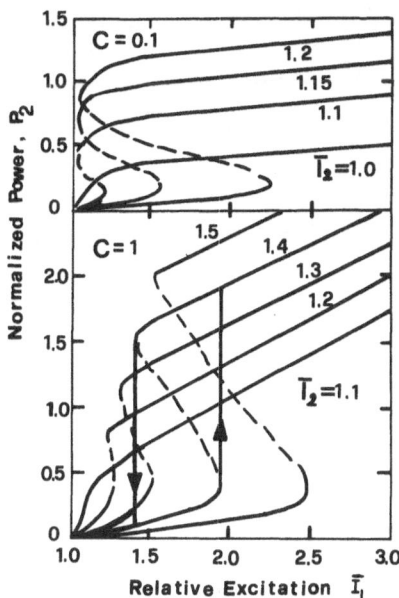

Fig. 3. Calculated output power P_2 vs. $\bar{I}_1$. The dashed portion of each curve indicates an unstable branch and is experimentally inaccessible.

To understand the origin of optical bistability in coupled-cavity lasers, we note that depending on the relative excitation of two sections, one cavity is optically pumped by the other; thus nonlinearity associated with gain saturation occurs. In Figs. 4 and 5 are shown the carrier density corresponding to the cases of Figs. 2 and 3. For instance, we see in Fig. 5 that the carrier density N_2 varies nonlinearly with $\bar{I}_1$. The carrier density N_2 decreases gradually and jumps down to a lower level, resulting from gain saturation in cavity 2 due to the increasing of optical pumping by the cavity 1 when I_1 increases.

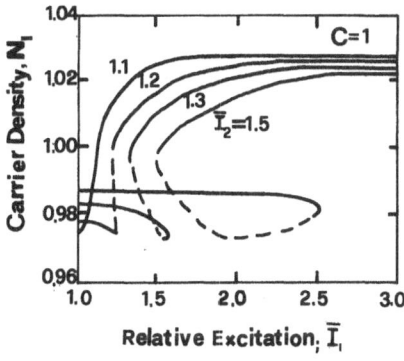

Fig. 4. Normalized carrier density N_1 vs $\bar{I}_1$. Dashed portion indicates unstable region.

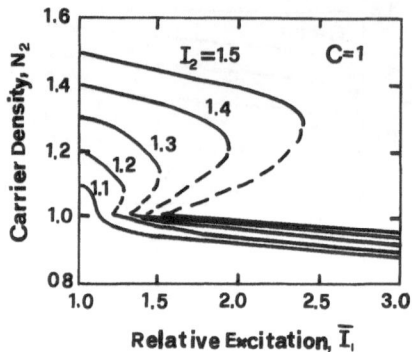

Fig. 5. Normalized carrier density N_2 vs $\bar{I}_1$.

IV. OBSERVATIONS OF OPTICAL BISTABILITY

IV.1. Bistability in L/I Characteristic of c^3 InGaAsP Laser

In our observations of bistability, a c^3 InGaAsP laser is used with the threshold current I_1^{th} = 14.5 mA and I_2^{th} = 18 mA, and is operating in a single transverse mode. The length for two cavities is about the same, L = = 250 μm, and the gap width d is less than 5 μm. In Figs. 6 and 7 the optical output power P_1 and P_2 from two outer facets are shown[13]. Fig. 6 is P_1 vs I_1 curves for I_2 = 0 and for I_2 = 21 mA (above threshold). The curve for I_2 = 0 is similar to the conventional L/I characteristic of lasers. However, the L/I curve exhibits a bysteresis loop for I_2 = 21 mA. Note the hysteresis loop in the P_2 vs I_1 curve in Fig. 7, which shows two stable states of the coupled-cavity laser as I_1 is varied. Similar behaviour is observed with higher values of I_2 except for an increase in the hysteresis width ΔI_1 and the power jump ΔP_2. A comparison of Fig. 6 and Fig. 7 with Figs. 2 and 3 shows that the observed bistable behaviour is in qualitative agreement with the calculated results.

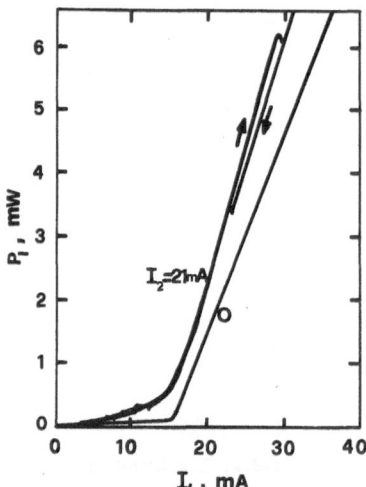

Fig. 6. Observed output power P_1 as a function of I_1 for two fixed values of I_2. Bistability occurs when I_2 is above threshold.

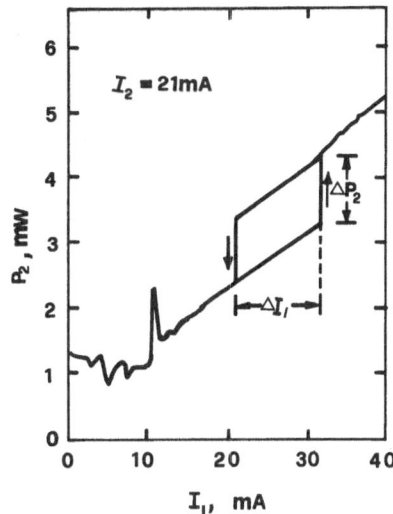

Fig. 7. Observed output power P_2 as a function of I_1 for I_2 = 21 mA, corresponding to the above-threshold operation of cavity 2.

IV.2. <u>Spectral Bistability in c³ InGaAsP Lasers</u>

Optical bistability in c³ InGaAsP lasers has been observed in the spectral domain[14]. The experimental setutp is shown in Fig. 8. The spectral output of this c³ laser is in a pure single longitudinal mode even under high speed modulation. The individual threshold currents of the laser and modulator sections are 20 and 25 mA, respectively. The spectral hysteresis is observed when the laser current is held constant and the modulator current is varied. When the laser current is set at 40 mA, the spectrally resolved output power vs modulator current is as shown in Fig. 9. The hysteresis loop in Fig. 9a is for the output at λ = 1.5122 μm, and 9b for the output at λ = = 1.5102 μm. In fact, the c³ laser is now operating in a tuning mode, and in a certain region of modulator current the c³ laser shows spectral bistability, as shown in Fig. 10. In this bistable region the c³ laser can lase in either of two longitudinal modes but never in both simultaneously. The variation in total output power between two states has been found to be less than 5%.

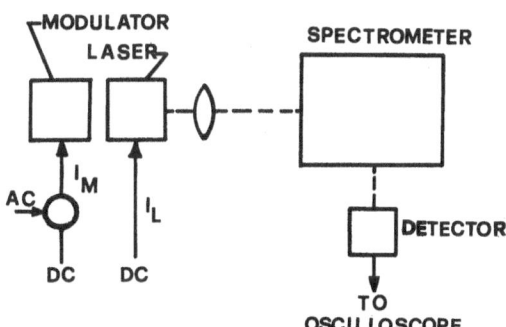

Fig. 8. Experimental arrangement for observing spectral bistability.

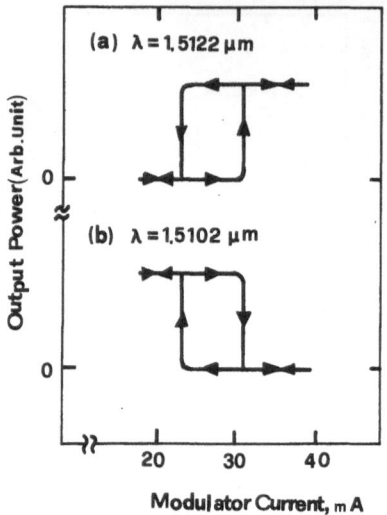

Fig. 9. Spectrally resolved output power: (a) output at $\lambda = 1.5122$ μm;
(b) output at $\lambda = 1.5102$ μm.

Because the bistable c^3 laser operates above threshold for both the
'ON' and 'OFF' states, this device has very short switching time. To measure
the switching speed the modulator section is dc biased inside the hysteresis
loop, then a train of short alternating positive and negative current pulses
is superimposed on the dc bias to make the device switch between two output
states. The switching time is found to be less than 1 ns, very fast compared
with typical switching times of 10 ns in absorptive bistable semiconductor
lasers[5].

The bistable behaviour described in Section IV.1 is different from
that of Section IV.2. In the latter case, bistability corresponds to the
two different longitudinal modes which can be excited for the same current
I_1 and I_2. However, in the case of Section IV.1, only a single longitudi-

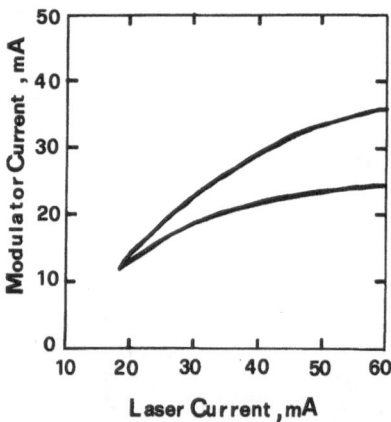

Fig. 10. Extent of the bistable region. Spectral bistability is
observed inside the wedge-shaped region.

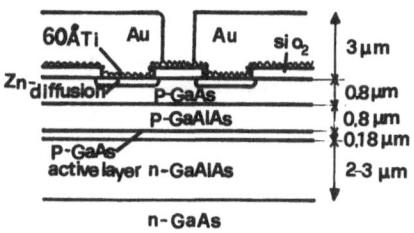

Fig. 11. Schematic of twin-stripe laser.

nal mode is considered and the optical bistability is related to the power jumps experienced by this mode as I_1 varies for a given I_2. Of course, the wavelength of this mode varies somewhat with the current I_1, mainly due to carrier-induced index changes. Further study has been performed concerning the spectral behaviour of bistable semiconductor lasers with an oxide-stripe structure[18].

IV.3. Optical Bistability in Twin-Stripe Semiconductor Lasers

A closely coupled twin-stripe laser is shown in Fig. 11, and the L/I characteristics are shown in Fig. 12 (see Ref. 11). From Fig. 12, at certain values of I_1 the same value of I_2 can correspond to two or more values of light output or vice versa. With both stripes driven together with similar currents, the light output from the twin-stripe laser versus drive current shows hysteresis loops which depend on the measuring points on the far-field distribution, as shown in Fig. 13. In addition, the magnitude of the hysteresis loop changes as the position of the photo-detector is varied across the near field, indicating that a shift occurs in the near-field distribution as the current alters. This bistable behaviour is associated both with a change in the dominant waveguiding mechanisms in the laser and a shift of near-field distribution along the laser junction[11].

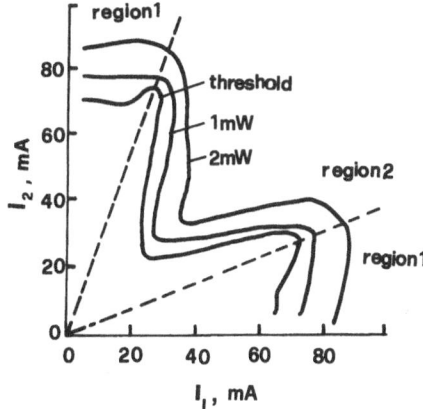

Fig. 12. Series of plots of currents I_1 and I_2 required to set the laser at threshold and with 1 mW and 2 mW light output.

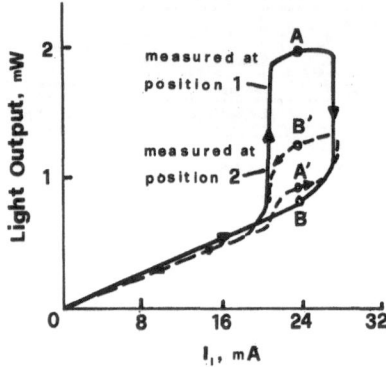

Fig. 13. Light output from the twin-stripe laser as a function of drive
current, showing hysteresis loops measured at two points in
the far-field distribution.

V. SUMMARY

Recent progress in the study of optical bistability in coupled-cavity
semiconductor lasers has been discussed. Experimental observations and theoretical analysis show that these kinds of bistable devices are attractive
because of their outstanding features; the bistable behaviour can be controlled both electronically and optically, and allows very short switching
times as well as high output power due to the above threshold operation.

APPENDIX

In cavity 1, E_1 after the round trip becomes E_1^{rt}

$$E_1^{rt} = r_m r_c e^{((\Gamma_1 g_1 - \alpha_1^{int})L_1 + i2k_1 L_1)} E_1 \tag{A1}$$

where r_m is the mirror reflection coefficient. Using the facet loss $\alpha_1^m =$
$= -(1/L_1)\ln r_m r_c$, Eq. (A1) can be rewritten as

$$E_1^{rt} = E_1 e^{((\Gamma_1 g_1 - \alpha_1)L_1 + i2k_1 L_1)}. \tag{A2}$$

The effective index $\bar{\eta}_1$ can be expanded around its threshold value

$$\bar{\eta}_1 = \eta_1 + (\partial \eta_1/\partial g_1)(g_1 - \alpha_1). \tag{A3}$$

By using the antiguiding parameter $R = -2(\omega/c)(\partial \eta_1/\partial g_1)$ and adding a phase
constant $2m_1\pi$ to the exponential in Eq. (A2), Eq. (A2) can be written as

$$E_1^{rt} = E_1 e^{((\Gamma_1 g_1 - \alpha_1)(1 - iR)L_1 + i2 \eta_1 L_1(\omega - \Omega_1)/c}. \tag{A4}$$

If $\Gamma_1 g_1 - \alpha_1 \ll 1/L_1$, $|\omega - \Omega_1| \ll c/2\eta_1 L_1$, the exponential in Eq. (A4) can be
expanded in a power series. Neglecting the quadratic and high-order terms,
we obtain

$$E_1^{rt} = E_1((\Gamma_1 g_1 - \alpha_1)(1-iR)L_1 + i2\eta_1 L_1(\omega-\Omega_1)/c + 1. \tag{A5}$$

The difference $E_1^{rt}-E_1$ can be considered as the contribution from cavity 1 to the accumulation of the optical field. Since this happens at a rate of $v_{g1}/2L_1 = c/(2\eta_{g1}L_1)$, the accumulation rate of the optical field contributed from cavity 1 should be $(v_{g1}/2L_1)(E_1^{rt}-E_1)$, i.e.

$$(G_1-v_1)(1-iR)E_1/2 + i(\eta_1/\eta_{g1})(\omega-\Omega_1)E_1, \tag{A6}$$

where $G_1 = \Gamma_1 v_{g1} g_1$ is the stimulated emission rate and $v_1 = v_{g1}\alpha_1$ is the photon--decay rate. A similar result can be obtained for E_2 in cavity 2. Taking the intercavity coupling into account, we obtain the following rate equation for the optical field:

$$\frac{dE_j}{dt} = \frac{i\eta_j}{\eta_{gj}}(\omega-\Omega_j)E_j + \frac{1}{2}(G_j-v_j)(1-iR)E_j + v_j \kappa_j E_{3-j}. \tag{A7}$$

REFERENCES

1. M. J. Adames, Int. J. Electron., 60:123 (1986)
2. J. A. Goldstone, Optical Bistability in: "Laser Handbook 4", M. L. Stich and M. Bass, ed., Elsevier Science Publisher BV, Amsterdam (1985)
3. H. M. Gibss, "Optical Bistability: Controlling Light with Light", Academic Press, New York (1985)
4. H. Kawaguchi and G. Iwane, Electron. Lett., 17:167 (1981)
5. Ch. Harder, K. Y. Lau, and A. Yariv, Appl. Phys. Lett., 39:382 (1981); 40:124 (1982)
6. H. Kawaguchi, IEE Proc. I, 129:141 (1982) Appl. Phys. Lett., 41:702 (1982)
7. Ch. Harder, K. Y. Lau, and A: Yariv, IEEE J. Quant. Electron., QE-18:1351 (1982)
8. P. Glas and R. Muller, Opt. Quant. Electron., 14:375 (1982)
9. V. Yu. Bazhenov, A. P. Bogatov, P. G. Eliseev, O. G. Okhotnikov, G. T. Pak, M. P. Rakhvalsky, M. S. Soskin, V. B. Taranenko, and K. A. Khairetdinov, IEEE Proc. I, 129:77 (1982)
10. W. A. Stallard and D. J. Bradley, Appl. Phys. Lett., 42:858 (1983)
11. I. H. White and J. E. Carroll, Electron. Lett., 19:338 (1983)
12. K. Otsuka and H. Iwamura, IEEE J. Quant. Electron., QE-19:1184 (1983)
13. N. K. Dutta, G. P. Agrawal, and M. W. Focht, Appl. Phys. Lett., 44:30 (1984)
14. N. A. Olsson, W. T. Tsang, and R. A. Logan, Appl. Phys. Lett., 44:375 (1984)
15. Y. C. Chen and J. M. Liu, Appl. Phys. Lett., 46:16 (1985)
16. C. M. Wang, L. D. Zhu, and Q. P. Chao, Acta Phys. Sinica, 34:1102 (1985)

17. D. M. Heffernan, J. McInerney, L. Reekie, and D. J. Bradley, IEEE J. Quant. Electron., QE-21:1505 (1985)

18. P. Phelan, L. Reekie, D. J. Bradley, and W. A. Stallard, Opt. Quant. Electron., 18:35 (1986)

19. S. W. Wang, C. M. Wang, and S. M. Lin, Acta Phys. Sinica, 35:1095 (1986)
    Chinese J. Semicond., 7:136 (1986)

20. H. Kawaguchi, Quantum Electron., 19:S1 (1987)

21. P. Glas and R. Muller, Opt. Quant. Electron., 19:S61 (1987)

22. D. Maclean, I. H. White, J. E. Carroll, C. J. Armistead, amd R. G. Plumb, Opt. Quant. Electron., 19:S103 (1987)

23. W. T. Tsang, N. A. Olsson, and R. A. Logan, Appl. Phys. Lett., 42:650 (1983)

24. L. A. Coldren, K. J. Ebeling, B. I. Miller, and J. A. Rentschler, IEEE J. Quant. Electron., QE-19:1057 (1984)

25. L. A. Coldren and T. L. Koch, IEEE J. Quant. Electron., QE-20:659 (1984)

26. C. H. Henry and R. F. Kazarinov, IEEE J. Quant. Electron., QE-20:735 (1984)

27. T. P. Lee, C. A. Burrus, P. L. Liu, W. B. Sessa, and R. A. Logan, IEEE J. Quant. Electron., QE-20:374 (1984)

28. H. K. Choi, K. L. Chen, and S. Wang, IEEE J. Quant. Electron., QE-20:385 (1984)

29. W. Streifer, D. Yevick, T. L. Paoli, and R. D. Burnham, IEEE J. Quant. Electron., QE-20:754 (1984)

30. G. P. Agrawal and N. K. Dutta, J. Appl. Phys., 56:664 (1984)

31. G. P. Agrawal, J. Appl. Phys., 56:3110 (1984)
    IEEE J. Quant. Electron., QE-21:255 (1985)

32. M. B. Spencer and W. E. Lamb, Phys. Rev. A, 5:893 (1972)

33. G. H. B. Thompson, "Physics of Semiconductor Laser Devices", Chap 7, Wiley, Chichester (1980)

34. H. Ishikawa, H. Imai, T. Tanahashi, Y. Nishitami, M. Takusayawa, and K. Takahei, Electron. Lett., 17:415 (1981)

SEMICONDUCTOR BISTABLE ETALONS FOR DIGITAL OPTICAL COMPUTING

A. Miller, I.T. Muirhead, K.L. Lewis, J. Staromlynska
and K.R. Welford

Royal Signals and Radar Establishment
Great Malvern, Worcestershire, WR14 3PS, U.K.

I. INTRODUCTION

A large number of semiconductors have now been shown to exhibit optical
bistability under appropriate conditions. Optical bistability, a process
in which an element illuminated with a given optical power can exist in
one of two possible stable transmission (or reflection) states, requires
two conditions, (a) an optical nonlinearity and (b) some form of feedback[1].
Bistable devices able to operate as short term memory elements and perform
the fundamental logic operations have been demonstrated at low power levels.
For semiconductors, large band gap resonant nonlinearities can be employed
which reduce the optical input power requirements to milliwatts or less.
The feedback may be produced by housing the nonlinear material within a
Fabry-Perot etalon or using the intrinsic feedback associated with the
dependence of optical absorption on a parameter such as temperature. Alternatively, the nonlinearity may be artificially created using electro-optic
effects and electrical feedback.

Semiconductors in the form of thin etalons are particularly useful
because of their compact size and band gap resonant nonlinearities. For
medium and high speed devices, nonlinear refraction of electronic origin
is of primary interest. Small gap semiconductors such as InSb, InAs and
CdHgTe give very large changes in refractive index due to optically generated free carriers and the blocking of interband transitions just above the
gap, while larger gap material such as GaAs, CdS, CuS, etc. may also exhibit
large nonlinearities associated with excitons. However, only InSb has provided a nonlinearity which is large enough to dominate over thermally induced changes in optical constants such that bistable devices can be held
in either upper of lower state for long periods with a c.w. laser.

An alternative approach is to make use of thermo-optic effects. These
effects are also resonant with the band gap energy in semiconductors. Optical interference filters based on II-VI semiconductors[2,3] are examples of
devices which are very attractive as test-beds for all-optical computing

schemes. They can be fabricated as large areas, room temperature, bistable
devices for visible light digital optical circuits[4]. These active filters,
have generally been fabricated by thermal evaporation of multi-layers of
ZnSe or ZnS and some low index material. Evaporated filters are limited in
thickness and tend to suffer from drift of operating point during use. We
have been assessing ZnSe Fabry-Perot filters fabricated by a new optical
coating technology of molecular beam deposition (MBD). In this technique,
entire filters are deposited in an ultra-high vacuum (UHV) system. UHV/MBD
techniques are capable of producing relatively thick, dense films, free of
porosity and with low contamination levels, which should improve the stability of thermo-optic devices[5]. An immediate advantage of the MBD technique for optical bistability is the ability it provides to produce much
thicker, mechanically stable layers. This is significant in terms of opti-
mising the structure for lower switching powers, reducing the required
temperature rise, controlling heat flow and the ability to achieve optical
bistability due to increasing absorption. Using these filters, we have
studied the dependence of critical switching power on wavelength, and
optimum operating conditions for a given filter construction, using a
single laser source, thus enabling a direct comparison with theoretical
predictions[6].

II. BACKGROUND

The first report of passive optical bistability in a semiconductor in
1978 employed ZnS as a spacer layer between dielectric multilayer reflectors. Although originally discussed by Karpushko and Sinitsyn[7] in terms
of an electronic nonlinearity, it has since been well characterized as a
thermal induced dispersive mechanism[2,3]. Most effort has involved ZnSe
since it is an established thin film coating material and provides a band
gap energy which is resonant with the dominant 514 nm green output of the
argon laser. It can be uniformly fabricated in large areas and therefore
offers a very high degree of parallel processing capability with room
temperature, visible light, operational convenience.

Interference filter etalons have the general structure, HLHL...
...(mHH)...LHLH where H indicates a quarter wavelength optical thickness
($\lambda/4$) layer of high index material (e.g. ZnSe, n=2.4) and L is a $\lambda/4$ layer
of low index (e.g. ThF_4, n=1.55). The integer, m, (typically 8 for thermally evaporated layers) gives the etalon spacer thickness. These etalons
can be self-tuned with a laser because of band edge absorption in the ZnSe
causing a rise in lattice temperature; the absorption edge shift to longer
wavelengths at higher temperatures results in an increase in the refractive index and therefore also in the round trip etalon phase change. When
the appropriate cavity conditions are met, optical bistability can occur
with incident powers typically in the milliwatt regime with focussed laser
beams. This power depends on such parameters as the effective optical
nonlinearity of the material, the design of the etalon, and wavelength.
The heatsinking of the device, normally on a glass substrate, will affect
both the switching power and the switching speed. Although the absorption
coefficient, α, varies rapidly with wavelength near the band gap, the
temperature dependence of the refractive index is much less sensitive to

wavelength. It may also be noted that thick polycrystalline samples have shown dispersive optical bistability at 476 nm, whereby a polished etalon of ZnSe provides feedback[8,9]; bistability due to increasing absorption in which the feedback is inherent in the thermally induced absorption edge shift was also achieved in this type of sample[10]. The complete MBD filters fabricated for the studies described below have 6 μm thick ZnSe spacers (m~120) with reflecting stacks constructed from layers of ZnSe and BaF_2. (Bistable filters have also been constructed with thick MBD ZnSe spacers with conventional, thermally evaporated $ZnSe/ThF_4$ mirrors[10]). Ultrahigh vacuum (UHV) film fabrication has produced ZnSe films with high laser damage thresholds[11]. The low index material chosen was BaF_2 (n=1.45); recent work on fluoride deposition has highlighted the potential of barium and lead fluorides as low index materials for use over a wide range of wavelengths. Techniques have been developed for the control of the micro-structure of the films offering the potential for controlling the proper-ties of different materials within novel structures.

Bistable interference filters have been used to demonstrate parallel digital loop circuits[4,12], (see Fig. 1), symbolic substitution[13] and full adder circuits[14]. Useful implementation of systems such as these impose testing requirements on the individual devices in terms of uniformity, stability, switching power, switching speed, signal gain, etc. To this end, bistable device research has already addressed these parameters in some detail. For instance, Wherrett et al[6] have theoretically analysed filter designs for minimun switching power in terms of mirror reflectiv-ities and spacer layer absorption and thickness, making comparison with experimental results on different filters. Optimization of the filter design in terms of heat sinking and element isolation bas also been discussed from an analysis of heat flow under conditions of bistabil-ity[6,10,15]. Switching times have been measured as a function of laser spot size by Mathew et al[16] and analysed in relation to the phenomenon of crit-ical slowing down. Bistable arrays have been operated using either fly's eye lens arrays[13] or holographic lenslets[17] to produce parallel address beams. High power pulses have been used to switch ZnSe bistable etalons through fast electronic nonlinearities[18].

ZnSe filters have now been operated over a wide range of wavelengths. The absorption edge of ZnSe films is generally less abrupt than bulk mate-rial, such that significant absorption is apparent well below the band gap energy. By optimizing mirror reflectivities for different wavelengths and introducing absorbing layers external to the etalon[19], bistable operation may be extended into the infrared.

The absolute stability of a given etalon structure is of some concern in determining whether the above objectives can be met. Although bistabil-ity effects can be readily produced at incident powers of 10-20 mW, this translates into power densities of about 10 kW/cm^2 for the typical spot widths used for etalon assessment. Since the repeated production of a bi-stable loop depends on a controlled excursion of refractive index over a small range, it is essential that the etalon structure is stable and free from effects that produce drift, e.g. desorption of water form pores of varying levels of stress. Earlier work on UHV film fabrication addressed

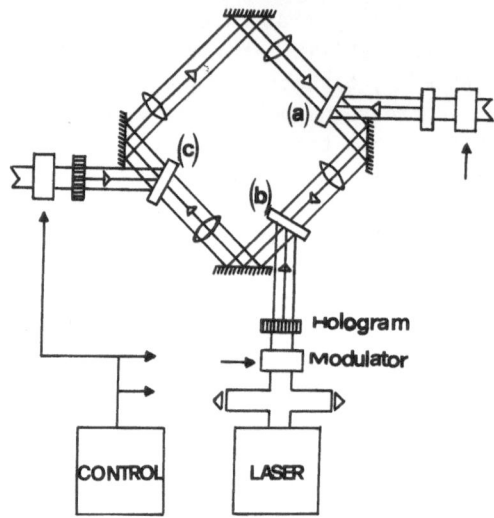

Fig. 1. An optical configuration used to demonstrate a 3 NOR-gate
loop incorporating parallel operation of 3 channels; see
Smith et al[4].

the feasibility of producing films with high resistance to damage induced
by pulsed laser irradiation[11]. Barium and lead fluorides are suitable low
index materials for use over a wide range of wavelengths. Techniques have
been developed for the control of the microstructure of the films offering
the potential of novel structure including graded index designs.

III. ETALONS FABRICATION AND ASSESSMENT

    The MBD layers were deposited in a Vacuum Generators load-locked UHV
system fitted with 3 Knudsen sources and in-situ surface diagnostics. The
source materials were contained within high purity graphite crucibles which
were carefully outgassed following baking of the entire deposition chamber
at 180°C. The ZnSe source was ultra-high purity polycrystalline ZnSe which
had previously been prepared by chemical vapour deposition from a mixture
of zinc vapour and hydrogen selenide. The barium fluoride source was high
purity crystalline optical grade material. Source temperature were adjusted
to give deposition rates of 0.9μm/h for ZnSe (896°C) and 0.7 μm/h for BaF_2
(1185°C) at pressures of 10^{-8}mbar. The glass substrate temperature during
growth for the entire MBD deposited layer was 300°C. Filters were built up
by first depositing a two-period stack comprising $\lambda/4$ thick layers of ZnSe
and BaF_2 centred at 633nm, followed by the ZnSe spacer and finally a matching two period mirror to complete the etalon. The multilayer were exception-
ally smooth with no evidence of surface texture visible by Normanski interference microscopy. A small built-in wedge allowed tuning of the etalon.

    Transmission spectra were recorded at a number of sample tempera-
tures, in order to determine the dispersion in both the refractive index
and the index change with temperature, dn/dT. The room temperature transmission is shown in Fig. 2. The dispersion results are shown in Fig. 3.

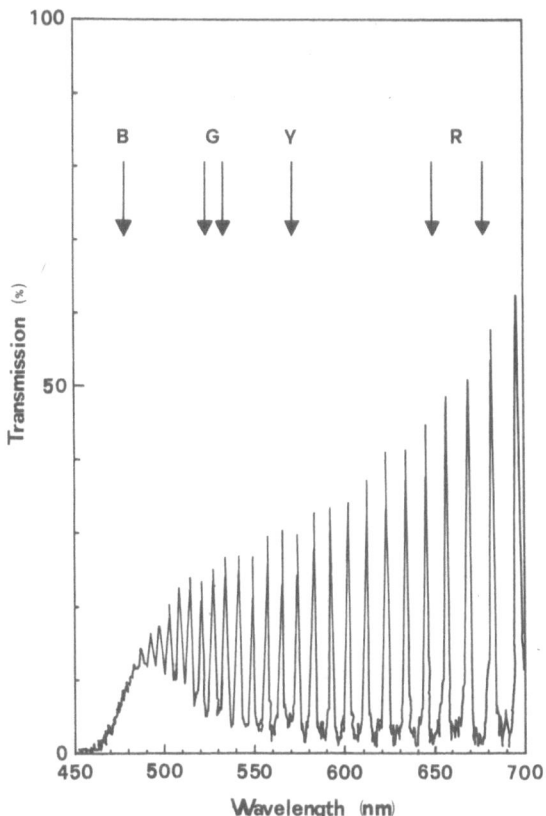

Fig. 2. Transmission of a complete MBD ZnSe etalon (sample AP245)
with mirror reflectivities centred at 633 nm. Arrows show
output wavelengths of a krypton laser. Note that fringes are
not fully spatially resolved.

We note that dn/dT increases by a factor of two between 676 and 521 nm.
Transmission measurements were also carried out at five wavelengths using
low power, focused krypton laser radiation in order to determine precise
maximum and minimum values by fully resolving the fringes[5]. From these
results peak mirror reflectivities of about 70% were deduced assuming
equal values of reflectivity for front (R_F) and back (R_B) mirrors. These
reflectivities may be compared with the theoretically predicted values of
R_F=81% and R_B=74% assuming no absorption in the stacks. Values of αd were
also estimated where α is the absorption coefficient and d is the etalon
spacer layer thickness. The ZnSe absorption coefficient varies from about
100 to 1000 cm^{-1} between 676 and 521 nm. The absorption edge is less
abrupt than for bulk polycrystalline ZnSe, as found for conventional,
thermally evaporated thin films[2].

IV. OPTICAL BISTABILITY

Spacer layer 6 μm thick produce etalon fringes with spacing at less
than 10 nm in visible. This allowed optical bistability to be observed at
five krypton laser wavelengths between 676 and 521 nm employing a spot

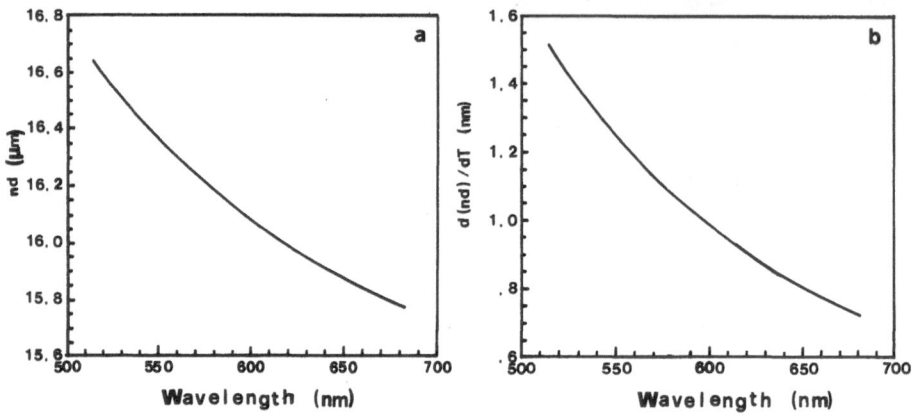

Fig. 3. Measured dispersion in (a) optical thickness, nd, and (b)
its temperature dependence, d(nd)/dT, (sample AP244).

size, $\omega_0 = 12$ μm (1/e^2 intensity radius). Input-output characteristics
measured at 531 nm are shown in Fig. 4a, with critical switching recorded
at 11 mW. Measurements of critical switching power at the five wavelegnths
are plotted in Fig. 4b; the minimum critical power observed was 9.5 mW at
568 nm. The optimum wavelength is a trade-off between the larger nonlinearity at shorter wavelengths due to higher absorption plus slightly larger
dn/dT, against the higher absorption causing a reduction in the finesse of
the etalon.

An expression giving the wavelength dependence of the critical
switching irradiance, I_c, for optothermal devices has been derived by
Wherrett et al[6]. Two regimes may be considered. For large spot sizes, such
that longitudinal heat dissipation into the substrate dominates cooling

$$I_c = \frac{\lambda \ \alpha \ x_s}{2\pi(\delta n/\delta T)\omega_0} \ \frac{f(R_F, R_B, \alpha d)}{\alpha d} \tag{1}$$

This is the product of material parameters, (x_s is the substrate
thermal conductivity) and an expression comprising a complicated function
of etalon parameters (reflectivities, absorption coefficient and thickness). In the limit of small spot sizes and thick films, cooling is
expected to be transverse with heat loss along the layers, in which case[8],

$$I_c = \frac{\lambda \ \ x_f}{(\delta n/\delta T)\omega_0^2} \ f(R_F, R_B, \alpha d) \tag{2}$$

where x_f is the film conductivity. In either case, for a given laser spot
size, the wavelength dependence of the critical switching power is given
by, $P_c \sim \lambda f/(\delta n/\delta T)$. This function is plotted in Fig. 4b for the etalon
parameters extracted from the transmission measurements. The line shown
has been scaled vertically to overlap the 568 nm result. Although the
dispersion in dn/dT is small inclusion results in an excellent fit to the
data.

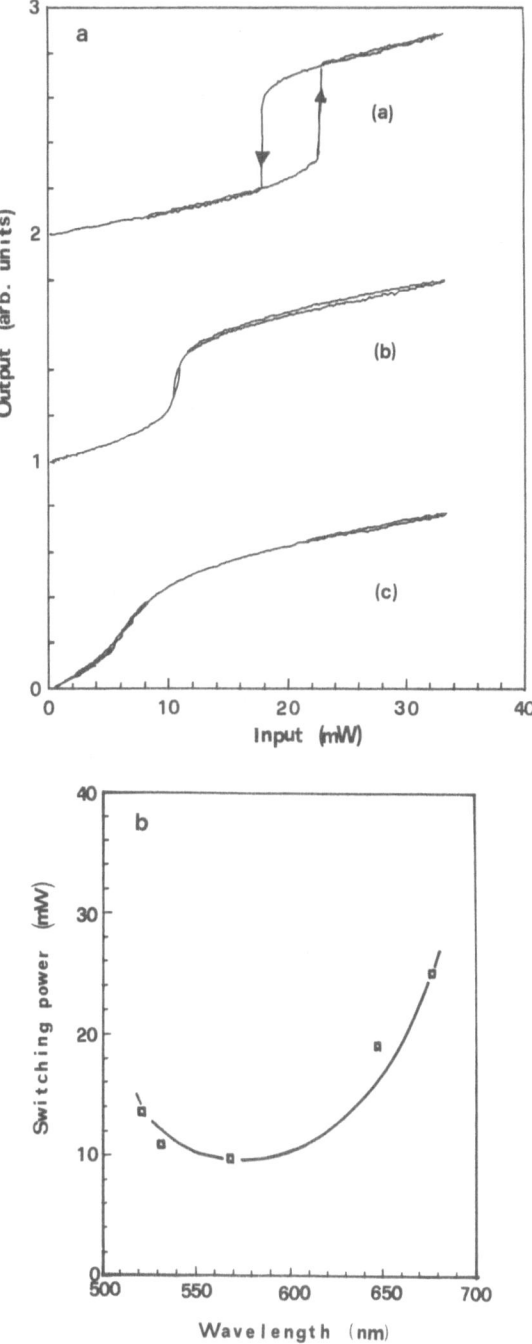

Fig. 4. (a) Input-output characteristics for AP245 at 531 nm;
(b) Measured critical switching powers for optical
bistability compared to theory (line).

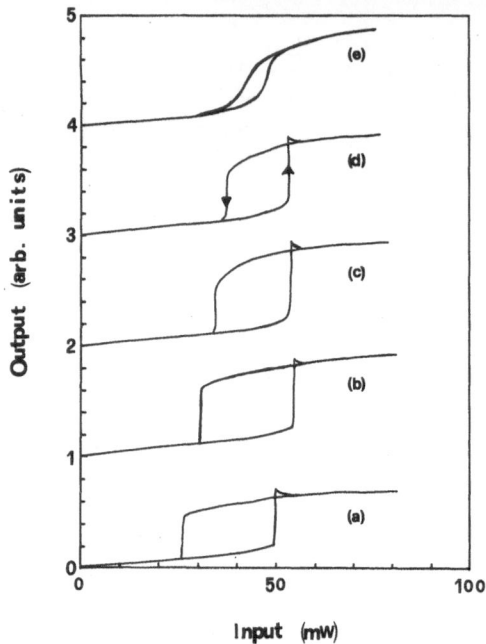

Fig. 5. Optically bistable loops at λ=568 nm for sample AP245 after
after (a) 0, (b) 1, (c) 2, (d) 3 and (e) 7 hours.

The size of the illuminated area and how it is heatsunk will effect
both the switching power and the switching speed[15]. Although the spot
diameter is four times the film thickness in these measurements, the ratio
of the thermal conductivity of ZnSe to that of glass is twenty. For the
measurements at 568 nm, equation 1 predicts a critical switching power of
1.5 mW, while equation 2 gives 110 mW for ZnSe on glass. Thus, significant
transverse cooling is implied by theory under the conditions used here,
and the higher value of critical switching power should apply. The measured
value of 9.5 mW would indicate however that cooling is significant both
along the layers and into the substrate.

Although some drift of the bistable loops was still apparent in these
initial MB structures, essentially stable operation could be maintaned for
many hours under certain conditions[5]. Fig. 5 shows an example of optical
bistability at a wavelength of 568 nm (yellow) in sample AP245; the input
power was continuosly ramped up and down at 0.5 Hz for seven hours with
the maximum incident power on each cycle reaching 80 mW. Single cycle,
input-output characteristics are shown at fixed times during this period.
Only small changes in the bistable loop are observed during the first
three hours and only after seven hours has the loop begun to deteriorate
significantly. At other wavelengths, faster deteriorations of the bistable
loop were recorded.

Optical bistability was observed in the same filter with an argon
laser operating at 514 nm, Fig. 6. Dispersive bistability is shown here at
much higher powers, partly because the filter design is far from optimised
at 514 nm, but also because a much larger spot size ($\sim$100μm) was employed

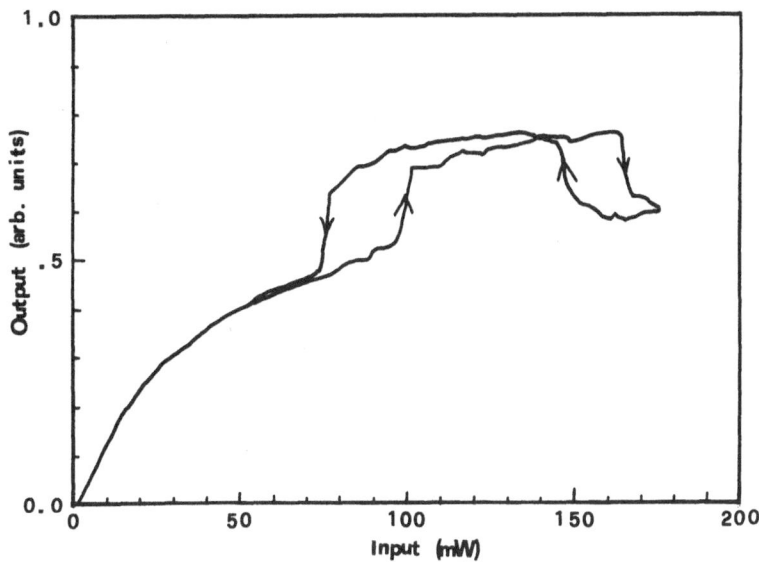

Fig. 6. Dispersive and absorptive bistability in sample AP245 at
λ=514 nm.

in this case. At higher powers a clockwise loop occurs due to absorptive
bistability. This effect, due to the inherent feedback between the absorption edge shift and the incident intensity, has been reported previously
in thick polycrystalline ZnSe samples at 476 nm[8,9], but not in evaporated
layers. The present samples satisfy the requirement, $\alpha d > 0.18$ at 514 nm[9]
for this form of bistability to occur.

V. CONCLUSIONS

We have demonstrated that high quality ZnSe etalons can be fabricated
entirely by ultra high vacuum, molecular beam techniques. These initial
filters are mechanically stable and exhibit optical bistability. The
wavelength dependence of the critical switching power is found to follow
the general theoretical predictions of Wherrett et al[6] as the absorption
close to the band gap energy increases with photon energy. Further improvements in mirror reflectivities and optimization of the design should allow
low power optical bistability with high throughput at the longer wavelengths.

Although some drift of the bistable loop was still apparent at high
incident power levels in these structures, stable conditions could be
maintained under continuous operation for periods of many hours at 568 nm
in these initial studies. This is already a substantial improvement over
conventional, thermally evaporated filters. Developments in this new
materials technology are likely to lead to further improvements in the
performance of opto-thermal devices and allow novel structures to be
fabricated because of the high degree of control a vailable with MB techniques. A large range of different materials can be deposited with a great
deal of flexibility possible in the choice of thin film structures.

We have also identified a permanent laser induced change in the nonlinear absorption and its associated temporal response[5]. These effects are observed under conditions of very high absorption, i. e. quite different to conditions for optical bistability, however this could prove to be a useful means of exploring more fully the general stability of these structures to laser irradiation and as a method of assessing different material multilayers combinations.

REFERENCES

1. H. M. Gibbs, "Optical bistability: controlling light with light", Academic Press, Orlando (1985)
2. S. D. Smith, J. G. H. Mathew, M. R. Taghizadeh, A. C. Walker, B. S. Wherrett and A. Hendry, Opt. Comm., 51:357 (1984)
3. G. R. Olbright, N. Peyghambarian, H. M. Gibbs, H. A. Macleod and F. Van Milligan, Appl. Phys. Lett., 45:1031 (1984)
4. S. D. Smith, A. C. Walker, F. A. P. Tooley and B. S. Wherrett, Nature, 325:27 (1987)
5. A. Miller, J. Staromlynska, I. T. Muirhead and K. L. Lewis, J. Mod. Opt., 35:529 (1988)
6. B. S. Wherrett, D. Hutchings and D. Russell, J. Opt. Soc. Am. B, 3:351 (1986)
7. F. V. Karpushko and G. V. Sinitsyn, J. Appl. Spectrosc., 29:1323 (1978)
8. M. R. Taghizadeh, I. Janossey and S. D. Smith, Appl. Phys. Lett., 46:331 (1985)
9. A. K. Kar and B. S. Wherrett, J. Opt. Soc. Am. B, 3:345 (1986)
10. Y. T. Chow, B. S. Wherrett, E. Van Stryland, B. T. McGuckin, D. Hutchings, J. G. H. Mathew, A. Miller and K. L. Lewis, J. Opt. Soc. Am. B, 3:1535 (1986)
11. K. L. Lewis, J. A. Savage, A. G. Cullis, N. G. Chew, L. Charlwood, and D. W. Craig, Assessment of optical coatings prepared by molecular beam techniques, in: National Bureau of Standards Special Publication (NBS, Washington, D.C.) 727:162 (1986)
12. B. S. Wherrett, Appl. Opt., 24:2876 (1985)
13. R. Jin, L. Wang, R. W. Sprague, H. M. Gibbs, G. C. Gilioli, H. Kulcke, H. A. Maclead, N. Peyghambarian, G. R. Olbright and M. Warren, Simultaneous optical bistable switching of adjacent pixels on ZnS and ZnSe interference filters, in: "Optical Bistability III" ed., H. M. Gibbs, P. Mandel, N. Peyghambarian, S. D. Smith, Springer-Verlag, Berlin, p. 61 (1986)
14. F. A. P. Tooley, N. C. Craft, S. D. Smith and B. S. Wherrett, Opt. Comm., 63:365 (1987)
15. E. Abraham and I. J. M. Ogilvy, Appl. Phys. B, 42:31 (1987)
16. J. G. H. Mathew, M. R. Taghizadeh, E. Abraham, I. Janossy, and S. D. Smith, Observation and analysis of critical slowing down in nonlinear visible interference filters, in: "Optical Bistability III", ed. H. M. Gibbs, P. Mandel, N. Peyghambarian, and S. D. Smith, Springer-Verlag, Berlin p. 57 (1986)
17. M. R. Taghizadeh, I. R. Redmond, A. C. Walker and S. D. Smith, Design and construction of holographic otpical elements for

photonic switching applications, <u>in</u>: "Photonic switching", T. K. Gufstafson and P. W. Smiths, eds., <u>Springer- Verlag</u>, Berlin, p. 111 (1987)

18. J. Y. Bigot, A. Daunois, R. Leonelli, M. Sence, J. G. H. Mathew, A. C. Walker and S. D. Smith, <u>Appl. Phys. Lett.</u>, 49:844 (1986)

19. A. C. Walker, <u>Optics. Commun.</u>, 59:145 (1986)

THREE PHOTON IONIZATION OF Na ATOMS

AND RELATED PLASMA PHENOMENA

F. Giammanco

Dipartimento di Fisica - Università di Pisa
Piazza Torricelli, 2, 56100 Pisa - Italy

I. INTRODUCTION

Laser induced ionization has become a popular technique for investigating the properties of laser-matter interactions (multiphoton ionization of
atoms and molecules, above threshold ionization, properties of Rydberg
levels, autoionizing levels and so on) as well for many applications (isotope separation, photodeposition, trace analysis, laser annealing and so
on). All such experiments exhibit common features, especially with regard to
the charge collection scheme. Specifically, ionized particles are collected
by externally applied electric and/or magnetic fields, and the induced
current is either detected by the optogalvanic technique, or mass- or
energy-analyzed by appropriate spectrometers.

In general, the response time of the collection apparatus is much
longer than the laser interaction time. Hence, in principle, during the
collection time, the initially laser-produced yields could be modified by
short-range interactions involving two or three body collisions, and by
collective long-range phenomena depending on the internal electric field
generated by the relative motion of both electrons and ions. Short-range
collisions affect the number density of the selectively ionized species, but
their influence can be reduced by decreasing the density owing to a quadratic
dependence. However, long-range interactions scale linearly as a function
of the charge density[1]; lowering the charge density, the density threshold
of collective behavior in many cases is limited by the signal-to-noise ratio
of the experimental set-up, and in other cases causes a disappearance of the
desired effects. Moreover, increasing the collection time, depending on the
screening of the external field induced by the self-generated electric
field, enhances the influence of binary collisions. In particular, in laser
isotope separation experiments, the latter effect constitutes a serious limit
in the collection of the selectively ionized isotope. In fact, in this kind
of experiment a fast extraction time is strickly required in order to avoid
charge-exchange collisions.

Nonlinear Optics and Optical Computing
Edited by S. Martellucci and A. N. Chester
Plenum Press, New York, 1990

Although many authors concerned with various studies or applications[2-8] have introduced so-called "space-charge" effects as a possible explanation of the observed anomalous dependences of electron and/or ion yields as a function of laser power, nevertheless up to now no experimental work has been especially devoted to the analysis of the above-mentioned phenomena and their range of influence.

The experiment described here was performed in order to increase the knowledge of those phenomena and to compare the observed time behavior of charged particles with the theoretical predictions of ref. 1. The Na ionization spectrum exhibits in the 5390-6000 Å region (corresponding to the tunability region of the Coumarin 153 dye solution)) several ionization pathways that can be classified as atomic or molecular depending on the resonance width. Thus it is possible to perform a detailed analysis of the charge behavior produced at different resonances as a function of the experimental parameters, that is, laser power, initial beam density and focusing conditions. Moreover, the region around the D lines has been extensively investigated[9-13], and this provides an important reference for the obtained results.

As pointed out in ref. 1, the behavior of the electron peak current as well the arrival time of the peak current as a function of the external electric field is a suitable measure of the collective effects. If the electrons behave as a single particle the peak current intensity and the arrival time exhibit respectively an $E^{1/2}$ and $E^{-1/2}$ dependence on the external electric field. In fact, in this case the density spatial profile simply translates; thus the peak current and arrival time depend only on the single particle velocity at the collection point.

Electron behavior has been analyzed as a function of the electric field with differing conditions of laser power, that is, different initial charge density and focusing conditions. As will be seen below, the focusing conditions introduce a further complication in analysing the data. In fact, the ion and electron signals result from the contribution of zones at different charge density, especially if a long focus lens is used. To separately analyze the effects of volume measurements have been performed with two different focusing aperture angles (10^{-1} and 10^{-2} rad).

II. EXPERIMENTAL SET-UP

A beam of Na vapor is ionized by an excimer pumped dye laser tunable in the 5390-6000 Å region (Coumarin 153), delivering at the peak of the gain profile a maximum unfocused power of 100 kW at a repetition rate of 10 Hz. The power spectrum of the dye laser is quite flat up to 5850 Å. In the region around the D lines the available power is a factor five less than at the peak.

To improve the transverse uniformity of the beam as well as to reduce the dimer density, the Na vapor flows through several cylindrical channels 3 mm long and 1 mm in diameter. The head of the oven is overheated to prevent the channels from clogging, thereby improving the beam density stability.

In order to obtain an accurate value for the beam density, two different techniques have been employed: the absorption of Na D line emitted by a
spectral lamp[14], and measurement of the apparent lifetime of the D lines
when laser excited[15]. Both measurements depend on the geometrical parameters
of the beam, but in differerent ways. In the first method the beam diver-
gence affects the width of the absorption line. Since the D line emitted by
the lamp is wider ($\cong 5$ GHz) than the absorption line ($\cong 1$ GHz), the value of
the density must be obtained by numerical integration, where the ratio
between emitted and absorbed line widths represents a critical parameter[14].
Moreover, a further uncertainty arises from the non-uniform spatial density
profile, which makes it difficult to derive the optical depth. The same
problem affects the measurement of the apparent lifetime, whose value
depends on absorption and re-emission of the vapor around the excitation
zone[15].

In any case, by combining the results of both methods it has been
possible to obtain the beam density to a precision of 10%. Furthermore, it
must be noted that the absorption technique, once calibrated, constitutes a
useful tool for continuously monitoring the beam density during the experiment, of course providing that the incoming light remains sufficiently weak
as to produce negligible population of the resonant levels.

The ions produced are collected by a cylindrical einzel lens at a fixed
potential (-20 V). The ion yields are focused in the field free region of
the lens by the radial field configuration at the entrance of the spectrometer and, after expansion in a 30 cm long time-of-flight drift tube, detected by a multichannel plate (MCP), whose active area is about 5 cm^2. A
plate at a fixed positive potential (+64 V) is placed in front of the
cylindrical collector to improve ion collection efficiency as well as to
measure the electron current. The diameter of the cylindrical lens and the
plate distance are chosen in such a way to reduce edge effects and hence to
obtain an uniform electric field along the direction of detection, whose
value is 14 V/cm. A test performed by moving the focal spot along the laser
axis did not show any significant change in the time of flight of ions along
the diameter of the cylinder. Therefore the ion spectrometer is able to
accept ions produced in a double conical volume around the focusing spot
whose length is approximately the diameter of the cylinder (1 cm).

The features of the ion spectrometer have been checked by irradiating a
metallic Na target at a very low laser intensity level to avoid plasma
effects. The measured time of fligth (TOF) for atomic ions is 32±2 μs and
the Na_2^+ TOF is in the right ratio of $2^{1/2}$. Both TOF's are in very good
agreement with the calculated values. It has also been verified that saturation effects inducing broadening and distortion of ion signal, due to mutual
ion repulsion in the drift tube as well as to the saturation of microchan-
nels in the ion detector, occur at a level of collected charge two orders of
magnitude greater than the maximum obtainable level in the experiment. As a
consequence, any change in the measured dependences can be ascribed to a
real physical effect.

The ion signal from the MCP is amplified by a 1 GHz bandpass circuit
and analyzed by boxcar triggered by the laser pulse. The electron current

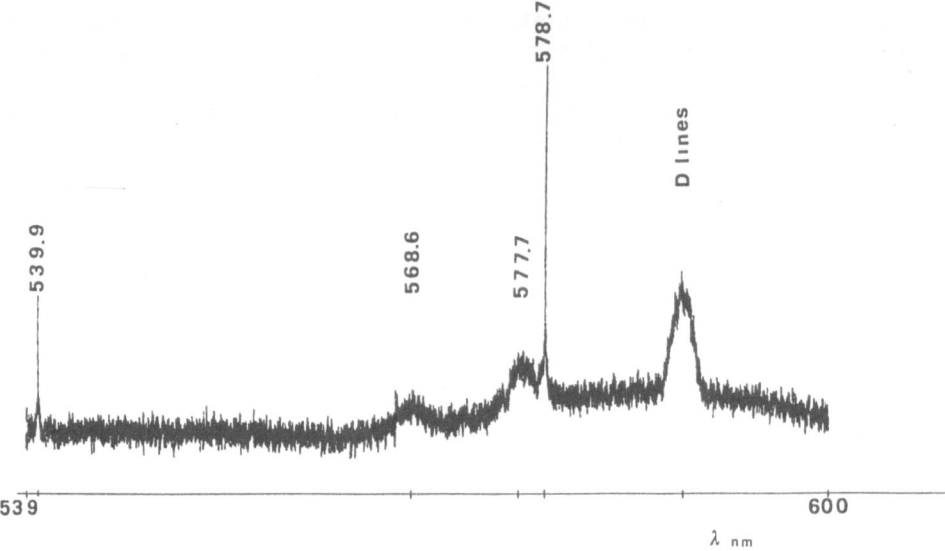

Fig. 1. Spectrum of the electron current peak as a function of laser
wavelength. Beam density 4.4×10^{11} atoms/cm^3, laser power
10^9 W/cm^2.

is collected by an AC optogalvanic technique on a 1 kΩ resistor between
plate and power supply, amplified by a 50 MHz bandpass circuit and analyzed by a boxcar. The bandpass of the electron current circuit is kept
lower than the corresponding ion circuit because the plate is more sensitive to the R.F. noise produced by the excimer laser discharge. Anyway, we
have verified that proportionality between the incoming and detected
signal is maintained in spite of slight rounding of the pulse shape.

   Finally, it must be noted that the above-mentioned values of the
potentials of both plate and cylindrical lens correspond to the best efficiency for ion detection. In particular the potential of the cylindrical
lens is critical within a few volts, since few lenses are employed in the
drift tube. Thus a small change in the first potential strongly affects
the ion trajectories inducing deflection out of the active MCP area. The
ion and electron current dependences as a function of laser power and
initial density have been obtained at the above-mentioned potentials; the
electron dependence on the applied electric field, independent of the ion
signal, has been checked only at the value corresponding to the most efficient collection of the total produced charges.

III. EXPERIMENTAL RESULTS

   Fig. 1 shows the resonances observed. The spectrum corresponds to the
peak electron current at maximum beam density and laser power to display
the overall resonances. The atomic ion spectrum is exactly the same. The
observed resonances may be classified as atomic or molecular according to
their resonant bandwidth. The observed atomic resonances occur at the following wavelengths: 5399 Å: 3S--5D (two photons)-continuum; 5787 Å: 3S--4D

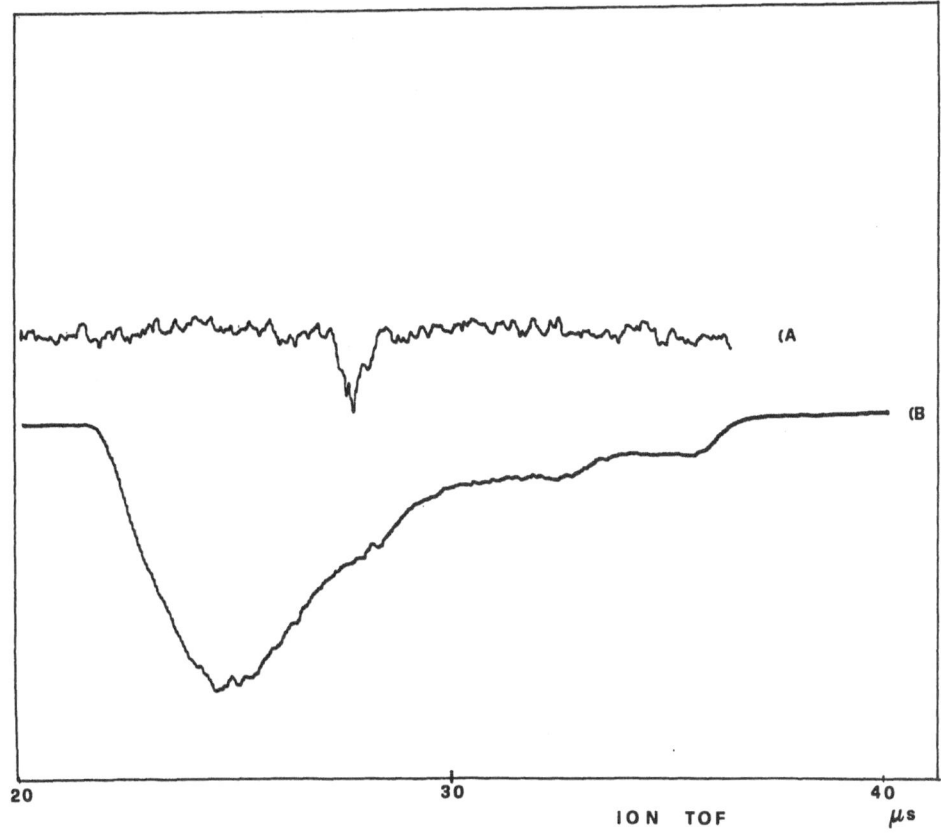

Fig. 2. Ion signal vs time after of flight spectrometer. Beam density
4.4 10^{11} atoms/cm^3. (a): laser power 10^8 W/cm^2; (b): laser
power 10^9 W/cm^2. λ =5787 Å.

(two photons)-continuum; 5890/96 Å: 3S--3P (saturated)-continuum (two
photons). The expected laser power dependences are I^3 for the 5399 Å and
5787 Å resonances, followed by a linear dependence after saturation of the
two-photon step, if achievable. On the other hand, multiphoton ionization
through the D-line pathway must exhibit an I^2 dependence owing to the strong
saturation of the resonant first step.

The wider resonances occur at the peak wavelengths 5777 Å, 5686 Å and
the well known band about the D lines. These resonances exhibit a struc-
tureless bell shape whose width is about 25 Å. In Fig. 1 the D line are
totally masked by the band, as already observed on increasing the laser
power. Owing to the third power law of the wider resonances, the atomic
resonances through the D-lines will again dominate if the laser power decreases[10-12].

Although a detailed analysis of the origin of the wider resonances is
beyond the scope of this work, it must be noted that all those resonances
exhibit common features, nearly the width of the tuning range around the
peak and the absence of any vibrational structure. Moreover, the 5777 Å
and 5686 Å pathways do not match any particular dimer band. On the contrary

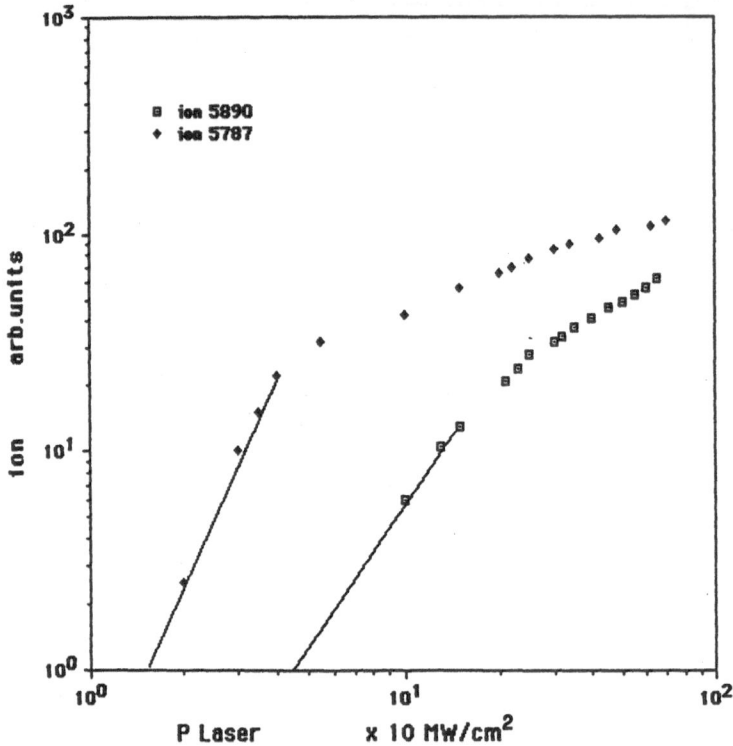

Fig. 3. Peak ion signal vs laser power at the resonances λ=5890 Å and λ=5787 Å. Beam density 1.6×10^{11} atoms/cm^3.

the 5686 Å wavelength matches the atomic transition 3P-4D. Thus other explanations are possible, such as a four photon atomic process involving the intermediate resonant transition 3P-4D, or a molecular dissociation in the 3P atomic state followed by two photon resonant absorption.

With regard to the type of ions produced, only the 5787 Å resonance and the band around the D lines (including the atomic pathways) are able to produce detectable Na_2^+ ions as the laser power and beam density increase (see below).

The behavior of the ion yields and of the electron current have been investigated as a function of laser power at the above-mentioned resonances and at three different beam densities, i.e., 1.6×10^{11} at/cm^3, 2.7×10^{11} at/cm^3 and 4.4×10^{11} at/cm^3. A common feature is observed immediately. Every slope changes more or less abruptly at a fixed value of the ion signal and of the electron current, independently of the laser power, beam density or the selected excitation pathway. Below that threshold every resonance follows the expected behavior and the ion signal is a symmetric peak positioned in time at 27 μs (Fig. 2a). As the charges produced increase up to the threshold, a new peak grows and eventually dominates the previous one (Fig. 2b). The position in time of the new peak corresponds to the TOF of the spectrometer as deduced from the calibration, i.e., 23 μs.

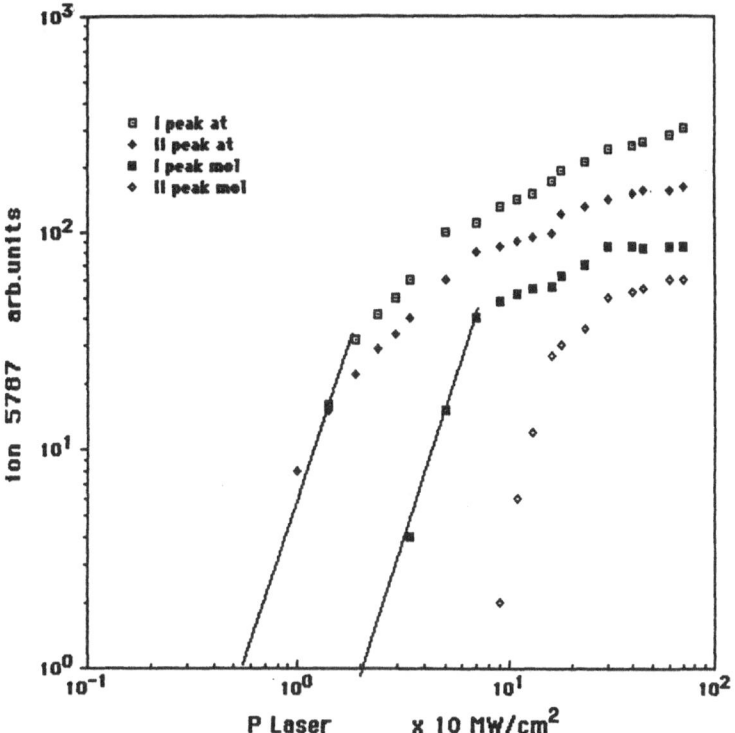

Fig. 4. Peak ion signal vs laser power λ=5787 Å. Beam density
2.7x10^{11} atoms/cm^3.

Figs. 3,4 and 5 show the behavior of the 5787 Å resonance at the selected initial beam densities, as a function of laser power. The labels indicate respectively the 23 μs and 27 μs peaks (I at. and II at.) and the molecular peaks, where detectable, whose arrival time is in the correct $2^{1/2}$ ratio with the corresponding atomic peak (I mol. and II mol.)

Figs. 6 and 7 show the corresponding behavior of the D line and its molecular band (labelled 5893 Å). The different slopes below charge threshold are evident, as well as the complete overlapping as the laser power increases (Fig. 6). At the maximum beam density both resonances exhibit the same slope below the charge threshold, and a molecular peak corresponding to the 23 μs atomic peak is observed (Fig. 7).

The main features as deduced from analysis of the ion yields behaviour are confirmed by the slopes of the electron current as a function of the laser power. Moreover, the time shape of the current changes as laser power increases up to the charge threshold; however, the time resolution, which depends on the distance between the plate and the interaction zone (4 cm), is not so high as to distinguish contributions from different zones to the ion signal. For a rough estimate, the ion density corresponding to the observed change of slope is about 10^{10} ions/cm^3 and the electron density about one order of magnitude less.

The unexpected time behavior of the ions, as well as the existence of

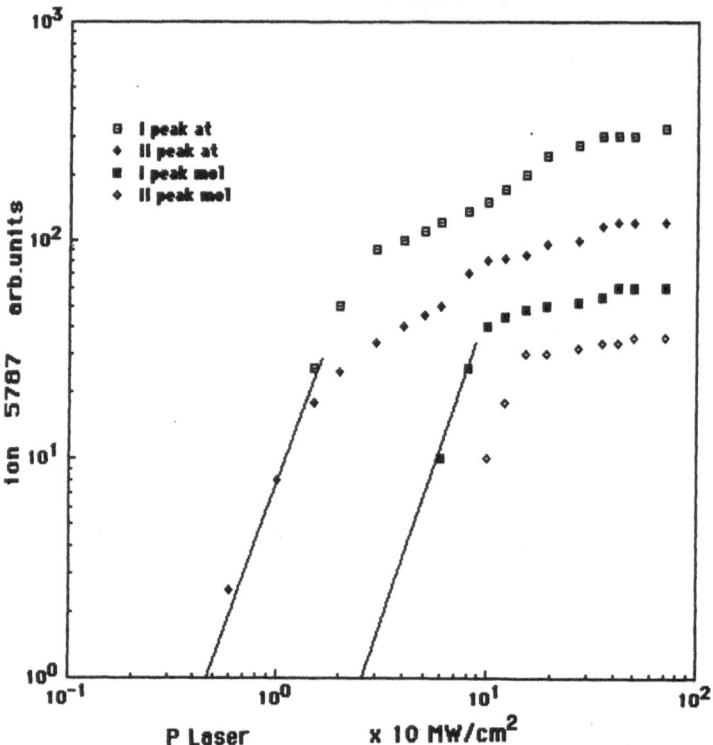

Fig. 5. Peak ion signal vs laser power λ=5787 Å. Beam density
4.4 x 10^{11} atoms/cm^3.

a charge threshold independent of the selected resonance, laser power and
beam density, suggests an explanation based on collective electron- ion
interactions[1]. As pointed out in the Introduction, the self-generated
electric field is a function of charge density and not of the total number
of changes. Therefore, the 23 µs peak could be ascribed to the unsaturated
region surrounding the focal spot which is sensed by the ion spectrometer;
the volume of this region is about 100 times greater than the volume of
the focal spot. Thus a very low charge density level might be sufficient
to obtain comparable intensities of the two peaks. Referring to Fig. 2b,
although the overlapping of peaks does not allow more than an estimate,
the density in the outer zones is about a factor 10 less than in the focal
focusing spot. The width of the first ion peak depends on the non-uniform
ion density in the focusing cone inducing a continuous variation of the
collection time between 23 µs (no screening effects) and 27 µs (maximum
screening effect). Of course, the decrease of the signal depends on the
decreasing contribution from the outer zones up to the focal spot[16].

In order to confirm the foregoing interpretation, the laser beam was
expanded and focused using a 10 cm converging lens. Fig. 8 shows the
behavior of the ion signal at the 5787 Å resonance, which is the resonance
much more affected by volume and collective effects.

The laser power density in the interaction zone is reduced compared with the previous measurements by an aperture at the entrance window. Only one peak is observed in this case, and its position in time varies from 23 μs to 26 μs as the laser power, and therefore the generated charge density, increases. The third power law is well verified up to the charge threshold, which occurs at a value comparable to the earlier measurements. The width of the signal shows a marked asymmetry towards more delayed arrival times as the laser power increases.

If the results of Fig. 8 are normalized taking into account the different resonance width, the ion behavior follows the third power law in the range of laser power investigated. In fact, plasma effects spread the initial charge profile but do not change the total number. If a similar normalization is applied to the results of Fig. 5, although this can only be approximate owing to the overlapping of the different signals, we still find a third power dependence above the charge threshold, but only if the ion molecular signal is negligible. As a consequence, we can surmise that the observed molecular ion signal depends more on collisions involving atomic ions (charge exchange between atomic ions and neutral molecules or direct molecular ion formation induced by collisions between atomic ions and neutral atoms) than on direct ionization of neutral molecules present in the beam. In fact, the ratio between atomic and molecular ions greatly exceeds the thermal equilibrium value in a beam ($\cong 10\%$ instead of $\cong 0.1$-1%). A simple estimate of the collision frequency suggests a negligible contribution

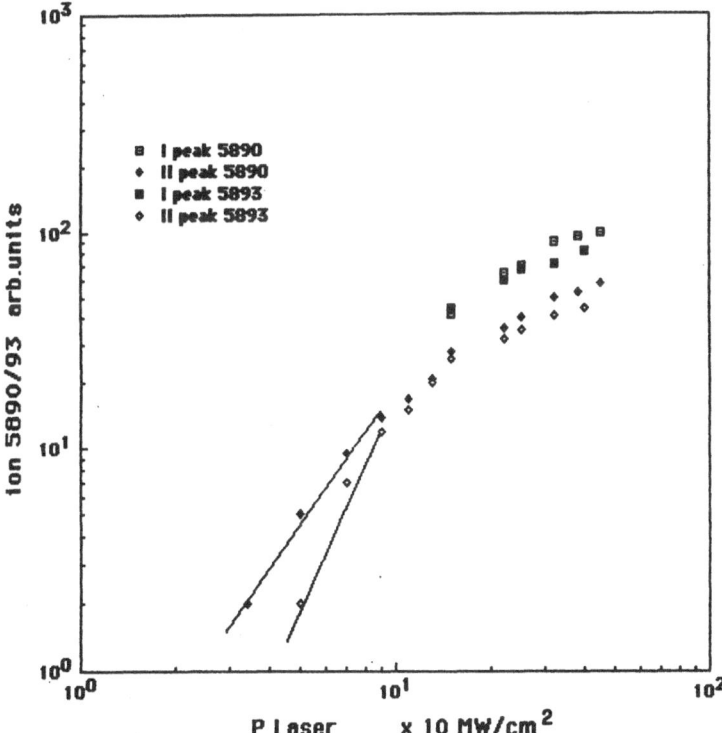

Fig. 6. Peak ion signal vs laser power at λ=5890 Å and λ=5983 Å (molecular band around D lines). Beam intensity 2.7×10^{11} atoms/cm^3.

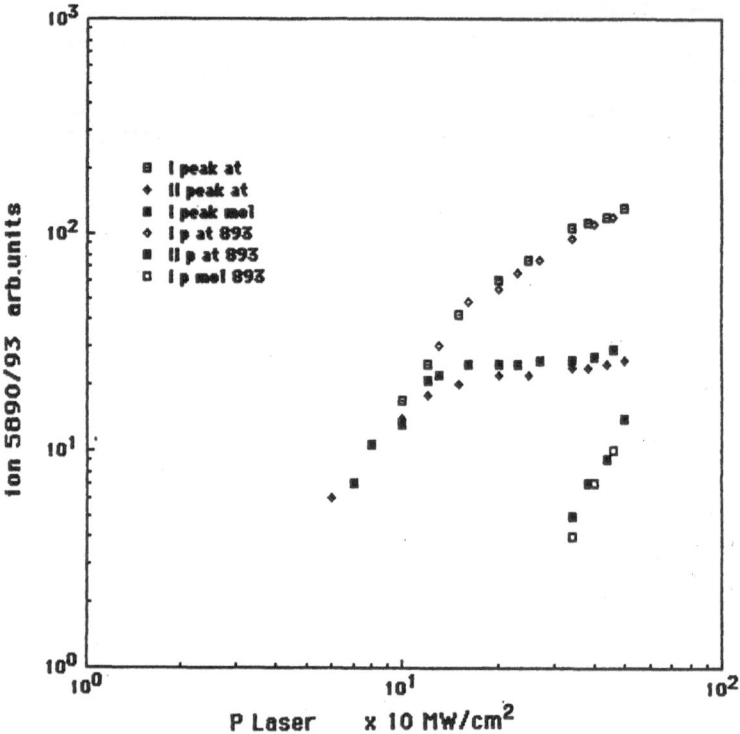

Fig. 7. Peak ion signal vs laser power at λ=5890 Å and λ=5893 Å.
Beam density 4.4×10^{11} atoms/cm^3.

during the collection time. Therefore the contribution of molecular ions may be considered as a further effect induced by plasma interactions.

As pointed out in the Introduction, the electron peak current and the arrival time of the current peak as a function of the external electric field give important information about the influence of collective effects.

In order to avoid complications induced by volume effects, i.e., con-tributions from zones of differing charge density, the analysis has been performed principally for a short focal length lens.

Fig. 9 shows the behavior of the current peak as a function of the external electric field at two different laser energies (1 mJ and 0.1 mJ), corresponding to $\cong 10^{10}$ and $\cong 10^8$ ions/cm^3 (Fig. 8) respectively. At high laser intensity, two peaks are observed. The second peak (labelled II peak 1 mJ) follows the 0.5 slope but exhibits an arrival time of some μs, whereas under the conditions of the experiment the arrival time for a single electron should be some ns. The arrival time of the second peak follows the single-particle -0.5 slope (Fig. 10). In contrast, the arrival time of the first peak is in quite good agreement with the expected value, although the integration of the electronic circuit does not permit an exact measurement, but its dependence on the electric field strongly deviates from the 0.5 slope for values of applied field less than 30 V/cm. At 0.1 mJ only one fast peak is observed (labelled I peak 0.1 mJ), whose dependence

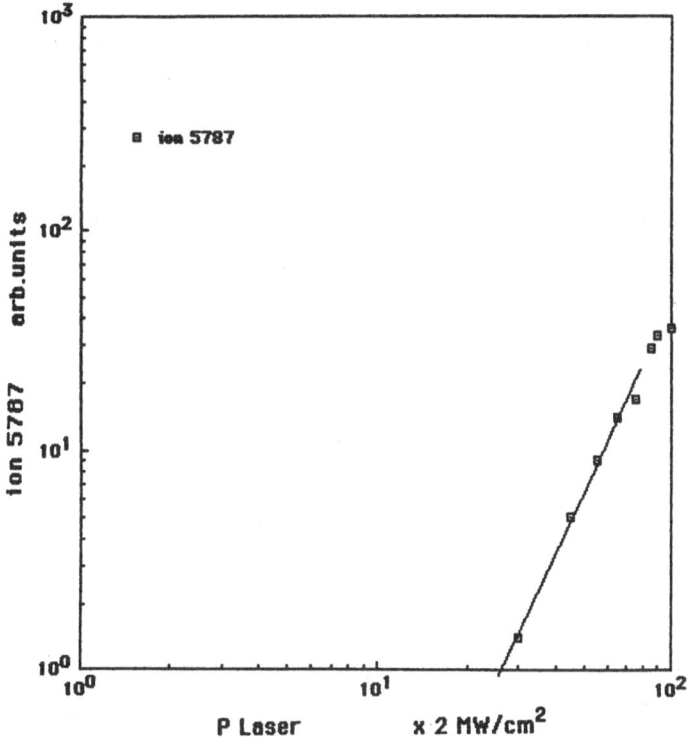

Fig. 8. Peak ion signal vs laser power. λ=5787 Å. Short focus lens. Beam density 4.4×10^{11} atoms/cm^3.

on the electric field is about the same as the dependence of the first fast peak at 1 mJ (Fig. 9). The ratio between fast peaks is very far ($\cong 5$) from the ion signal. The correct ratio of produced charges is recovered by taking into account the total number of collected charges, that is, by multipling both fast and delayed peaks by the corresponding widths.

IV. THEORETICAL APPROACH

A very simple explanation of both ion and electron anomalous behavior can be carried out following the theoretical approach of ref. 1, where the time-space evolution of both ions and electrons is analyzed from the point of view of the fluid equations. Although a detailed comparison between theory and experiment is still in progress, some important features may be pointed out.

For the sake of simplicity, a preliminary solution is obtained by assuming a one-dimensional expansion of a Gaussian profile, $n = n_0 f(t)^{1/2}$ · $\exp(-f(t)x^2)$, where $f(t)$ is an unknown function of time and x the spatial coordinate normalized to the initial half width. We initially use the single-fluid approximation, and neglect both internal and external electric fields. The fluid velocity, that is, the mean velocity of fluid referred to a fixed frame, can be related to the function $f(t)$ by using the continuity equation. In fact, dividing all terms by n, the continuity

73

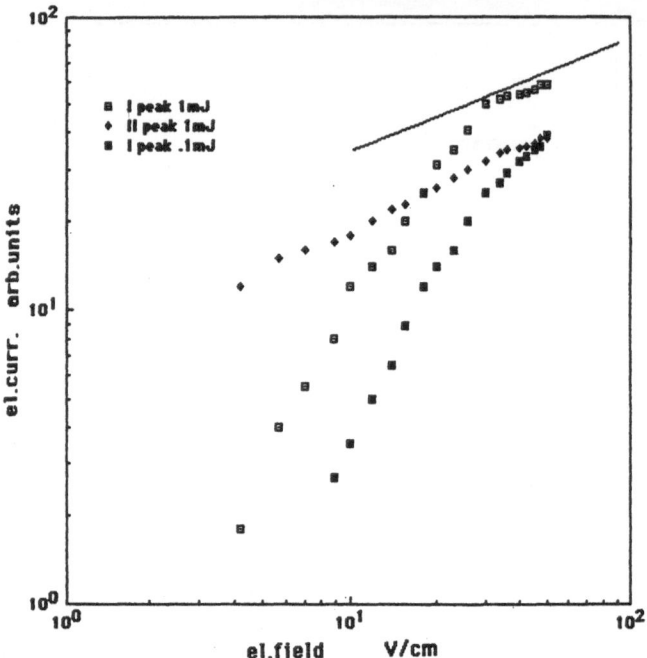

Fig. 9. Peak electron current as a function of the applied electric
field. Short focus lens. Beam density 4.4×10^{11} atoms/cm^3.
$\lambda = 5787$ Å. The labels I peak 1 mJ and II peak 1 mJ indicate
the two observed peaks (see text) at a laser power of 2×10^8
W/cm^2. I peak 0.1 mJ indicate the peak observed at 2×10^7 W/cm^2.

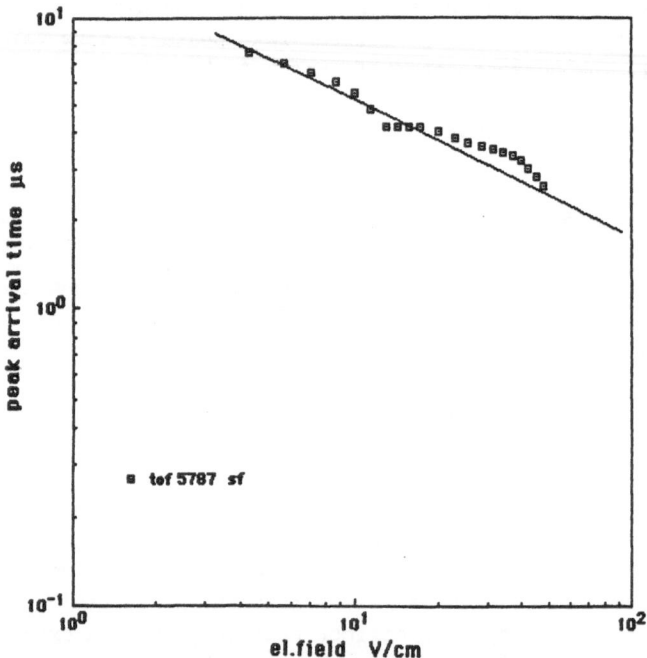

Fig. 10. Time of arrival of the II peak 1 mJ (see text) as a function
of the applied electric field. Same conditions as Fig. 9.

equation is transformed into a first order linear equation in x for variable v, whose coefficients are the time and space derivatives of $\ln(n)$. Thus, the dependence of v on the function $f(t)$ and x is given by v $= - (x/2)D\ln(f(t))$, where D represents the total time derivative. The equation of motion immediately leads to a one-variable second order differential equation for the function $f(t):D^2f_2(t)=(3/2) (Df(t))^2/f(t)-2B_jf(t)^2$, where B_j is given by $2kT/(m_jd^2)$; d and m_j represent respectively the half-width of the initial density profile and the mass of the charged particle. The initial conditions are $f(0)=1$ and $Df(t)\big|_{t=0}=0$.

In ref. 1 it is demonstrated that the form of the solution does not change if a time-varying electric field is included, provided that its spatial dependence is linear. In such a case, the time dependence of the field affects the differential equation for the $f(t)$ function as well as the fluid velocity. The solution of the single-particle like equation $D v_s=(e/m_j)E(t)$ must be added to the field-free solution for v and the spatial variable x must be replaced by $\int v_s dt-x$.

Using the method of solution just described, it is possible to analyze the two-fluid system coupled by the self-generated electric field. Assuming the single-fluid shapes for both ions and electrons and integrating the one-dimensional Poisson equation between $-\infty$ and x, introducing a variable change $\varepsilon_j=f_j(t)^{1/2}(\int v_{sj} dt-x)$ and using a theorem about the mean value of the integrals, the self-generated field is given by $E(t)=4\pi n\cdot$ $\cdot (\varepsilon_i-\varepsilon_e)\exp(-\varepsilon_*^2)$, where ε_* represents a value between ε_i and ε_e. If ε_* is negligible the electric field is a linear function of x and, by using the same technique of single-fluid solution, the complete time evolution is described by the equations

$$D^2f_j(t)=(3/2)(Df_j(t))^2/f(t)\pm\Omega_{pj}^2 \ f_j(t)(f_i(t)-f_e(t))-2B_jf_j(t)^2$$

$$D^2x_j(t) =\pm\left[(e/m_j)E_0+\Omega_{pj}^2 \ f_j(t)^{1/2}(x_e-x_i)\right]$$

(1)

where the index j indicates both the species and $x_j=\int v_{sj}$ dt. The upper sign refers to the ion behavior and Ω_{pj} is the plasma frequency at the maximum density n_0. E_0 represents the externally applied electric field.

A numerical calculation has verified the limits of the approximation assuming an initial charge density in the 10^{10} to 10^{11} cm^{-3} range. Fig. 11 shows as an example the spatial profile of the electron density at time t=100 ns for $E_0=0$ as determined by the numerical code, compared with the above-mentioned solution. In the region A, whose width is about the half-width of the Gaussian profile, the behavior given by equations 1) and by the numerical solution completely overlap. The evolution of the region C is well represented by the single-fluid solution, immediately obtained from the equations (1) if $\Omega_{pj}=0$, that is, if the self-generated electric field is neglected. The region C is related to the fraction of electrons able to quickly overcome the internal field.

The transition region B is very thin but exhibits unstable numerical evolution. That numerical instability could indicate the limit of the collisionless fluid approximation, preceding a particle motion that in the

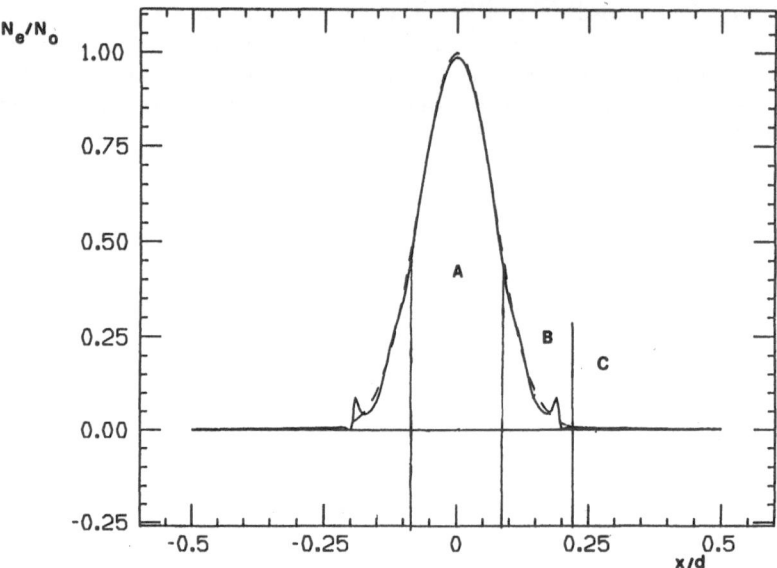

Fig. 11. Calculated electron spatial profile at t=100 ns. Initial
charge density 10^{10} el/cm^3. ---: numerical code, - - -: two-
fluid solution. a) strongly coupled region (two-fluid); b)
transition unstable region; c) free expansion (single-fluid
solution).

transition region surely exhibits chaotic features. In fact region B is
limited by two ordered motions of different kinds: fast oscillating motion
around the ions in region A until the coupling term depending on Ω_{pj} decreases more than the term depending on the thermal motion $(2B_j f_j(t)^2)$;
and fast translational motion depending on the electron and mass energy by
the term depending on the electron and mass energy by the term $2B_e f_e(t)^2$
(region C). Therefore a detailed analysis of the transition zone requires
the introduction into the fluid equations of collisional terms taking into
account the momentum exchange between ordered motions of electrons as well
as between electron and ion motion[17]. In any case, considering that the
transition region involves only a small fraction of the total produced
charge, the previous model seems a suitable approach to explain the
observed behavior.

Referring to Fig. 9, the first (fast-1 mJ) peak is produced by
electrons coming from a region not affected by the self-generated field
(region C of Fig. 11). The evolution of the peak intensity can be ascribed
to different effects, depending first on a non linear growth of the C
region (Fig. 11) as a function of the external electric field, and
secondly related to the characteristics of the single-fluid expansion[1].
Actually, considering that the behavior of the 0.1 mJ peak (Fig. 9), not
affected by trapping owing to too low a charge density, is quite similar
to the 1 mJ fast peak, the explanation based on the single-fluid model
seems more realistic. Moreover, the 1 mJ fast peak approximates the 0.5
slope as the electric field increases, that is, when the collection time
is less than the characteristic time of thermal expansion $(\cong (B_e)^{-1/2}$
$\cong (md^2/2kT)^{1/2})$.

In the conditions of the experiment, that corresponds to reasonable values of the applied electric fields (30 to 100 V/cm), in quite good agreement with the observed field (Fig. 9), whereas very strong values should be required to completely overcome the self-generated field ($\gtrsim$1 KV/cm).

The 1 mJ second peak exhibits completely different features. The arrival time and peak of current behavior exhibit the single-particle slopes (-0.5 and 0.5 respectively), but the absolute value of the arrival time corresponds to that expected for ions.

As previously pointed out, the electrons are trapped in the A zone (Fig. 11) until the internal electric field is greater than the external field. As an effect of ion thermal motion, the initial profile spreads out and thus the density decreases. When the density is low enough that the self-generated field assumes a value smaller than the external one, the electrons are affected by the external field. During the collection time, the further spread induced by the thermal motion of electrons is less effective. In fact that spread strongly depends on the initial width of the electron profile (d^{-2} see above). In order to reduce the self-generated field sufficiently, the initial width must increase by about one or two orders of magnitude. Therefore, the electron profile simply translates during the collection phase and that explains the observed dependence on the electric field. If this explanation is accurate, the delay of the second peak must correspond to a delay in ion collection. In fact, the screening of the external field must also induce an increase in the ion collection time. From Fig. 10 the measured delay of the electron arrival time at the same value of electric field used in ion collection (14 V/cm) is $\cong$4 μs.

Recalling the discussion of Sec. 3 concerning the behavior of the ion signal, the delay in time of flight of ions produced in the focusing spot, as strongly affected by electron-ion mutual interaction, is just about 4 μs.

V. CONCLUDING REMARKS

Although a detailed quantitative comparison between theory and experiment requires further improvement in the model, nevertheless the previous discussion gives a suitable explanation of the observed main features. Therefore it seems evident that the effects due to electron-ion mutual interaction can represent a serious limit in a wide range of laser induced ionization experiments. The principal effects on the ions are spread of the initial profile, delay of the peak and enhancement of the collision rate affecting the total number (see the previous discussion concerning the production of molecular ions). The last mentioned effect depends on the cross section of the collisional processes, and therefore on the kinds of interacting bodies.

For instance, in experiments performed in noble gases[4-7] the influence of collisions is strongly reduced and thus the total number of collected ions can be a suitable measure of laser-matter interactions. On the other hand, if the final objective of the experiment is the collection of selectively produced species, the spread of the profile makes it difficult to separate atoms or molecules whose atomic weight differs by few mass units.

The features of electron behavior are more complex and make very hard to carry out a quantitative correlation between collected and produced charges, especially if different initial energy electrons are produced by the laser interaction as in the above threshold ionization experiments [4-7]. As predicted theoretically[1] and observed experimentally (see Sec. 3 and 4), the electrons deviate from single-particle motion at such a density level that space-charge effects are negligible (single-fluid motion)). As the charge density increases, a growing fraction of electrons is trapped by the ions and its collection is strongly delayed by the internal field; in this case also, an analysis in terms of totally produced electrons seems unreliable and in many experiments does not give the required informations.

Considering that the previously described experiment has been performed under conditions of most experiments, it may be asserted that there is no way to avoid collective effects, especially with regard to the electron evolution. Hence it seems important to improve the experimental investigations as well as the theoretical treatment of the time-space evolution of the yield produced by laser interactions.

REFERENCES

1. F. Giammanco, Phys. Rev A-36, 5658 (1987)
2. M. Crance, J. Phys. B-19, L267 (1986)
3. B. Carre', F. Roussel, G. Spiess, J. M. Bizeau, P. Gerard and F. Wuilleumier, Z. Phys. D-1, 79 (1986)
4. P. H. Bucksbaum, M. Bashkansky, R. R. Freeman and T. J. McIlrath, Phys. Rev. Lett. 56, 24 (1986)
5. P. Kruit, J. Kimman, H. G. Muller and M. J. van der Wiel, Phys. Rev. A-28, 248 (1983)
6. A. Lompre', A. L'Huillier, G. Mainfray and C. Manus, J. Soc. Am. B-2, 1906 (1985)
7. F. Yergeau, G. Petite and P. Agostini, J. Phys. B-19, L663 (1986)
8. For a review of space-charge problems in different fields of application and investigation see:
   "Proc. of the Third International Symposium on Resonance Ionization Spectroscopy and its Applications", Hilger, London (1986) and also "Proc. of Fourth RIS Symposium", NBS Gaithersburg, MD-Hilger (1988)
9. M. Allegrini, W. P. Garver, V. S. Kushawaha and J. J. Leventhal, Phys. Rev.A-28, 199 (1983)
10. J. Boulmer and J. Weiner, Phys. Rev. A-27, 2817 (1983)
11. C. E. Burkhardt, W. P. Garver and J. J. Leventhal, Phys. Rev. A-31, 35 (1984)
12. J. Keller and J. Weiner, Phys. Rev. A-30, 213 (1984)

13. C. Y. R. Wu, F. Roussel, B. Carre', P. Breger and G. Spiess, J. Phys. B-18, 239 (1985)

14. Mitchell and Zemansky, "Resonant Radiation and Excited Atoms", Cambridge U.P. (1961)

15. W. P. Garver, M. R. Pierce and J. J. Leventhal, J. Chem. Phys. 77, 1201 (1982)

16. F. Giammanco and S. Gozzini, Nuovo Cimento B-66, 1 (1981)

17. L. Spitzer, "Physics of Fully Ionized Gases" - Interscience, New York (1956)

QUANTUM WELLS AND FAST NONLINEARITIES

ULTRAFAST DYNAMICAL NONLINEARITIES

IN III-V SEMICONDUCTORS

J.L. Oudar

Centre National d'Etudes des Télécommunications
196, avenue Henri Ravera F-92220 Bagneux, France

I. INTRODUCTION

The application of optical effects to digital information processing[1,2] requires proper understanding of the nonlinear mechanisms, and the development of appropriate materials.

The III-V semiconductors are particularly interesting for digital processing, because they possess large optical nonlinearities and are therefore useful for optical logic gates, phase conjugation and optical bistability. Moreover their use in optoelectronic devices such as diode lasers, modulators and photodetectors has led to the development of sophisticated epitaxial growth techniques, which allow the fabrication of synthetic microstructures such as multiple quantum wells with thickness control down to the atomic level[3]. This results in an increased flexibility to design structures with new electronic and optical properties[4,5].

As is well known, the optical nonlinearities of a material are enhanced when light interacts resonantly with its excited states. When several optical frequencies come into play in this nonlinear interaction, quantum mechanical perturbation theory predicts that such resonant enhancements can occur not only at each of these frequencies, but also at their combinations, e.g. when the sum or the difference of two photon energies comes close to the energy separation of two eigenstates of the material. This has given rise in recent years to a wealth of experimental and theoretical studies, and has been widely used for spectroscopic applications[6].

There are two classes of nonlinear optical processes which depend upon resonance[7]. In the first class, "parametric processes", the material only acts as an intermediate medium for the interaction between the light beams. There is no net energy exchange between light and matter after the nonlinear interaction, and hence the material system is in the same energy state as before the interaction. The material response is reactive, and light induces only virtual transitions between energy eigenstates. In the second class, there is some energy exchange, because real transitions are involved, giving rise to dissipative processes. In this case, the material response time depends on the relaxation dynamics of its excited states, hence these non-parametric processes are called "dynamic nonlinearities". For obvious reasons of speed and dissipation, one would prefer to use only parametric effects in practical devices, however in many cases the enhancement of optical nonlinearity through non-parametric effects is so large that they are more advantageous in terms of optical power requirements. In addition, the speed limitations are not such a problem if the nonlinear response is governed by ultrafast relaxation processes.

II. THEORETICAL DESCRIPTION OF DYNAMIC OPTICAL NONLINEARITIES

While parametric processes are typically used for frequency mixing, second harmonic generation and parametric amplification, dynamic nonlinearities are applied to optical gates, phase conjugation and optical bistability. The latter may be of the third order type described by a third order nonlinear susceptibility $\chi^{(3)}$, or they may involve higher (odd) order nonlinear terms $\chi^{(5)}, \chi^{(7)}$, etc. in a strongly nonlinear regime. In particular, in the case of absorption saturation, and its associated nonlinear refractive index, a $\chi^{(3)}$ description is only valid in the low intensity limit, and all the higher order terms must be considered in the strongly saturated regime.

It is useful to recall the range of $\chi^{(3)}$ values in various materials, so as to appreciate the enormous differences in orders of magnitude that one finds between purely nonresonant parametric processes, and multiresonant dynamic nonlinearities. For third harmonic generation in crystals, the typical $\chi^{(3)}$ values[8] extend from 2.10^{-15} esu for LiF to 1.10^{-10} esu for Ge, involving parametric processes only. For band-gap resonant non-linear refraction, an effective value of $\chi^{(3)}$ as large as 1 esu has been quoted for InSb[9], which illustrates the gigantic enhancement of dynamic nonlinearities due to photoexcitation of electron-hole pairs in semiconductors. At this point it is worth recalling that such steady-state effective $\chi^{(3)}$ is directly proportional to the lifetime of electron-hole pairs in the material. Therefore an effective $\chi^{(3)}$ of 1 esu observed with a carrier lifetime of 400 ns is comparable to an effective $\chi^{(3)}$ of 10^{-6} esu observed with pulsed of 400 fs duration, or in steady-state with a carrier lifetime of this latter value. In the latter case a corresponding 10^{6} increase in light intensity is then needed to achieve the same change in optical properties. However, 10^{-6} esu is still quite large compared to the values quoted above for pure parametric processes. Wherett has discussed in a comprehensive way[10] why the effective $\chi^{(3)}$ may differ by so many orders of magnitude from one material to the other. Finally, let us mention that in addition to the saturation of band-to-band transitions discussed here, other mechanisms contribute to optical nonlinearities in semiconductors, such as the plasma contribution in small gap semiconductors[11], the screening of excitons[12], the nonlinear response of biexcitons[13] and the saturation of bound-exciton absorption[14]. These other mechanisms are beyond the scope of the present discussion and we refer the reader to the referenced literature.

II.1. Nonlinear response functions

The appropriate framework for describing the nonlinear optical response of a material phenomenologically, is to consider the "constitutive relations", which relate the polarization P to the electric field E applied to the material. When a perturbation approach can be used, these equations are expanded in a power series in E^{15}.

When the fields are expressed as a function of time, one may define nonlinear response functions at the various orders of perturbation. At order r, the polarization $\bar{P}^{(r)}(t)$ depends on the r-th power of $E(t)$, and the most general response function is written as:

$$\bar{P}^{(r)}(t) = \int dt_1 \int dt_2 \cdots \int dt_r \bar{R}^{(r)}(t-t_1, t-t_2, \cdots t-t_r)|$$

$$|\bar{E}(t_1)\bar{E}(t_2)\cdots\bar{E}(t_r) \tag{1}$$

where $R^{(r)}(t)$ is a tensor of rank r+1 and the | notation stands for the r-fold dot product of $R^{(r)}(t)$ with the E-field products. Here we have used the fact that $R^{(r)}(t)$ describes a permanent property of the material, so it is invariant with respect to a time translation[7,15].

It is interesting to note that the nonlinear response functions can be expressed quantum mechanically in a very compact form, using the formal expansion of Dyson's series[16]. In the electric dipole approximation, the perturbation term in the hamiltonian is -p.E(t), and it is useful to transform the electric dipole operator p with components p_1 into the interaction representation

$$\tilde{p}_1(t) = e^{iH_0t/\hbar} p_1 e^{-iH_0t/\hbar} \tag{2}$$

where H_0 is the unperturbed hamiltonian of the material. One then obtains directly for the r-th order term of polarization a multiple integral expression in the form of eq. (1), with

$$R^{(r)}_{ij\ldots n}(\tau_1, \tau_2, \ldots \tau_r) = (\tfrac{i}{\hbar})^{(r)} \langle [[\cdots [\tilde{p}_i(0), \tilde{p}_j(-\tau_1)], \ldots], \tilde{p}_n(-\tau_r)] \rangle \cdot$$

$$\cdot \; \Gamma(\tau_r > \ldots > \tau_2 > \tau_1 > 0) \tag{3}$$

where $\langle \; \rangle$ is the ensemble average of the r-fold commutator products ([A,B] means AB-BA) and Γ is a generalized Heaviside function, equal to 1 when the time ordering is as indicated in the argument of Γ, and null otherwise.

If one prefers to describe the fields in the frequency domain, one works with the Fourier transform $E(\omega)$ expressed as

$$\bar{E}(\omega) = \int dt e^{i\omega t} \bar{E}(t) \tag{4}$$

and one describes the response of the material by its nonlinear susceptibilities. Then the polarization $\bar{P}^{(r)}(t)$ is expressed as

85

$$\overline{P}^{(r)}(t) = \int d\omega_1 \int d\omega_2 \ldots \int d\omega_r \overline{\chi}^{(r)}(\omega_1,\omega_2,\ldots\omega_r) \mid \overline{E}(\omega_1)\overline{E}(\omega_2)\ldots\overline{E}(\omega_r) \cdot$$
$$\cdot e^{-i(\omega_1+\omega_2+\ldots+\omega_r)t} \qquad (5)$$

where the nonlinear susceptibility $\chi^{(r)}$ is related to the nonlinear response function $R^{(r)}(t)$ through the generalized Fourier transform[13]:

$$\overline{\chi}^{(r)}(\omega_1,\omega_2,\ldots\omega_r) = \int d\tau_1 \int d\tau_2 \ldots \int d\tau_r \overline{R}^{(r)}(\tau_1,\tau_2,\ldots\tau_r) \cdot$$
$$\cdot e^{-i(\omega_1+\omega_2+\ldots+\omega_r)t} \qquad (6)$$

Although these two descriptions are equivalent in principle, the non-linear susceptibilities are more convenient to use in the case of steady-state monochromatic light beams, while the nonlinear response functions seem more adapted to the case of ultrashort light pulses. With monochromatic beams the Fourier components $E(\omega)$ are in the form of δ-functions and the integrals in eq. (5) are straightforward to evaluate. Similarly, $R^{(r)}(t)$ in eq. (1) is simply the material polarization at time t, in response to a succession of δ-function electrical pulses occurring at earlier times $t_1,t_2,\ldots t_r$. In practice, however, most light pulses have a duration of at least several optical cycles, so their central frequency can be on resonance with some excited states of the material. Therefore even when using ultrashort light pulses one cannot completely forget the frequency domain description. More specifically, these possible resonances appear when one develops the operators $p_1(t)$ of eq. (3) on the basis of the energy eigenstates $|n>$ of energy E_n, of the unperturbed hamiltonian H_0. From eq. (2) one obtains for the corresponding matrix element

$$<m|\tilde{p}_1(t)|n>=<m|p_1|n>e^{i(E_m-E_n)t/\hbar} \qquad (7)$$

A convenient way to keep track of the various terms in eq. (3) is to use the diagrammatic approach of Yee and Gustavson[16]. It can be used in the frequency domain for quickly selecting the relevant terms in a given situation[17], or in the time domain for displaying the various time-ordered interactions[18].

II.2. Some specific cases

The formalism discussed above is very general and can be applied in principle to every physical situation where the perturbation approach is meaningful. In order to be more specific, we need to consider the various characteristic times relevant to a given instance of transient nonlinear optical effects. Such characteristic times for a light pulse are the pulse duration t_r and the coherence time t_c, the latter being roughly equal to the inverse of the light frequency bandwidth. A particular excited state of the material has in general a finite population lifetime T_1, while the coherent superposition of two states induced by resonant light is charac-terized by a dephasing time T_2. This terminology is standard in the context of two-level systems[19] that will shortly be discussed below. In addition, at very short time scales in condensed matter, the scattering events are not independent of each other and one may need to consider the correlation time T_c of the many-particle systems that interact with a given excited state. Finally, the light-matter interaction is charac-terized by the offset from resonance Δ and the Rabi frequency Ω, propor-tional to the electric field amplitude and to the transition dipole moment.

In order to classify this rather complex situation we shall now discuss some limiting cases in which one characteristic time is much shorter than the others, thus allowing some simplifications in the theo-retical description.

When the optical frequencies are all far away from any resonance, the material is transparent, and the polarization response is essentially instantaneous with respect to the driving fields. In this limit the non-linear interaction only involves parametric processes, appropriately described in the frequency domain by the nonlinear susceptibilities.

In the opposite limit, that of absorbing materials, the nonlinear response is delayed, so that a time domain description is appropriate in the pulsed regime. The various limiting cases for the temporal behaviour of resonant light-matter interactions are more clearly apparent when the material system can be approximated by an ensemble of 2-level systems. The time evolution of 2-level systems is most conveniently described by the so-called Bloch equations for the population difference $\Delta\rho$ between ground state $|a\rangle$ and excited state $|b\rangle$, and for the off-diagonal elements $\rho_{ab} = \rho_{ba}^*$ of the density matrix describing this statistical ensemble. Calling ω_0 the resonance frequency and V_{ab} the light-matter interaction part of the Hamiltonian, these equations read

$$\frac{d}{dt}(\Delta\rho) = \frac{2i}{\hbar}(V_{ba}\rho_{ab} - V_{ab}\rho_{ba}) - \frac{(\Delta\rho - \Delta\rho^{eq})}{T_1} \tag{8}$$

$$\frac{d}{dt}(\rho_{ab}) = i\omega_0\rho_{ab} + \frac{i}{\hbar}V_{ab}\Delta\rho - \frac{\rho_{ab}}{T_2} \tag{9}$$

where the relaxation time T_1 governs the decay of $\Delta\rho$ towards its equi-librium value $\Delta\rho^{eq}$ and T_2 governs that of ρ_{ab} towards 0.

First let us consider the case of a "transform-limited" pulse, with $t_p = t_c$, interacting with a 2-level system, with $T_2 \ll T_1$. Three limiting cases can be distinguished.

(a) In the ultrashort pulse limit, $t_p \ll T_2$ we are in the situation of coherent transient phenomena, such as photon echoes, superradiance, self-induced transparency, etc. These phenomena are best described with the help of the vector model of Feynman et. al.[20], that allows an elegant geo-metrical interpretation of the solutions of the Bloch equations.

(b) When the characteristic times are such that

$$T_2 \ll t_p < T_1 \qquad (10)$$

we are in the regime of dinamic nonlinearities, where the coherent response ρ_{ab} is quasi-stationary with respect to the driving electric field, while the temporal evolution of the nonlinear response is mostly governed by that of the population term $\Delta\rho$. If V_{ab} is written in terms of a slowly varying amplitude $\Omega(t)=(W/h)E_0(t)$, as $V_{ab}=(\hbar/2)\Omega(t)\exp(i\omega t)$, then the coherent response takes the form $\rho_{ab}=(1/2)\rho_0\exp(i\omega t)$, so that the solutions of eqs. (8,9) are written as

$$\Delta\rho(t) = \Delta\rho^{eq} - \frac{T_2}{1+(\Delta T_2)^2} \int_{-\infty}^{t} dt' \; \exp(-\frac{t-t'}{T_1})\Omega^2(t')\Delta\rho(t') \qquad (11)$$

$$\rho_0(t) = \frac{\Omega(t)\Delta\rho(t)}{\Delta-i/T_2} e^{i\omega t} \qquad (12)$$

where $\Delta = \omega-\omega_0$.

We note that, due to the approximation of a very short T_2, eqs.(8) and (9) have been decoupled, so that eq. (11) can first be solved for a given excitation pulse; then the polarization is directly deduced from eq. (12) when $\Delta\rho$ is known as a function of time.

Another way of writing eq. (12) is to describe it as renormalized (population dependent) linear response

$$P(t) = \chi_{eff}(t)E(t) = \chi^{(1)} \frac{\Delta\rho(t)}{\Delta\rho^{eq}} E(t) \qquad (13)$$

so that the time variations of $\chi_{eff}(t)$ are a direct measure of the population kinetics $\Delta\rho(t)$.

(c) A third limiting case occurs when t_p is much larger than both relaxation times T_1 and T_2. This corresponds to a quasi-stationary situation where the population can follow the time evolution of the light intensity (proportional to $\Omega^2(t)$). Then one obtains the well-known steady-state solution of the 2-level system

$$\Delta\rho(t) = \frac{\Delta\rho^{eq}|1 + (\Delta T_2)^2|}{1 + (\Delta T_2)^2 + T_1 T_2 \Omega^2(t)} \qquad (14)$$

$$\rho_0(t) = \frac{T_2(i + \Delta T_2)\Omega(t)\Delta\rho^{eq}}{1 + (\Delta T_2)^2 + T_1 T_2 \Omega^2(t)} \qquad (15)$$

which displays at exact resonance $\Delta=0$, a nonlinear response P_{NL} characteristic of absorption saturation

$$P_{NL} = \frac{P_L}{(1+I/I_s)} \qquad (16)$$

where $I/I_s = T_1 T_2 \Omega^2$ is the ratio of light intensity I to the saturation intensity I_s, and P_L is the linear response in the limit $I \rightarrow 0$. Away from resonance, the spectral behaviour has the form of a "power-broadened" Lorentzian lineshape.

Comparing these three limiting cases, we note that the description occurs in the time-domain for the coherent transient case (a), in the frequency-domain of the quasi-stationary case (c), while in the case (b) of dynamic nonlinearities, an intermediate description is needed, incorporating frequency and time variables as in eqs. (11,12).

Finally, if the excitation pulse has a coherence time t_c significantly shorter than t_p, such as the case of a white light continuum, one can have more complex situations, including the case where $t_c \ll T_2$ but $t_p \gg T_1$, which implies that the coherent response $\rho_{ab}(t)$ is transient, while the population kinetics is quasi-stationary[21,22], or the case where the applied electric field is the sum of two (or more) light beams at different frequencies, resulting in frequency mixing processes, resonantly enhanced in the quasi-degenerate case.

III. ULTRAFAST DYNAMICS OF ABSORPTION SATURATION IN GaAs, BULK AND MULTIPLE QUANTUM WELLS

III.1. Saturation of the band-to-band transitions in semiconductors

Although the 2-level system (2LS) approach developed in the previous section is most usually applied to the case of very weakly interacting atoms, such as in an atomic vapor, this model is general enough to provide a good starting point for the discussion of direct optical transitions in a semiconductor. This point if view must be taken with some precautions however, as will be discussed in the following.

It is well known that in a semiconductor the atoms are so strongly interacting that their energy states are distributed over energy bands, the electronic wavefunctions being in the form of delocalized Bloch waves labeled by a "momentum vector k". It is therefore impossible to identify each of the atoms with a particular 2LS, and instead one has to deal with 2-band systems. However, direct optical transitions between the valence and conduction bands in a semiconductor obey the k vector selection rule, which means that the optical interaction only couples states with the same k vector[23]. Therefore, to the extent that the various k states within a band can be considered as independent of each other, one may single out a particular vector k within the 2-band system, and call this subset a 2LS. If the interband transitions are induced by monochromatic light at frequency ω, the optical field is in resonance with only a small number of such 2LS's, those which obey the condition

$$(E_c(k) - E_v(k))/\hbar = \omega_{cv}(k) = \omega, \tag{17}$$

the other states belonging to other (off-resonance) 2LS's. We are thus led to describe our 2-band system simply as a continuous distribution of

2LS's, with a certain density of states (number of "LS's per unit energy and per unit volume, sometimes called the "joint density of states")[23]. Upon photoexcitation these 2LS's are partially populated, leading to absorption saturation and nonlinear refraction. The optical response of this continuous distribution of partially populated 2LS's is the sum of the responses of each individual 2LS. The population-dependent renormalized linear susceptibility (in the sense of eq. (13)) is then expressed as

$$\chi_{eff}(t) \propto \sum_{\vec{k}} \frac{|\hat{e} \cdot \vec{p}_{cv}(\vec{k})|^2}{\omega_{cv}(\vec{k}) - \omega - i/T_2(\vec{k})} [1 - f_e(\vec{k},t) - f_h(\vec{k},t)] \qquad (18)$$

where the sum is within the Brillouin zone, f_e and f_h are the electron and hole occupation factors, $\hat{e}$ is the light field polarization, and $\vec{p}_{cv}$ is the transition momentum matrix element between the conduction and valence states.

The last bracket in eq.(18), which governs the time evolution of the effective susceptibility, is analogous to the term Δ_p in eqs.(8)-(15), simply because the electron occupation factor in the valence band is $1-f_h$. In many cases, the electron and hole occupation factors correspond to thermal equilibrium, and are described by Fermi-Dirac distributions which depend only on the energy E_i of the particular state within band i:

$$f_i(E_i) = \frac{1}{1+\exp[(E_i-E_i^F)/kT_i]}, \quad \text{with } i = e,h. \qquad (19)$$

where E_i^F and T_i are the Fermi energy and the electronic temperature in the band i.

The absorption coefficient, given by the imaginary part of eq.(18), is then a relatively direct measure of these occupation factors, especially when they vary more slowly with energy than the width of the Lorentzian lineshapes in eq.(18). In that case, the Lorentzian linefunctions can be approximated by δ-functions, so that the saturated absorption coefficient can be written as

$$\alpha(\omega,t) = \alpha_0(\omega)[1-f_e(E_e,t)-(f_h(E_h,t)] \qquad (20)$$

where $\alpha_0(\omega)$ is the initial (unsaturated) absorption coefficient, E_e and E_h are the energies of the optically-coupled states. The absorption saturation in this thermalized regime gives rise essentially to a blue shift of the absorption edge, sometimes called the "dynamic Burstein shift"[24,25]. Eq.(20) is currently used for analyzing the saturated absorption spectra of samples excited by picosecond light pulses. On the time scale of a few picoseconds an internal equilibrium is reached by the electron-hole (e-h) plasma through the interelectronic collisions which allows us to define an electronic temperature T_e. In most experiments the initial carrier energy is relatively high, so that T_e is initially larger than the lattice temperature T_L, and one can use the transient absorption spectra to monitor the cooling of this hot e-h plasma towards towards T_L[26]. We note that

eq.(20) is generally not valid in the case of the non-thermalized distributions generated by femtosecond pulses, as will be discussed in the next
section. In that case it is necessary to go back to the more general
eq.(18).

The assumption that the various one-electron states are independent
of each other is of course only an approximation, due to several additional effects:

(a) excitonic effects, which arise from the Coulomb interaction between
an electron in the conduction band and the hole left in the conduction band
after the photoexcitation process. These excitonic effects modify the
absorption spectra of pure unexcited samples, leading to additional resonances (excitonic lines) and to an enhancement of the light-matter coupling at photon energies above the band-gap[27];

(b) many-body effects, which arise from the direct and exchange Coulomb
interactions between the various carriers already present in the material.
They cause a broadening of the excitonic lines and, at higher values of
plasma density, their disappearance due to the screening of the electron-
hole attraction; simultaneously the band-gap renormalization[28] causes a
red-shift of the photoluminescence spectra at high plasma densities.

Part of the many-body effects are taken into account in the simple 2LS
picture through the introduction of a phenomenological dephasing time for
the e-h pair transition, which accounts for the finite lifetime of the
electron and hole states, due to carrier-carrier collisions, as well as
other collisions with phonons, impurities, etc. This finite lifetime can
be calculated as the imaginary part of the quasi-particle self-energy[29],
while its real part corresponds to a renormalization of the 2LS's energy
separation.

The remainder of these additional effects, in particular the interplay between excitonic and many-body effects, need a much more sophisticated theory[30,31]. This theory requires the self-consistent calculation
of the many-body problem, only accessible through heavy formalisms, e.g.
involving Green's functions, the results of which are usually difficult to
translate into physical concepts.

III.2. Femtosecond studies of carrier momentum relaxation

The recent development of tunable femtosecond light pulses has made it
possible to study directly in the time domain carrier-carrier collisions and
the resulting momentum relaxation of the electrons and holes within their
bands. This study is performed by creating a highly non-equilibrium electron-
hole distribution with an intense ultrashort light pulse. Such a non-equilibrium situation actually results from the fact that selection rules normally
restrict the number of optically-coupled states to a relatively small domain
in k-space. As a result, the dynamics of absorption saturation (see Fig. 1)
depend directly upon the carrier scattering rates out of (and back into)

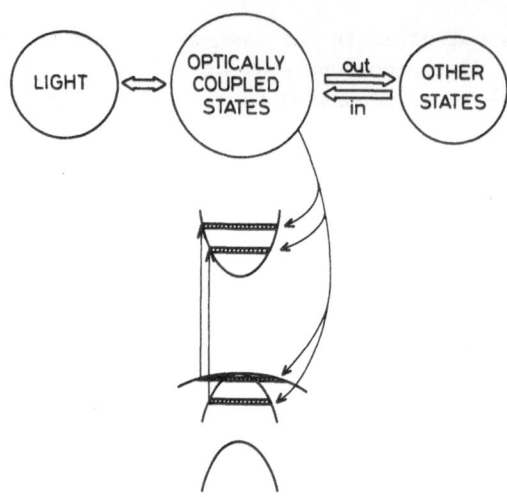

Fig. 1. Schematic representation of the processes controlling the
dynamics of band-to-band absorption saturation in a direct-
gap semiconductors.

these optically-coupled states, through carrier-phonon and carrier-carrier
collisions.

Absorption saturation experiments with femtosecond light pulses
provide the possibility of observing the orientational relaxation of an
initially anisotropic momentum distribution of electron-hole (e-h) pairs[32].
This effect occurs when e-h pairs are excited by linearly polarized light,
since this polarization imposes a preferential direction in the k-space
distribution of electrons and holes. With intense ultrashort pulses the
generation rate can become larger than the collision rate, making it
possible to preferentially saturate the e-h pairs whose electron (or hole)
wavevector points in a given direction. This anisotropic momentum distribution can be revealed by a corresponding anisotropy in the transmission
coefficients of a weak probe pulse polarized parallel or perpendicular to
the pump pulse polarization. With a mixed probe polarization (e.g. at 45°
with respect to the pump), this differential attenuation induces a rotation of the probe polarization, which can be measured as a function of
probe delay. From the decay of this polarization signal (see Fig. 2), one
can directly infer the orientational relaxation time of photo-excited e-h
pairs.

Another example is the observation of a dynamical spectral hole-
burning feature centered at the pump frequency in pump-probe experiments
with femtosecond pulse[33]. This effect also arises from a preferential
saturation of the optically-coupled states, but this time it is observed
in the spectral domain with broadband femtosecond pulses.

Provided that the pump frequency spectrum is narrow enough - close to
the Fourier-transform limit - a quasi-monoenergetic distribution of
carriers is initially produced in the immediate neighbourhood of the pump
frequency, as qualitatively described in Fig. 3. The spectral width of
this distribution is somewhat broadened, due the finite width of the pump

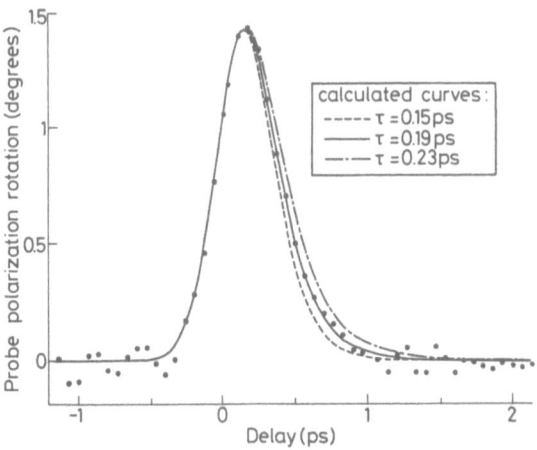

Fig. 2. Temporal evolution of the polarization signal, corresponding
to an induced anistropy of absorption saturation produced
by linearly polarized pump pulses (from ref. 32).

spectrum, but also due to the collision-broadening contribution discussed
in the previous section. At later stages, this highly non-equilibrium
(non-thermal) situation evolves towards a thermalized regime, due to
carrier-carrier collisions, and then electron-phonon interactions bring
the plasma into equilibrium with the lattice, which results in a cooling
of the electron-hole plasma down to the lattice temperature.

This qualitative picture is supported by the actual observation of a
preferential bleaching around the pump frequency, as displayed in Fig. 4.
This feature is characteristic of nonthermal carrier distributions, and
provides information on the primary steps of carrier relaxation, i.e., the
first scattering events out of the optically-coupled states. In bulk GaAs,
and in GaAs Multiple Quantum Wells (MQW's) the effect is more easily

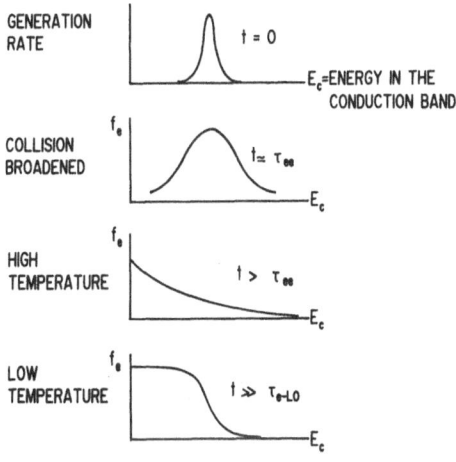

Fig. 3. Schematics of the time evolution of photo-excited carriers.

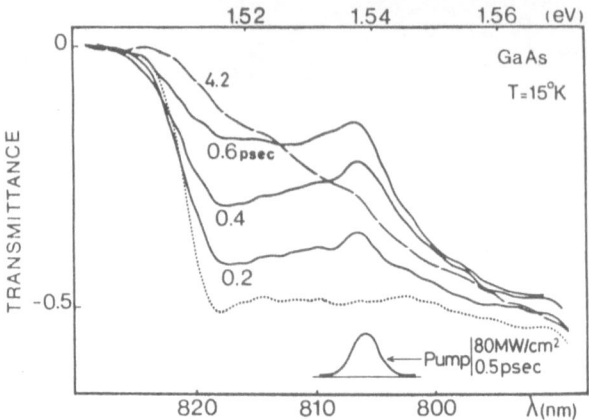

Fig. 4. Time-resolved transmittance spectra showing a spectral hole-
burning feature around the pump wavelength, due to non-
thermal carrier distributions (from ref. 33).

observed when the photon excess energy above the band-gap is smaller than
the energy of an optical phonon. This prevents electron relaxation by
optical phonon emission, and leaves carrier-carrier interactions as the
main scattering processes. A lineshape analysis allows us to separate the
ultrafast contribution of nonthermal carriers from the thermalized distribution of carriers generated at earlier times, and permits determination
of the e-h pair dephasing time T_2. With bulk GaAs at 15°K[33], T_2 is found
equal to 300 fs at 19 meV above the band-gap.

Calculations of plasma dynamics in the femtosecond regime[34] are found
in quite good agreement with the experimental curves (see Fig. 5). These
calculations show that the carrier dephasing time first increases with the
plasma density, but then stabilizes or even decreases at densities higher
than 10^{17} cm^{-3}, due to plasma degeneracy and the screening of Coulomb
interaction.

This effect is also observed in GaAs MQW's[35], and our experimental
spectra (shown in Fig. 6) obtained under the same excitation conditions as
in our previous experiment, allowed us to determine T_2 = 150 fs in the MQW
sample[36], a value significantly shorter than for bulk GaAs. This indicates
larger carrier-carrier scattering rates in quantum wells, which is mainly
attributed to the reduced screening efficiency of the e-h plasma, a characteristic feature of two-dimensional structures.

III.3. <u>Effect of stimulated recombination in multiple quantum wells</u>

In low temperature MQW's, another interesting effect was observed which
relates to the ultrafast dynamics of absorption saturation. While interband recombination processes usually lie in the ns range, we have observed
in fairly narrow well samples (75 Å well width) a surprisingly fast recovery of absorption that could only be attributed to an ultrafast recombination of the photoexcited carriers[37].

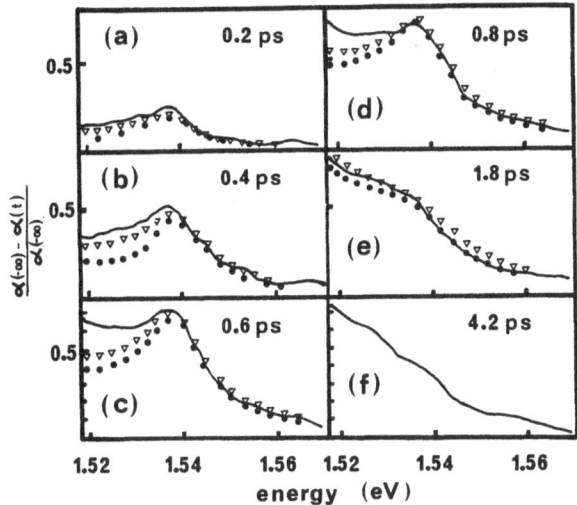

Fig. 5. Comparison between the experimental results of Fig. 4 (solid
line) and a model calculation (dots and triangles) of the
plasma dynamics with different screening efficiencies (from
ref. 34).

As shown in Fig. 7, the carrier density was found to decrease by more
than a factor of 2 within the first 10 ps when the sample was cooled at
15°K, while room temperature results displayed a much slower recombination.
This is due to the large stimulated recombination rate caused by amplified
spontaneous luminescence guided along the MQW structure, as confirmed by
time-resolved studies of the edge-emitted luminescence[37,38]. The temperature dependence of this effect, shown in Fig. 8, is particularly strong
between 80 and 140°K. Such a strong temperature dependence cannot be
explained by the conventional theories of gain in quantum wells[39], which

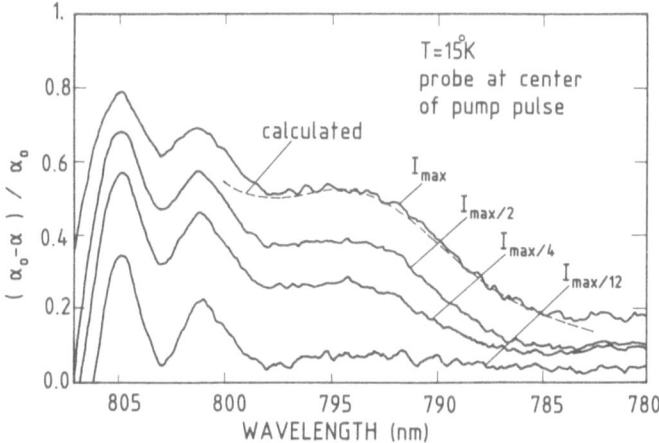

Fig. 6. Normalized difference spectra showing dynamic spectral hole-
burning in 114 Å thick GaAs quantum wells (from ref. 36). The
pump wavelength is 792 nm.

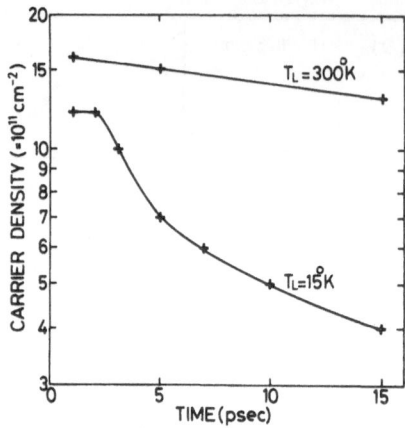

Fig. 7. Time evolution of carrier density in a photo-excited multiple
quantum well structure, for two values of sample temperature
T_L (from ref. 37).

points to a strong excitonic enhancement of stimulated recombination at
temperatures below 120°K[40]. The transmission spectra[41] observed under
conditions of small but non-zero gain (a few hundred cm^{-1}), show a clear
transition at 120°K between a high temperature regime in which both the
excitions and the bands are bleached, to a low temperature regime in which
absorption saturation only appears in the excitonic region.

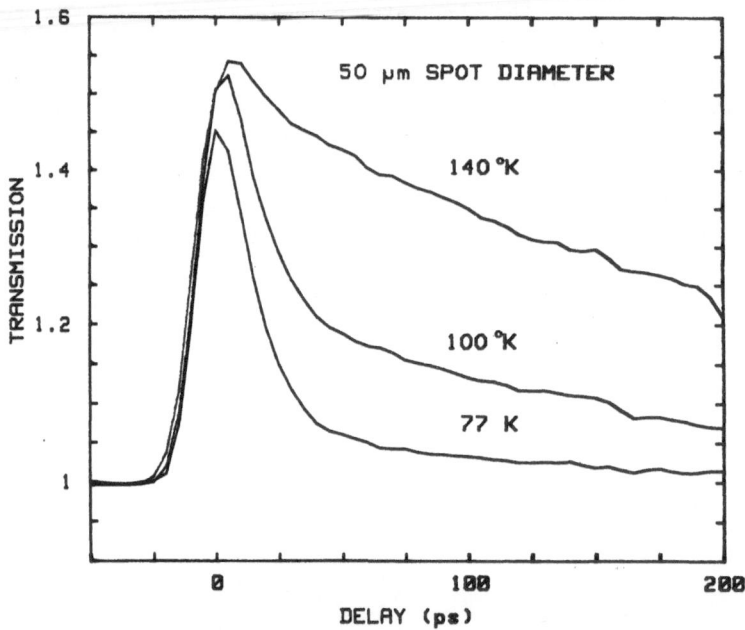

Fig. 8. Influence of the sample temperature on the absorption
recovery of 75 Å thick GaAs quantum wells.

IV. CONCLUSION

The preceding overview of dynamic optical nonlinearitis in III-V semiconductors has been used for discussion the general framework of time-dependent and frequency-dependent nonlinear optical response functions. The simple case of two-level systems has been useful for discussing the various limiting cases usually encountered in the transient regime of optical excitation. Specializing the discussion to the case of absorption saturation in direct-gap semiconductors, it has been shown that this effect provides a convenient tool for investigating the kinetics of photo-excited carriers at the ultimate temporal resolution allowed by femto-second light pulses. Finally, some specific examples of ultrafast non-linear optical response in GaAs-based epitaxial structures have been briefly reviewed. It is hoped that the insights provided by these funda-mental studies will be useful in improving the design of optical logic gates for optical computing.

ACKNOWLEDGEMENTS

The subjects briefly described here are the results of a close col-laboration with my coworkers, both at the Centre National d'Etudes des Telecomunications, Bagneux, and at the Laboratoire d'Optique Appliquée, Palaiseau. In particular I would like to thank Izo Abram and Rama Raj for fruitful discussions, as well as Danièle Hulin and Alain Migus for their pleasant cooperation.

REFERENCES

1. H. M. Gibbs, "Optical Bistability: Controlling Light by Light", Academic, New York (1985)
2. S. D. Smith, Appl. Opt. 25, 1550 (1986)
3. R. Dingle, in: "Festkörperprobleme XV", p.21, H. J. Queisser ed., Pergamon/Vieweg Braunschweig (1975)
4. L. Esaki, J. Phys., (Paris), C 5, 1 (1987)
5. D. S. Chemla and D. A. B. Miller, JOSA B 2, 1155 (1985)
6. Y. R. Shen, "The principles of nonlinear optics", Wiley, New York (1984)
7. J. L. Oudar, in: "Nonlinear Optics: materials and devices", p. 91, C. Flytzanis and J. L. Oudar, eds., Springer, Berlin (1986)
8. J. Jerphagnon, S. K. Kurtz and J. L. Oudar, "Nonlinear Dielectric Susceptibilities", Landoldt-Börnstein, New Series, Group III, 18, 456 (1984)
9. D. A. B. Miller, C. T. Seaton, M. E. Prise, and S. D. Smith, Phys. Rev. Lett., 47, 197 (1981)
10. B. S. Wherrett, in: "Nonlinear optics: materials and devices", p. 180, C. Flytzanis and J. L. Oudar, eds., Springer, Berlin (1986)
11. R. K. Jain and M. B. Klein in: "Optical Phase Conjugation", p. 307, R. A. Fisher, ed., Academic, New York (1982)
12. J. Shah, R. F. Leheny, and W. Wiegmann, Phys. Rev. B 16, 1577 (1977)

13. A. Maruani, J. L. Oudar, E. Batifol and D. S. Chemla, Phys. Rev. Lett. 41, 1372 (1978)

14. M. Dagenais and W. F. Sharfin, Appl. Phys. 46, 230 (1985)

15. P. N. Butcher, "Nonlinear Optical Phenomena", Columbus (1965)

16. T. K. Yee and T. K. Gustavson, Phys. Rev. A 18, 1597 (1978)

17. J. L. Oudar and Y. R. Shen, Phys. Rev. A 22, 1141 (1980)

18. P. X. Ye and Y. R. Shen, Phys. Rev. A 25, 2083 (1983)

19. L. Allen and J. H. Eberly, "Optical Resonance and Two-level Atoms", Wiley, New York (1975)

20. R. P. Feynman, F. L. Vernon, and R. W. Hellwarth, J. Appl. Phys. 28, 49 (1957)

21. N. Morita and T. Yajima, Phys. Rev. A 30, 2525 (1984)

22. T. Hattori and T. Kobayashi, Chem. Phys. Lett. 133, 230 (1987)

23. G. Harbeke, in: "Optical Properties of Solids", pp. 28-43, edited by F. Abeles, North Holland (1972)

24. R. N. Zitter, Appl. Phys. Lett. 14, 73 (1969)

25. J. Shah, R. F. Leheny and C. Lin, Solid State Commun. 18, 1035 (1976)

26. C. V. Shank, R. L. Fork, R. F. Leheny and J. Shah. Phys. Rev. Lett. 42, 112 (1979)

27. R. J. Elliott, Phys. Rev. 108, 1384 (1957)

28. T. M. Rice, Solid State Physics 32, pp. 1-86, edited by H. Ehrenreich, F. Seitz, and D. Turnbull, Academic, New York (1977)

29. H. Haug and D. B. Tran Thoai, Phys. Status Solidi B 98, 581 (1980)

30. H. Haug and S. Schmitt-Rink, Prog. in Quantum Electronics 9, 3 (1984)

31. R. Zimmermann, Phys. Status Solidi B 146, 371 (1988)

32. J. L. Oudar, A. Migus, D. Hulin, G. Grillon. J. Etchepare and A. Antonetti, Phys. Rev. Lett. 53, 384 (1984)

33. J. L. Oudar, D. Hulin, A. Migus, A. Antonetti, and F. Alexandre, Phys. Rev. Lett. 55, 2074 (1985)

34. J. Collet, J. L. Oudar and T. Amand, Phys. Rev. B34, 5443 (1986)

35. W. H. Knox, C. Hirlimann, D. A. B. Miller, J. Shah, D. S. Chemla, and C. V. Shank, Phys. Rev. Lett. 56, 1191 (1986)

36. J. L. Oudar, J. Dubard, F. Alexandre, D. Hulin, A. Migus and A. Antonetti, J. Phys. (Paris), C5, 511 (1987)

37. J. Dubard, J. L. Oudar, F. Alexandre, D. Hulin and A. Orszag, Appl. Phys. Lett., 50, 821 (1987); erratum ibid. 50, 1969 (1987)

38. D. Hulin, M. Joffre, A. Migus, J. L. Oudar, J. Dubard and F. Alexandre, J. Phys. (Paris), C5, 267 (1987)

39. N. K. Dutta, Electron. Lett. 18, 451 (1982)

40. S. Schmitt-Rink, C. Ell and H. Haug, Phys. Rev. B 33, 1183 (1986)

41. J. L. Oudar and J. A. Levenson, XVIth Internat. Conf. on Quantum Electron., paper ThB7, Tokyo (July 1988)

NONLINEAR OPTICAL MATERIALS AND DEVICES

N. Peyghambarian, S.W. Koch, H.M. Gibbs, and H. Haug*

Optical Sciences Center - University of Arizona
*Institut f. Theoretische Physik - Univ. of Frankfurt, FRG

I. INTRODUCTION

Nonlinear optical materials in which the index of refraction depends on the excitation density have emerged at the forefront of research because of their importance to optical communication, signal processing and computing[1-3]. These materials are the major components of nonlinear waveguides, bistable devices and optical logic gates. For high speed otpical signal processing and switching, devices should be fast with response times in the picosecond or femtosecond time domain.

The research on nonlinear materials is focused on both inorganic semiconductors and organic thin films. In direct-gap inorganic semiconductors, the absorption of photons excites electrons from the valence band into the conduction band. These electrically charged quasi-particles interact through the Coulomb potential. The interaction is attractive for oppositely charged particles, while it is repulsive for charges of the same sign. The attractive interband interaction causes a strong correlation between electrons and holes, and may lead to formation of bound states of excitons. In the so-called Wannier approximation, excitons may be regarded as hydrogen-atom-like quasi-particles, which are characterized both by a Bohr radius, a binding energy, and an effective mass. Organic semiconductors with π-conjugated electrons also exhibit interesting optical nonlinearities. Organic polymers have rapid response times which make them suitable for high speed switching applications.

The optical nonlinearities in these materials may be employed to realize devices such as an all-optical nonlinear directional coupler or so-called SEED, i.e., (self electro optic effect) devices. The optical nonlinearities are also a major contributing factor in the operation of semiconductor diode lasers. Here, we summarize optical nonlinearities of GaAs/AlGaAs multiple quantum wells, GaAs n-i-p-i structures, CdSe quantum dots, and organic compounds. Recent femtosecond experiments that resolve the time dynamics of the nonlinearities are mentioned. In particular, we describe the

Nonlinear Optics and Optical Computing
Edited by S. Martellucci and A. N. Chester
Plenum Press, New York, 1990

response time measurements of flouro-aluminum phthalocyanine thin films. The application of the nonlinearities to photonic switching is emphasized.

II. SEMICONDUCTOR OPTICAL NONLINEARITIES

II.1. GaAs-AlGaAs Multiple-Quantum Wells

Quantum confinement effects in optically excited semiconductor micro-structures arise if at least one spatial dimension of the material becomes comparable to or smaller than the exciton Bohr radius which is the charac-teristic length scale of an electron-hole pair. Well-known examples of such semiconductor systems are the multiple-quantum-well structures made of al-terning layers of active and transparent materials. Laser excitation in the appropriate frequency regime generates electron-hole pairs within the quasi-two dimensional active layers. These layers provide confinement in one space dimension which is already sufficient to largely enhance excitonic effects.

Modern crystal growth techniques like molecular beam epiataxy (MBE) and metalorganic chemical vapor deposition (MOCVD) make it possible to manufac-ture high quality multiple-quantum-well (MQW) structures with precise layer thickness and high uniformity[4]. The excitonic effects can be easily seen at room temperature in MQWs as a result of the confinement effect (which results in smaller exciton Bohr radius and consequently larger exciton binding energy, leading to an enhanced exciton oscillator strength) and small LO-phonon broadening.

Even though the physical processes causing renormalization of the opti-cal spectra near the band edge of bulk and MQW systems are somewhat differ-ent, the overall features of the nonlinearities themselves are quite simi-lar. The magnitude of the nonlinear refractive index has been measured for MQWs using degenerate four-wave mixing[5,6], pump-probe spectroscopy[7,8] and nonlinear interferometry[9,10]. The nonlinear and refraction have been measured as a function of well size at room-temperature. As an example, Fig. 1 shows the nonlinear absorption spectra for bulk GaAs and three MQWs. The well size in the quantum wells were 76 Å, 152 Å and 229 Å, respectively. The exciton peak which is barely resolvable in bulk material, is more prominent in the 299 Å sample as shown in Fig. 1b. For low pump intensity, exciton saturation dominates. For high intensity, the nonlinear behavior in the 299 Å sample is similar to that of bulk GaAs because of the relatively weak confinement. Figs. 1c and 1d exhibit similar curves for the 152 and 76 Å samples. The increase in the excitonic absorption is now quite apparent. Notice that separation of the heavy- and light-hole excitons is very clear in the 76 Å sample in contrast to the 152 Å sample. The corresponding variation in the refractive index, Δn, is calculated using a Kramers-Kronig transformation of absorption changes.

Curves of Δn versus intensity for all four samples are plotted in Fig. 2a. Note that for a given Δn, a lower pump intensity is required for the 76 Å MQW sample than for the bulk sample. These results indicate that the smaller MQWs have larger nonlinear refractive indices. However, we note that the comparison of intensity-dependent index changes is somewhat misleading. For

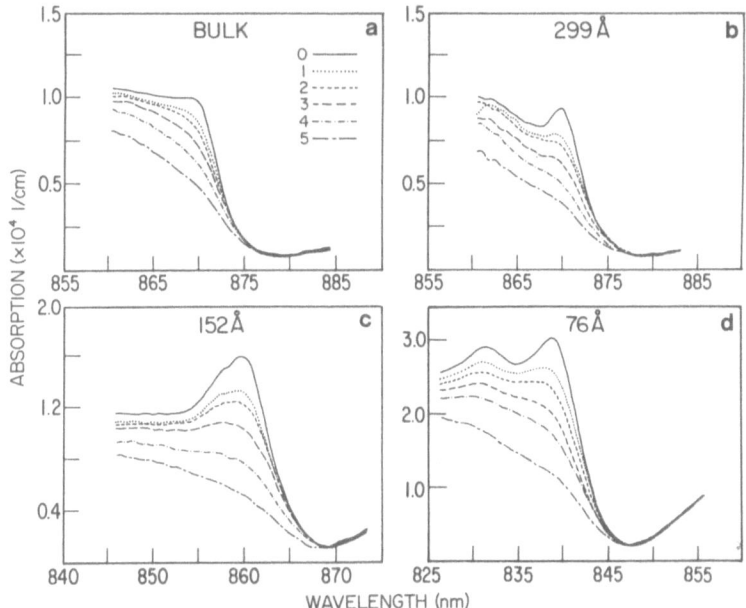

Fig. 1. Experimental room-temperature absorption spectra for: (a) Bulk GaAs; (b) 299 Å MQW; (c) 152 Å MQW; (d) 76 Å MQW. The curves labeled in the figure represent pump beam intensities (in W/cm^2) of : (0) 0; (1) 670; (2) 1270; (3) 2650; (4) 5400; (5) 11.700.

resonant excitation and sufficiently long pulses ($\geq$ ps), changes of the optical material properties depend on the excitation intensity only through the carrier density. Moreover, the various samples have a different absorption at the pump frequency so that the same excitation intensity generates a different number of carriers. Therefore, to obtain a more fundamental comparison of the material nonlinearities themselves the index change per excited carrier, $\Delta n/N$ is compared for these samples as shown in Fig. 2b. $\Delta n/N$ increases by a maximum factor of 3 for $N \cong 10^{17}$ cm^{-3} as the MQW well size decreases from bulk to 76 Å.

As can be seen form Fig. 1, the excitons bleach and broaden with increasing carrier concentration. Several models were employed in order to determine the excitonic absorption saturation with carrier concentration for the MQWs. A reasonable fit to the experimental data was obtained using the simple saturation model:

$$\alpha(N) = \frac{\alpha_0}{1 + N/N_s}$$

where α_0 is the linear absorption coefficient at the heavy-hole exciton peak and N_s is an empirical "saturation carrier concentration". The N_s and α_s for the MQW samples obtained from fits of the experimental results are given in Table I. The larger α_0 for the smaller-well MQWs result from the decrease in their exciton Bohr radii, a_B which causes an increased oscillator strength. These results indicate that the saturation density N_s is nearly well-size

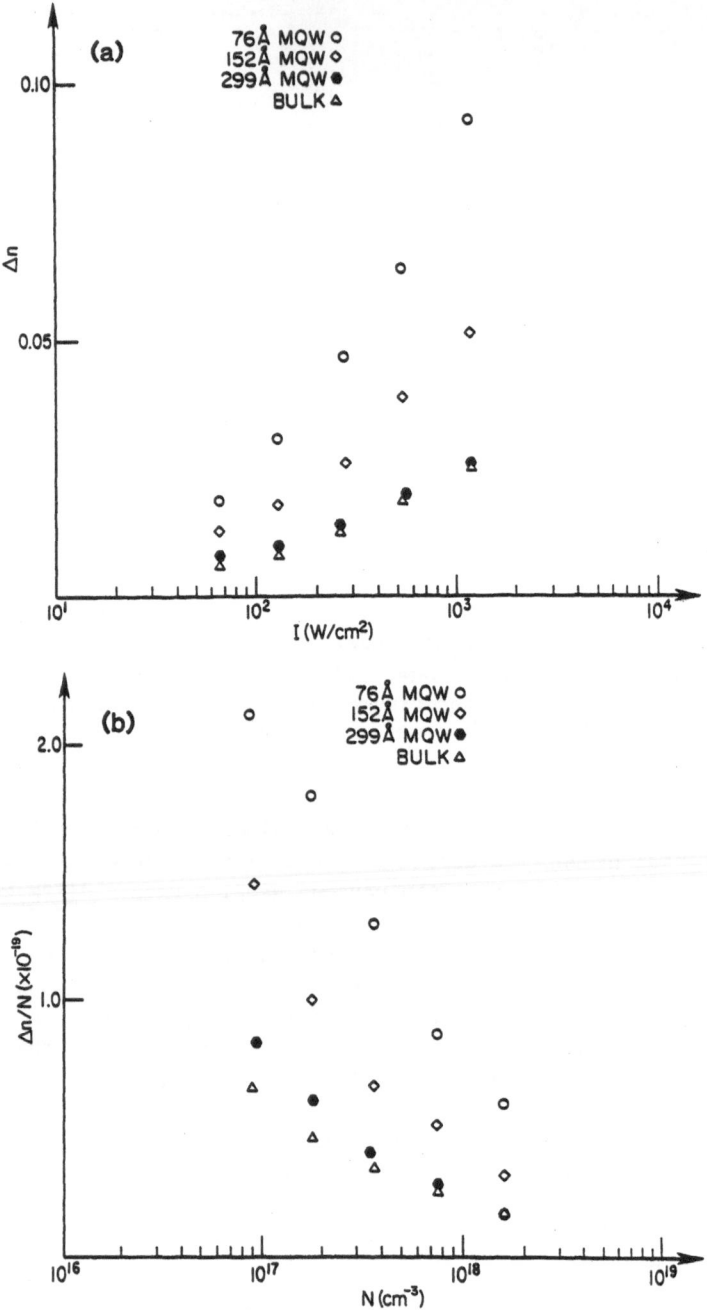

Fig. 2. (a) The maximum change in refractive index vs. pump beam
intensity; (b) the maximum change in index per carrier concentration vs. the carrier concentration.

independent and that the factor of three increase in the optical nonlinearities is mainly due to the factor of three increase in the exciton absorption.

Table I. Values of α_0 and N_s determined by least squares fit.

Sample	$\alpha_0 (\times 10^4 cm^{-1})$	$N_s (\times 10^{17} cm^{-3})$
76 Å	2.7	9.0
152 Å	1.4	7.2
299 Å	0.8	11.1

II.2. Semiconductor Doping Superlattices (n-i-p-i structures)

Doping superlattices, also called n-i-p-i crystals, are periodic arrays
of n- and p- doped layers separated by undoped intrinsic layers. These structures display novel characteristics such as a tunable absorption coefficient[11], tunable photoluminescence spectrum[12], and variable carrier lifetime[13]. Most significantly, the intensity levels required to obtain nonlinear transmission changes in GaAs n-i-p-i are orders of magnitude smaller
than in bulk or multiple-quantum-well GaAs. Therefore, these materials are
more suitable for low-speed optical computing and applications that require
a large number of switching devices to be operated simultaneously. In the
n-i-p-i structure, a space charge potential of ionized impurities in the
doping layers periodically modulates the bands. The ability to change this
built-in potential via charge carrier injection accounts for the novel properties. We have directly measured the nonlinear transmission spectrum as a
function of intensity as well as the intensity-dependent carrier build-up
and decay lifetimes[14].

The structure we used was grown by MOCVD and consisted of 10 periods of
alternating layers of n- and p- doped GaAs 130 nm thick with a doping concentration of 2×10^{17}; no intrinsic layers were included. With no excess
carriers injected, the built-in potential is approximately 1.32 eV corresponding to a maximum electric field of approximate3ly 1.95×10^5 V/cm in
the depletion regions. For the optical transmission experiments, the GaAs
substrate was removed.

In the experiment we directly measured the nonlinear transmission
changes through the sample by a pump-and-probe technique. The pump beam
from a cw dye laser was tuned to above the bandgap of the GaAs (λ=800 nm)
for efficient generation of free carriers. The pump and a broadband probe
were synchronously modulated by acousto-optic modulators; the pump and probe
pulse widths were 100 and 1 microseconds, respectively. The separation time
between pump pulses (typically 300 microseconds)) was chosen so that the
excess carriers have a chance to decay. The arrival time of the probe pulse
relative to the rising edge of the pump pulse could be varied continuosly,
even probing times after the pump is turned off.

The differential transmission spectrum $\Delta T/T$ is shown in Fig. 3 at
various pump intensities. As photocarriers accumulate and neutralize the
space charge, the internal electric fields are suppressed and the absorp-

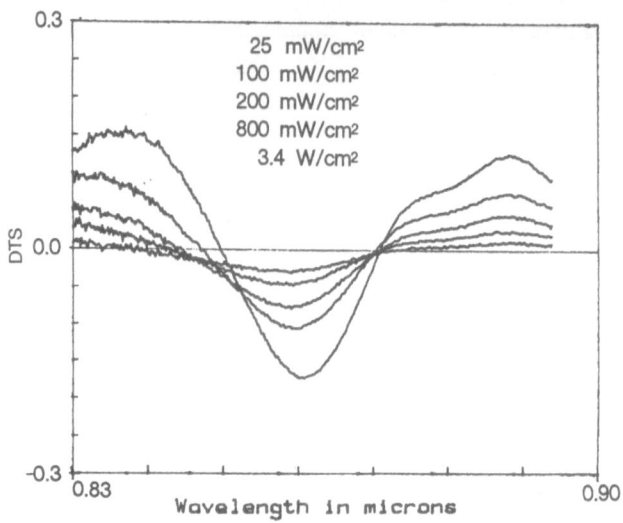

Fig. 3. Differential transmission spectra ΔT/T at room temperature for
different pump intensity values. The intensity values from
highest to lowest are 3.4 W/cm^2, 800 mW/cm^2, 200 mW/cm^2, 100
MW/cm^2, 25 mW/cm^2.

tion edge pivots to sharper slopes. This enhances the transmission at long
wavelengths while increasing the absorption in the band-edge region. Fig. 4
shows the build-up time required to obtain the maximum modulation for a
given intensity. For the higher intensity (12 W/cm^2) the maximum modulation reached steady state in 3 microseconds. However, this build-up time was
increased to 40 microseconds for a lower intensity (1.2 W/cm^2) pump. In
general, increasing the injection rate by orders of magnitude did not change
the carrier density proportionally, or at the same rate. In Fig. 5 we measured the intensity-dependent decay dynamics by delaying the probe relative
to the falling edge of the pump pulse. On a log-log scale the decay of the
maximum modulation is linear, indicating a decay time that increases exponentially with decreasing carrier density. Both the build-up and decay dynamics can be effectively modelled with a rate equation whose carrier decay
time depends exponentially on carrier density, N as shown in Fig. 4b. This
density-dependent carrier lifetime dependence is quite different from pure
bulk or multiple-quantum-well GaAs and is part of the reason why n-i-p-i
structures can show sizeable nonlinear effects with vey low intensities.

Even larger transmission changes may be obtained in hetero n-i-p-i
superlattices, where there is a compositional variation in addition to doping
modulation. An example of hetero-n-i-p-i superlattice uses a larger gap
material (such as AlGaAs) as the doping layers and a smaller gap material
(such as GaAs) as the intrinsic layer between the doped layers. In these
structures, there is a potential associated with the spatial variation of
band structure of the two different materials in addition to the space charge
potential. The hetero-n-i-p-i structures combine the unique features
of the compositional and doping modulated superlattices.

Recently, interesting optical nonlinearities have been observed in a

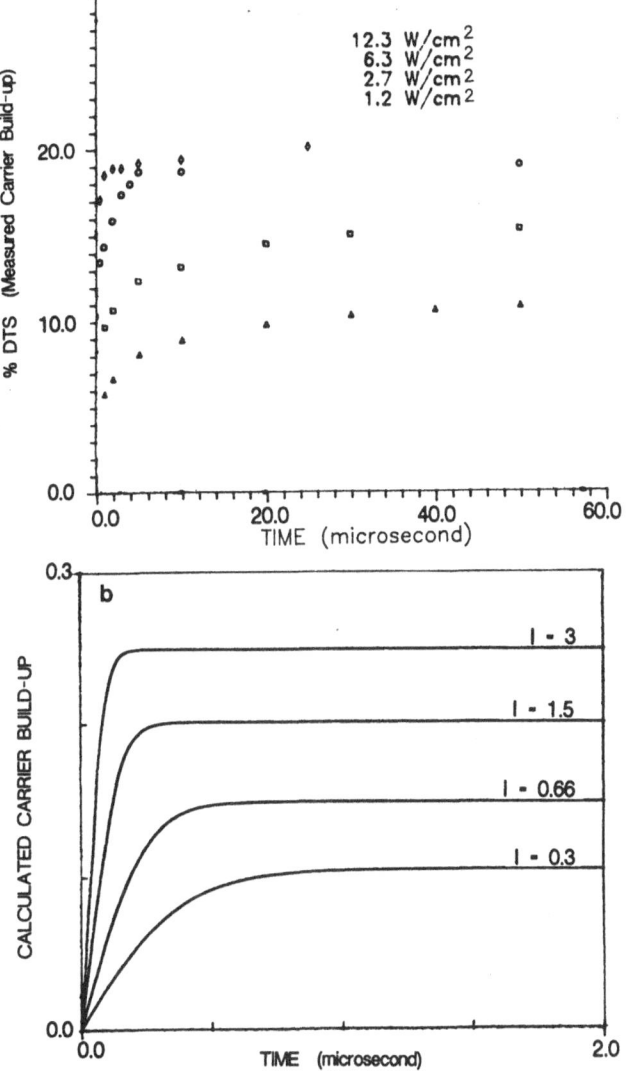

Fig. 4. Intensity-dependent carrier build-up dynamics. (a) The
maximum modulation $\Delta T/T$ available at the given intensity is
plotted as a function of time; (b) the results from a rate
equation model using a carrier-dependent lifetime are plotted.
$dN/dt = -N/\tau(N) = I$; $\tau(N) = \exp(-kN)$, $k = 10$, and $I = (3, 1.5,
0.66, 0.3)$.

GaAs-AlGaAs hetero-n-i-p-i superlattice[16]. The reported structure that was
grown by MOCVD used a MQW between doped layers of $Al_{0.32}Ga_{0.68}As$ with dopant
concentrations of $p \cong n \cong x \ 10^{18} cm^{-3}$. Absorption changes of more than 2000
cm^{-1} were observed with low intensities of 375 mW/cm^2. This corresponded to
an index change of $\cong 0.02$ or a value of $\Delta n/I$ of 50 cm^2/kW which is more than
two order of magnitude larger than that in a GaAs-AlGaAs MQW. Again the
larger nonlinear index had been obtained at the expense of the slower speed.

II.3. CdSe Quantum Dots

Stimulated by the interest in quantum-well, researchers are currently trying to further reduce the dimensionality of the semiconductor systems. It has become customary to use the term "quantum wires" for structures where the electron-hole pairs are confined in two space dimensions and "quantum dots" (QD) where three dimensional confinement occurs. Thus, as quantum dots we loosely refer to all those semiconductor micro-structures which confine the laser-excited electron-hole pairs in all three space dimensions. Examples of such systems are colloids[17,18] (small semicon-ductor particles suspended in liquid) and special glasses doped with CdS, CdSe, CuCl, or CuBr crystallites that clearly exhibit quantum confinement[19,20]. The microcrystallites in these glasses form out of the super-saturated solid solution of the basic constituents originally brought into the glass melt. The crystallites are more or less randomly distributed in the glass matrix. Under some growth conditions, the average crystallite size follows the Lifshitz-Slyozov law[21] $R \cong t^{1/3}$, where R is the crystallite size and it is the duration of the heat treatment during which the crystallites actually grow.

The theoretical investigations of quantum confinement in semiconductor microcrystallites have been pioneered by Efros and Efros[22] and by Brus[17,18]. Various regimes of quantum confinement have been introduced, depending on the ratio of the crystallite radius R to the Bohr radius of the electronhole pairs, $a_B = h^2 \varepsilon_2 / \mu e^2$, holes, $a_h = h^2 \varepsilon_2 / m_h e^2$, respectively, where $1/\mu = 1/m_e + 1/m_h$ and ε_2 is the background dielectric constant of the semiconductor material. Efros and Efros[22] attributed to these regimes quantization of the exciton, $R > a_B$ (weak quantization regime), quantization of the electron, $a_e > R > a_h$ (intermediate or moderate quantization regime), and quantization of electron and hole, a_e, $a_h > R$ (strong quantization regime), respectively. Using the effective mass approximation these authors showed that the increasing kinetic energy of the confined quasi-particles leads to a blue shift of the electron-hole-pair ground state energy. This blue shift is always proportional to $1/R^2$, but the prefactors are different in the different confinement regimes.

We have made a comprehensive study of steady-state nonlinear optical properties of specially-prepared quantum-confined CdSe microcrystallites suspended in a transparent borosilicate glass matrix. Three samples were investigated. the average crystallite diameters of these samples were measured using transmission electron microscopy to be 30 Å, 44 Å, and 79 Å, respectively. For bulk CdSe, the exciton Bohr radius (a_{ex}) is $\cong$ 56 Å therefore, our samples with smaller crystallite sizes, fall within the so-called intermediate confinement regime.

The room-temperature absorption spectrum of the samples is shown in Fig. 6. As reported in Ref. 20, a lower heat treatment temperature decreases the microcrystallite size and results in a confinement-induced energy shift of the lowest electron-hole transition (absorption edge) and the formation of discrete states. Evident from the figure, are three distinguishable peaks for the 30 Å sample. the lowest energy absorption peak is assigned to the transition from the highest hole subband level to the lowest electron subband level, $E_{1S,A}$, the intermediate peak, $E_{1S,B}$, to the transition from the

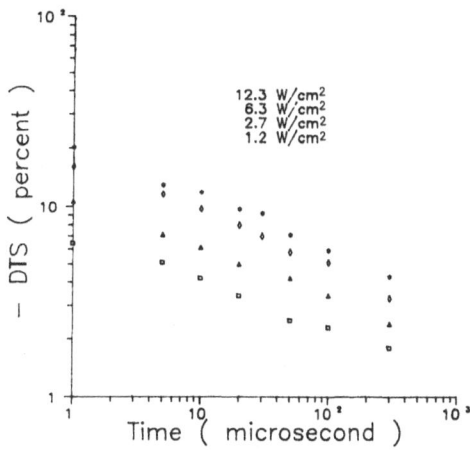

Fig. 5. Intensity-dependent carrier decay dynamics. The time decay of
the maximum modulation $\Delta T/T$ is plotted for different initial
pumping intensities.

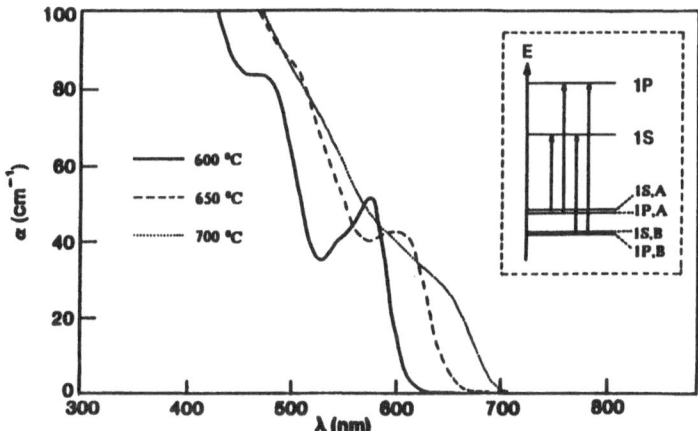

Fig. 6. Room temperature absorption spectrum of the CdSe QD sample.
Inset shows the simplified representation of the QD energy
levels.

hole level in the B-valence band to the lowest electron level, the highest
energy absorption peak to the second quantum-confined transition, $E_{1P,A}$.
These are shown schematically in the inset of Fig. 6.

To compare the QD-size dependence of the optical nonlinearities for
the three samples, we performed a simple self-saturation experiment, using
a nitrogen-pumped dye laser operating at a 20-Hz-repetition rate. However,
the photodarkening effect was observed at room temperature. When the irradiation increased above a certain threshold level, the transmission decreased
with time, i.e. number of pulses. At low temperature, photodarkening effects
decreased considerably.

The absorption vs. intensity at 10°K for the samples is shown in Fig. 7.
The wavelength was chosen at the $E_{1S,A}$ - transition peak for all three

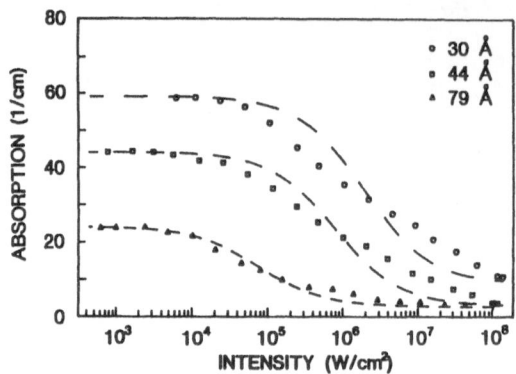

Fig. 7. Absorption at the peak of the first electronic transition
vs. intensity for the samples at 10°K.

samples. This data was fit to an absorption model with a nonsaturable (background) absorption. Taking data at large enough intensities (before the
onset of photodarkening) permitted us to almost completely bleach the samples; the dashed curves of Fig. 7 are plotted using least-square-fit parameters obtained from the above mentioned saturation model. These results
tend to indicate that smaller QDs exhibit larger saturation intensities.
However, the decay time of the nonlinearity should be included for a proper
comparison. Room temperature measurements have shown that the decay time is
shorter for the smaller QDs, as expected due to the increase in QD surface
to volume ratio.

Next, in order to better understand the behavior of the 30 Å QD optical
nonlinearities, our nanosecond laser system was used in a number of differential transmission spectroscopy (DTS) experiments. Fig. 8 shows the DTS for
various pumping intensities and their corresponding linear absorption spectra
(inset) at 300°K. Different pump wavelengths were used (indicated by the
vertical arrows).

Much can be learned from the results of Fig. 8: i) at a given pump
intensity, the DTS near $E_{1S,A}$ is more pronounced than that near $E_{1S,B}$; ii)
Pumping near $E_{1S,A}$ produces a larger DTS than pumping near $E_{1S,B}$ (the DTS is
larger in Fig. 8a than in 8b). These observations are consistent with the
above peak assignments: i) filling the 1S-electron state bleaches both the
$E_{1S,A}$ - and $E_{1S,B}$ - transitions, simultaneously, the more electrons we
excite into the 1-S - conduction band stae, the more the 1S-hole state transitions (from all subbands) are blocked; ii) the $E_{1S,A}$ - absorption is
greater than the $E_{1S,B}$ - absorption (see linear spectrum of Fig. 6); therefore, $E_{1S,A}$ - pumping bleaches more efficiently than $E_{1S,B}$ - pumping.

We have also measured the population relaxation time in the samples. A
300 fs decay was obtained for the 30 Å sample while 500 fs was measured for
the 44 Å sample[23]. The short time scales are attributed to scattering and
recombination involving surface states. These results indicate that the main
advantage of the quantum dots is their rapid response time. Thus, the
quantum dots may be employed for high speed switching application.

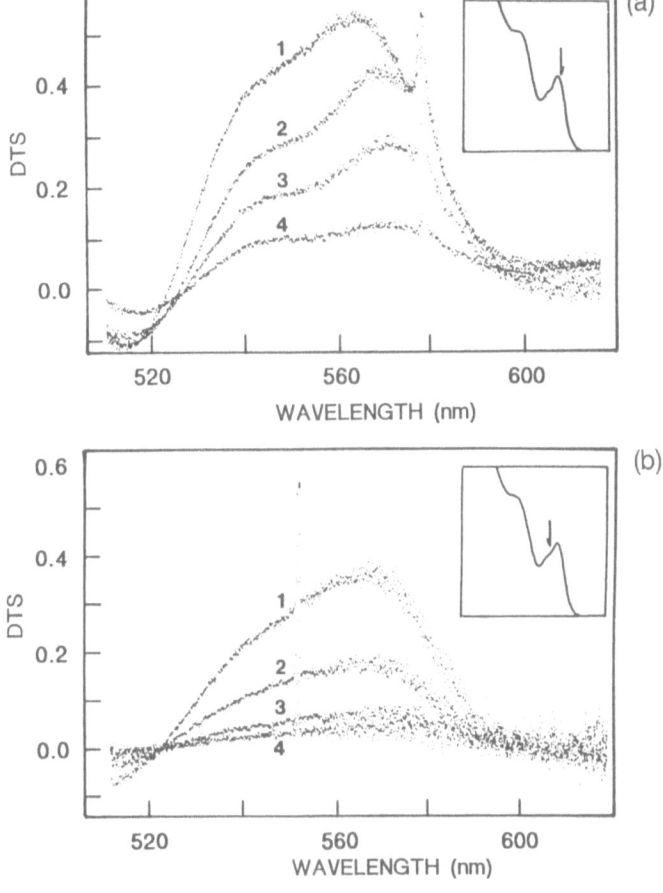

Fig. 8. DTS for the 30 Å QD sample at 300 K. (a) λ_p = 581 nm, the curves labeled in the figure represent pump beam intensities (in MW/cm^2) of (1) 4.7; (2) 2.4; (3) 1.3; and (4) 0.6; (b) λ_p = 552 nm, pump beam intensities: (1) 5.81; (2) 2.93; (3) 1.40; (4) 0.72.

III. NONLINEAR ORGANICS

Another class of nonlinear materials with ultrafast response time are the organics. It is well known that organic semiconductors with π-conjugated electrons may exhibit relatively large nonlinearities. Furthermore, the structural design flexibility of organics that can in principle be utilized to synthesize "engineered" materials, make the nonlinear optical properties of organics very interesting.

Here as an example, we describe our femtosecond experiments on thin films of flouro-aluminum phthalocyanine (AlPc-F). Phthalocyanines (Pc), which are thermally stable up to 400°C contain a two-dimensional π-electron system with or without a central metal atom or ion. Because of their strong absorption coefficient (α_{max} =1.5x10^4 cm^{-1}) in the visible region and their conductive properties, Pc thin films have been used in photovoltaic, photo-conductive and photoelectrochemical applications[24,25].

From third-harmonic generation measurements using the fundamental (1.06 μm) of a Nd: YAG laser, $\chi^{(3)}$ in polycrystalline phthalocyanine films was found to be 5×10^{-1} esu[26].

Whe have directly measured the electronic dynamics of AlPc-F thin films after excitation by femtosecond laser pulses[27]. AlPc-F polycrystalline thin films with thickness of 0.8 μm were grown by vapor sublimation at 150° C and 10^{-6} Torr pressure. The films were deposited on fused silica optical flats. The thin film surfaces have characterized by scanning electron micrographs and exhibit crystallites less than 2 μm long. AlPc-F exhibits electronic transitions in the visible (Q band at 600-800 nm) and in the near UV (B or Soret band at 300-400 nm) as shown in the linear absorption spectrum of our films in the inset of Fig. 9. Both Q and B bands, with E_u symmetry, arise from $\pi-\pi^*$ excitations.

In the experiment a more energetic pump pulse with wavelength of 6200 Å is first used to create and excited state population, then the probe beam is monitored at the electronic band edge between 7400 and 8600 Å. The differential transmission spectrum (DTS), is detected. Fig. 9 shows the DTS of the probe pulse at various time delays between the pump and probe pulses. The spectrum labeled - 150 fs was taken when the probe proceeded the pump by 150 fs. The signal is very small in this case because the probe pulse was monitoring the unexcited sample before the major portion of the pump pulse had arrived at the sample. At t = -100 fs, a noticeable increase in the transmission is observed as the pump and probe pulses were beginning to overlap. The DTS reached its maximum value at t = +100 fs.

Fig. 10 shows the evolution of the DTS signal at longer time delays up to 15 ps. The DTS signal diminishes as the ground state recovery completes.

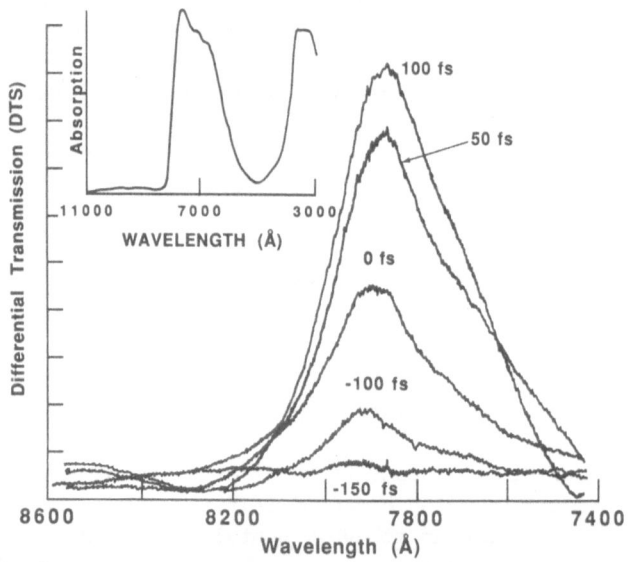

Fig. 9. The differential transmission spectra (DTS) at various delay
times. Inset: absorption spectrum of an 0.8 μm thick AlPc-F
polycristalline thin film.

The 15 ps spectrum shows a small detectable signal. The pump intensity dependence of the DTS signal is displayed in the inset of Fig. 10. This signal is detected at the peak of DTS, i.e. 7850 Å, for a fixed time delay of 100 fs. At lower pump intensities the DTS signal is linear to the applied power density. At an average power of 80 μW, which corresponds to a power density of $4x10^9$ W/cm^2, the signal reaches a plateau and a saturation effect is observed. The absorption bleaching (increase DTS) is 45% for $5x10^9$ W/cm^2 pump intensity. Fig. 11 exhibits the decay dynamics of the DTS signal as a function of time for different pumping intensities at 7850 Å. It is apparent that the excited population relaxes in a couple of picoseconds. Thus, the material has a much faster relaxation than the electron-hole recombination in inorganic semiconductors. However, the faster response time is obtained at the expense of smaller nonlinearity. The rapid decay time of organics such as AlPc-F may be used for high speed switching and signal processing applications.

IV. NONLINEAR OPTICAL DEVICES

Optical devices may be constructed using the nonlinearities described in the previous section. Nonlinear devices may be divided into etalons, where the light travels prependicular to the thin material film or wave-guides where the light propagates parallel to the surface. The material non-linearities also affect the operation of active device such as semiconductor diode lasers. Here we will give examples of each category of optical devices.

IV.1 Etalon Devices

A semiconductor etalon is formed when the material is sandwiched be-tween two partially reflecting mirrors. Nonlinear etalons may perform opti-cal bistability and logic operations. In such devices, the transmission of the device is abruptly altered as a result of the application of a switching laser beam. Optical bistability is a one-wavelength operation while optical

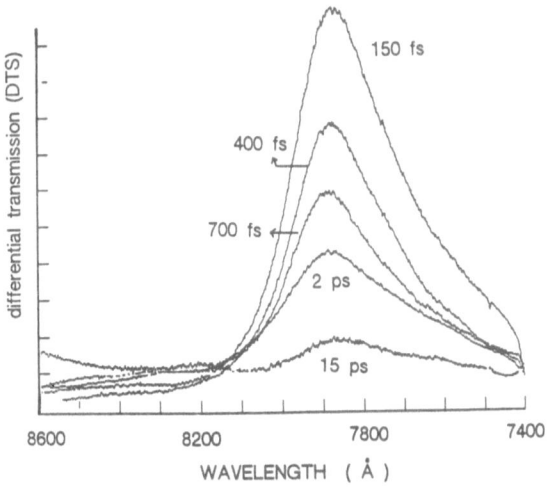

Fig. 10. The DTS at various delay times. The DTS intensity at 100 fs for λ = 7850 Å with different pumping powers.

111

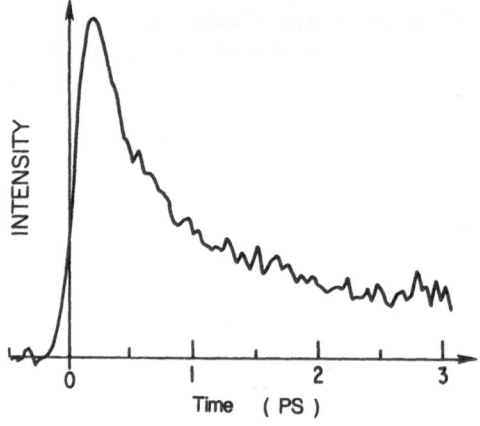

Fig. 11. The decay dynamics of AlPc-F for λ = 7850 Å.

logic gates usually employ two wavelengths. Various logic functions such as
AND, OR and NOR have been demonstrated using electronic nonlinearity GaAs-
-AlGaAs MQWs[28]. 1-ps switching-on and off times have also been demonstrated
in these materials using optical Stark effect[29].

IV.2. Waveguide Devices

There are several possible designs for semiconductor waveguides, such
as planar guides, channel structures, ridge waveguides and strip-loaded
structures, to name only a few examples[30-34]. The merits of the respective
designs depend to a large extent on the desired operation and on the ease
of fabrication. Most of the reported bistability in waveguides have been
of thermal origin. For the case of a MQW strip-loaded channel waveguide
electronic optical bistability has been reported recently .

Semiconductor waveguide devices can perform useful operations such as
nonlinear directional coupling. The nonlinear directional coupler (NLDC)
consists of two closely spaced parallel guides. The light is inserted into
only one of the guides but it tunnels into the neighboring guide after a
characteristic length which is the so-called half-beat length. The beat
length is a very sensitive function not only of the design parameters like
separation between the guides but also of the refractive indices of the
guides. Since the refractive index changes with changing carrier density,
the beat length changes too. Hence, for a directional coupler which has just
half the linear beat length, the light injected into one guide is totally
coupled into the other guide for low light intensity; whereas it might stay
in the original guide for sufficiently high intensities. Such a nonlinear
directional coupler has been realized[36] in a metal-strip-loaded GaAs
structure, where transverse light- and probably also carrier-confinement is
introduced by the strain field at the edges of the metal stripes. Nonlinear
directional coupling on picosecond time scale has also been realized recently in single-mode GaAs/AlGaAs strip-loaded waveguides[37]. The experiment was
performed by varying the intensity of a single input light beam to change
the relative transmission of the two channels. This single-beam method has

some undesirable features: the total absolute output of the NLDC in the bar-state is much higher than in the cross-state and the large detuning from the band edge needed for high transmission results in a decrease in nonlinearity.

Another technique was proposed and demonstrated that uses the output of a single laser, allows a GaAs/AlGaAs NLDC to be operated at a wavelength where its transmission is reasonably high, and still maintains efficient switching. This technique is based on the anisotropic optical properties of quantum well (QW) structures, as measured by Weiner et al.[39]. It shows that there is an effective blue shift of the band edge for the light polarized perpendicular to the QW layers due to the absence of the heavy-hole-exciton absorption feature. Thus using a control beam whose polarization is parallel to the QW layers, a large change in index of refraction can be induced. The signal beam can have a significantly larger transmission than the control beam because of its orthogonal polarization that gives it an effectively larger detuning from the band edge.

To test this idea, 10-ps pulses in the 830-890 nm wavelength region were used. The control and signal pulses were endfire-coupled into the NLDC. The guiding region consisted of a 100 A GaAs/AlGaAs MQW. The concept was first shown using a NLDC with 0.82 μm etch depth, 1 μm channel spacing, and 3.1 mm interaction length. Since the length of this NLDC is longer than one coupling length, light is first cross-coupled, and then coupled back into the input guide. In the absence of a control beam, a signal pulse at 880.5 nm wavelength, even with a high input energy (about 230 pJ), is unable to induce switching. However, when this amount of energy is put into a control pulse having the same wavelength but an orthogonal polarization, efficient

(a)

(b)

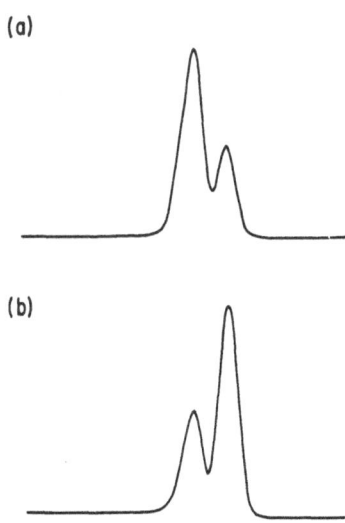

Fig. 12. Two-beam switching of a NLDC at 867 nm. Only the signal-pulse transmission is shown. The coupled-in energy of the control pulse is about 20-30 pJ, resulting in a 2:1 contrast in the signal-beam transmission. (a) un-switched (cross-state); (b) switched (bar-state). The two traces have the same scale.

switching of the signal pulse does occur. Another NLDC with 1.7-nm length and 1 -m channel spacing has shown a 2:1 contrast at 867-nm wavelength with only 20-30 pJ in the control pulse coupled into one channel (Fig. 12). The total absolute transmission of the signal pulse for both output states was nearly unchanged.

IV.3. Active Devices

Optical nonlinearities also affect the operation of active devices such as the semiconductor diode lasers. We have recently developed a many-body theory for semiconductor lasers which is valid both for three dimensional bulk materials and for quasi-two dimensional quantum-well structures[40]. Included in this theory are effects such as the reduction of the bandgap with increasing carrier density, the electron-hole plasma screening of the Coulomb potential, as well as the enhancement of the optical interband transitions due to the attractive electron-hole interaction. We start from the equation of motion for the interband polarization $P(k)$ between the dipole-coupled valence and conduction band[41-43].

$$(i\frac{\partial}{\partial t} - \Delta\varepsilon_{eff}(k) + i\gamma) \, P(k) = - \, (1 - f_h(k) - f_e(k)) \, (d_k E(t) +$$

$$+ \sum_{k' \neq 0} \cdot V_s(k - k') \, P(k')). \tag{1}$$

Here $f_{e/h}(k)$ are the distribution functions of electrons/holes in their bands, d_k is the dipole matrix element, $E(t)$ is the light field, V_s is the screened Coulomb potential, and γ^{-1} is the dephasing time. The transition energy $\Delta\varepsilon_{eff}$

$$\Delta\varepsilon_{eff}(k) = \varepsilon_e^0(k) + \sum_{k' \neq 0}(V_s(k')) - \sum_{k' \neq 0} V_s(k - k') \, (f_e(k') +$$

$$+ f_h(k')). \tag{2}$$

is renormalized due to screening and intraband exchange effects, and $\varepsilon_{e/h}^0$ is the unrenormalized (low excitation) energy of electrons and holes, respectively. Eq. (1) contains as limiting cases both the Wannier equation of electron-hole excitations in semiconductors and the quantum optical Bloch equation of the off-diagonal element of the two-level density matrix. Restricting ourselves to the time scale of picoseconds or longer, we can assume that the electron/hole distribution functions $f_{e/h}$ are given by Fermi functions with quasi-chemical potentials defined within the respective bands. Using a Pade approximation, we obtain the optical susceptibility $\chi(\omega)$ as[40]

$$\chi(\omega) = \frac{1}{V} \sum_k \frac{\chi_0(k,\omega)}{1-q_1(k,\omega)} \, , \tag{3}$$

where

$$q_1(k,\omega) = \frac{1}{d_k} \frac{1}{V} \sum_{k'} V_s(k - k') \, \chi_0(k',\omega). \tag{4}$$

The factor $1/(1-q_1)$ in Eq. (3) describes the Coulomb enhancement of

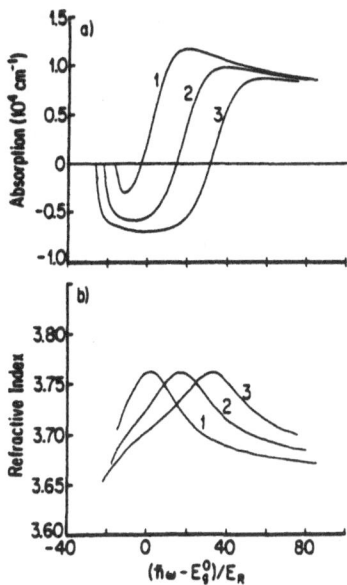

Fig. 13. Gain/absorption (upper part, a) and refractive index (lower
part, b) spectra for quasi-2d GaAs at room-temperature for
the plasma densities: (1) $2 \cdot 10^{12}$ cm^{-2}; (2) $4 \cdot 10^{12}$ cm^{-2}; (3)
$6 \cdot 10^{12}$ cm^{-2}.

the optical transitions caused by the interband attraction of electrons
and holes.

As usual, we obtain the spectra of absorption (gain) and index of
refraction from the real and imaginary part of susceptibility χ. Results for
the cases of bulk GaAs and quasi 2-d GaAs quantum-well material have been
reported in Ref. 44. As representative example, we reproduce in Fig. 13 the
gain/absorption and refractive index spectra for a quasi-2d GaAs quantum-
well at T=300°K and various plasma densities. The figure clearly demonstrates the expansion of the gain spectrum with increasing plasma density as
consequence of the increasing chemical potential (on the high energy side)
and the decreasing bandgap (on the low energy side), respectively. As is

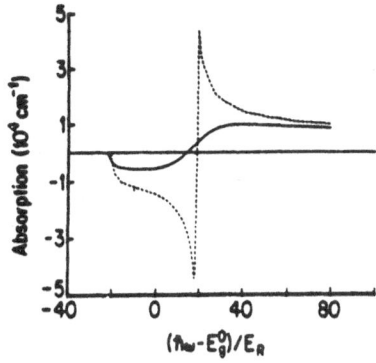

Fig. 14. Gain/absorption in spectra for quasi-2d GaAs for a density of
$4 \cdot 10^{12}$ cm^{-2} . T=10°K (dotted lines) and T=300°K (full lines).

discussed in Ref. 44, the shape of the gain spectra deviates significantly from that of bulk material, expressing the difference in the density of states in two and three dimensions and the stronger electron-hole Coulomb correlation in 2d. Fig. 14 demonstrates the dramatic influence of the temperature on the shape of the gain/absorption spectrum. At low temperatures the Coulomb enhancement factor has a pronounced maximum leading to sharp peaks in the gain/absorption spectra just below and above the chemical potential.

ACKNOWLEDGMENTS

The authors would like to acknowledge support from the Optical Circuitry Cooperative of the University of Arizona, the National Science Foundation (grant numbers EET8610170, and INT8713068 travel grant), JSOP, NATO (travel grant numbers 86/0747 and 87/0736), ONR/SDIO (grant number N00014-86-K-0719), DARPA/RADC (grant number F30602-87-C-0009), and the John von Neumann Computer Center for the CPU time. S. W. Koch is also with Physics Department at University of Arizona. This work was conducted in collaboration with S. H. Park, B. Fluegel, R. Morgan, J. Sokoloff, P. Harten, Z. Z. Ho, V. Williams, A. Jeffery, M. Lindberg, Y. Z. Hu, R. Jin, C. L. Chaung, M. Warren, G. Khitrova, A. Chavez-Pirson, J. Morhange, W. Gibbons, K. Komatsu, D. Sarid, D. Hendricks, D. Hulin, A. Migus, M. Joffre, A. Antonetti, D. W. Hall, N. F. Borrelli, J. N. Polky, A. C. Gossard, J. English, W. Wiegmann, M. Derstine, J. Lehman, P. Ruden, M. M. Hibbs-Brenner, G. A. Pubanz, and C. Ell.

REFERENCES

1. H. M. Gibbs, "Optical Bistability - Controlling Light with Light", Academic, New York (1985)
2. N. Peyghambarian and H. M. Gibbs, J. Opt. Soc. Am. B2, 1215 (1985)
3. S. W. Koch, N. Peyghambarian, and H. M. Gibbs, J. Appl. Phys. 63, R1 (1988)
4. R. Dingle, W. Wiegmann, and C. H. Henry, Phys. Rev. Lett. 33, 827 (1974)
5. D. A. B. Miller, D. S. Chemla. D. J. Eilenberg, P. W. Smith, A. C. Gossard, and W. Eiegmann, Appl. Phys. Lett. 42, 925 (1983)
6. M. W. Derstine, D. E. Grider. J. A. Lehman, P. P. Ruden, and N. Peyghambarian, proceedings of SPIE conference on "Optical Computing and Nonlinear Materials", ed. by N. Peyghambarian, vol. 881, p. 131 (SPIE, Los Angeles, 1988)
7. H. C. Lee, A. Hariz, P. D. Kapkus, A. Kost, M. Kawase, and E. Garmire, Appl. Phys. Lett. 50, 1182 (1987)
8. S. H. Park, J. F. Morhange, A. D. Jeffery, R. A. Morgan, A. Chavez-Pirson, H. M. Gibbs, S. W. Koch, N. Peyghambarian, M. Derstine, A. C. Gossard, J. H. English, and W. Weigmann, Appl. Phys. Lett. 52, 1201 (1988)
9. Y. H. Lee, A. Chavez-Pirson, S. W. Koch, H. M. Gibbs, S. H. Park, J. Morhange, A. Jeffrey, N. Peyghambarian, L. Banyai, A. C. Gossard, and W. Wiegmann, Phys. Rev. Lett. 57, 2446 (1986)

10. G. R. Olbright, N. Peyghambarian, Appl. Phys. Lett. 48, 1184 (1986)
11. G. H. Dohler, H. Kunzel, K. Ploog. Phys. Rev. B, 25, (4) (1982)
12. G. H. Dohler, H. Kunzel, D. Olego, K. Ploog, P. Ruden, H. J. Stolz, G. Abstreiter, Phys. Rev. Lett. 47, 864 (1981)
13. W. Rehm, P. Ruden, G. H. Dohler, K. Ploog, Phys. Rev. B28, (10) (1983)
14. A. Chavez-Pirson, S. H. Park, M. Pereira, N. Peyghambarian, J. A. Lehman, P. P. Ruden, and M. K. Hibbs-Brenner, proc. of CLEO, 1989 (Optical Society of America, Washington, DC, 1989)
15. G. H. Dohler, J. Vac. Sci. Technol. B1, 278 (1983)
16. A. Kost, E. Garmire, A. Danner, P. Dapkus, Appl. Phys. Lett. 52, 637 (1988)
17. L. E. Brus, J. Chem. Phys. 80, 4403 (1984)
18. L. E. Brus, IEEE J. Quant. Electron. QE-22, 1909 (1986)
19. A. I. Ekimov, A. L. Efros, and A. A. Onuschenko, Solid State Comm. 56, 921 (1985)
20. N. F. Borrelli, D. W. Hall, H. J. Holland, and D. W. Smith, J. Appl. Phys. 61, 5399 (1987); D. W. Hall and N. F. Borrelli, JOSA B5, 1650 (1988)
21. I. M. Lifshitz and V. V. Slyozov, J. Phys. Chem. Solid. 19, 35 (1961)
22. Al. L. and A. L. Efros, Sov. Phys. Semicond. 16, 772 (1982)
23. N. Peyghambarian, S. H. Park, R. A. Morgan, B. Fluegel, Y. Z. Hu, M. Lindberg, S. W. Koch, D. Hulin, A. Migus, J. Etchepare, M. Joffre, G. Grillon, D. H. Wall, and N. F. Borrelli, in: "Optical Switching in Low Dimensional Systems", ed. H. Haug and L. Banyai (Academic Press, 1989) in press.
24. W. J. Buttner, P. C. Rieke, and N. R. Armstrong, J. Phys. Chem. 89, 1116 (1985)
25. T. J. Klofta, P. C. Rieke, C. A. Linkous, W. J. Buttner, A. Nanthakumar, T. D. Mewborn, and N. R. Armstrong, J. Electrochem. Soc. 132, 2134 (1985)
26. Z. Z. Ho, C. Y. Ju, and W. M. Hetherington, J. Appl. Phys. 62
27. Z. Z. Ho and N. Peyghambarian, Chem. Phys. Lett. 148, 108 (1988)
28. J. L. Jewell, Y. H. Lee, M. Warren, H. M. Gibbs, N. Peyghambarian, A. C. Gossard, and W. Wiegmann, Appl. Phys. Lett. 46, 918 (1985)
29. D. Hulin, A. Mysyrowicz, A. Antonetti, A. Migus, W. T. Masselink, H. Morkoc, H. M. Gibbs, and N. Peyghambarian, Appl. Phys. Lett. 49, 749 (1986)
30. H. Kogelnik, Topics in Appl. Phys. 7: Integrated Optics (Springer Verlag, Berlin 1985)
31. V. Ramaswamy, Bell. Sys. Tech. J53, 697 (1974)
32. A. Yariv and P. Yeh, "Optical Waves in Couplets", Wiley & Sons, New York (1984)
33. G. I. Stegeman and R. H. Stolen ed., in: "Nonlinear Guided Wave Phenomena", special issue of JOSA B 5, 264-574 (1988) and references therein
34. P. R. Berger, Y. Chen, P. Bhattacharya, and J. Pamulapati, Appl. Phys. Lett. 52, 1125 (1988)
35. M. Warren, W. Gibbons, K. Komatsu, D. Sarid, D. Hendricks, H. M. Gibbs, and M. Sugimuto, Appl. Phys. Lett. 51, 1209 (1987)
36. P. LiKamwa, J. E. Sitch, N. J. Mason, J. S. Roberts, and P. N. Robson, Electron. Lett. 22, 1129 (1986)

37. R. Jin, C. L. Chuang, H. M. Gibbs, S. W. Koch, J. N. Polky, and G. A.
     Pubanz, <u>Appl. Phys. Lett.</u> 53, 1791 (1988)
38. R. Jin, C. L. Chuang, H. M. Gibbs, M. Warren, J. Sokoloff, P. Harten,
     N. Peyghambarian, J. N. Polky, and G. A. Pubanz, proceeding of
     CLEO, 1989 (Optical Society of America, Washington, DC, 1989)
39. J. S. Weiner, D. S. Chemla, D. A. B. Miller, H. A. Haus, A. C.
     Gossard, W. Wiegmann, and C. A. Burrus, <u>Appl. Phys. Rev.</u> A39,
     (February 1989)
40. H. Haug and S. W. Koch, <u>Phys. Rev. Lett.</u> A39, (February 1989)
41. see, e.g. "Optical Nonlinearities and Instabilities in Semicon-
     ductors, ed. H. Haug (Academic Press, N. Y. 1988);
     L. Banyai and S. W. Koch, Z. Physik B63, 283 (1986);
     S. Schmitt-Rink, C. Ell, and H. Haug, <u>Phys. Rev.</u> B33, 1183 (1986)
42. J. F. Muller, R. Mewis and H. Haug, <u>Z. Physik</u> B69, 231 (1987)
43. M. Lindberg and S. W. Koch, <u>Phys. Rev.</u> B38, 3342 (1988)
44. C. Ell, H. Haug and S. W. Koch, <u>Opt. Lett.</u>, 7, 356 (1989)

EXCITONIC OPTICAL NONLINEARITIES, FOUR WAVE MIXING AND

OPTICAL BISTABILITY IN MULTIPLE QUANTUM WELL STRUCTURES

A. Miller, P. K. Milsom and R. J. Manning

Royal Signals and Radar Establishment
Great Malvern, Worcestershire, WR14 3PS, U.K.

I. INTRODUCTION

Spatial confinement of optically generated free carriers and excitons in low dimensional semiconductors can result in enhanced nonlinear optical phenomena[1]. This is of practical consequence for optical switching devices[2], mode-locking of semiconductor lasers[3] and phase conjugation. For instance, confinement of excitons in GaAs/AlGaAs multiple quantum well (MQW) structures leads to clearly resolved absorption features at room temperature; nonlinear absorption and refraction can be observed at wavelengths close to these features at less than a milliwatt of incident c.w. optical power[1].

Although these are extremaly sensitive nonlinearities, several criteria must be met before they can be usefully employed in devices. An important requirement for excitonic nonlinearities is that sufficient change of optical constant be induced before the exciton absorption is saturated. Larger absorption and refractive index changes can be achieved as a result of band filling and band gap renormalisation, but usually at the expense of higher input powers[4].

The absolute magnitude and the detailed optical response of band gap resonant effects in semiconductors also depend on excess carrier time constants. Carrier diffusion in MQWs is highly anisotropic; this can effect carrier lifetimes and four wave mixing time constants determined by carrier diffusion.

In this paper, we discuss excitonic nonlinear refraction in GaAs /AlGaAs MOWs at room temperature to illustrate the implications of the saturation of the nonlinearity and temporal effects. We describe detailed measurements of the magnitude of the excitonic nonlinear refraction as a function of input power, and associated carrier dynamics, using the transient grating technique with sub-picosecond time resolution. These four wave mixing experiments are a development of the two beam, self-diffrac-

Nonlinear Optics and Optical Computing
Edited by S. Martellucci and A. N. Chester
Plenum Press, New York, 1990

tion studies by D. A. B. Miller et al[5], but allow the spatial dynamics of
the optically excited carriers to be monitored.

II. NONLINEAR REFRACTION

The physical processes responsible for nonlinear refraction associated with the excitonic absorption resonance in GaAs have been described
by Schmitt-Rink et al[6]. In quantum wells, the exciton binding energy is
enhanced to a degree where it exceeds the room temperature linewidth, the
thermal contribution to which is similar to bulk material, and excitonic
features are therefore clearly resolved, (see Fig. 2). Confinement also
raises the degeneracy at the top of the valence band such that an exciton
doublet is observed and may be associated with light and heavy hole bands
with the latter at lower photon energy. Any exciton created by optical
excitation is rapidly ionized (within a time, t < 400 fs) by LO phonon
collisions to form an electron-hole plasma[5]. Thus on the timescales encountered in experiments to be described here (t > 1 ps), excited carriers
fill the states at the bottom (top) of the conduction (valence) band,
inhibiting the creation of more excitons by phase space filling and coulomb
screening[7]. Nonlinear refraction results as a causal consequence of absorption saturation and is resonant with the exciton wavelength.

III. EXPERIMENTAL DETAILS

Four wave mixing results are presented for two MQW samples. Both were
fabricated by molecular beam epitaxy by Philips Research Laboratories,
Redhill, U.K., and consisted of 120 periods of 65Å thick GaAs quantum
wells with 212A $Al_xGa_{1-x}As$ (x~0.4) barriers. The substrates were removed
by selective etching, anti-reflection coatings deposited on both surfaces
and the samples mounted on sapphire[8].

A tunable, dye laser produced pulses of ~600fs duration between 810
and 870nm. A cavity dumper reduced the pulse separation to ~65ns at an
average power output of ~35mW. The output was split into two excite and
one probe beam for transient grating measurements; the probe pulses were
delayed via a stepper motor driven, variable path length stage. Detection
of first order diffracted light from the probe beams gave efficiencies and
lifetimes. From excite-probe measurements, the carrier lifetimes were
found to be 4 and 70ns for the two samples, KLB257 and KLB269 respectively[8]. The very different lifetimes were a result of different growth conditions.

Two separate four wave mixing configurations allowed for the creation
of free carrier gratings either in the plane of the wells of across the
wells. In the forward travelling geometry shown in Fig. 1a, two excite
pulses, incident at an angle ϕ, either side of the normal to the sample
surface, were coincident in space and time on the sample (spot size, ω_0 =
= 25 µm). In so doing, they interfere to form a sinusoidal intensity
pattern of period, λ, along the wells, determined by ϕ,

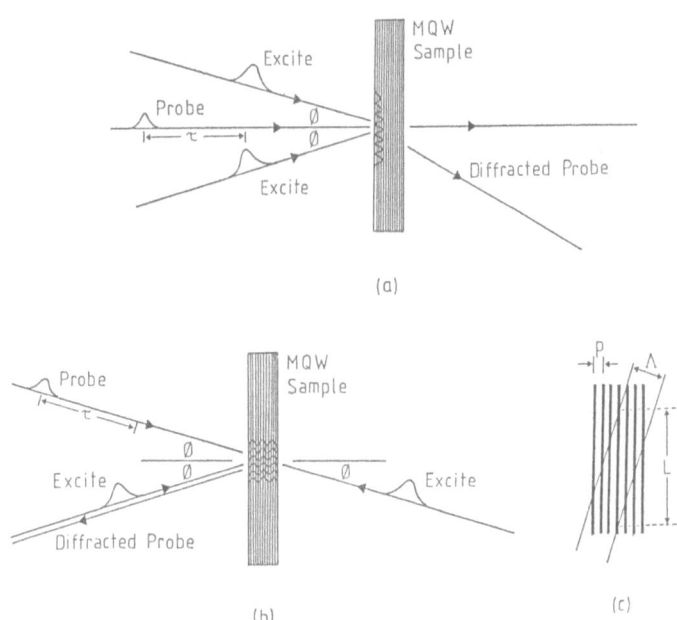

(a)

(b)   (c)

Fig. 1. Transient grating configurations, (a) forward travelling, and
(b) counter-propagating, and (c) an illustration of how the
quantum wells are positioned relative to the standing wave
intensity maxima at some rotation of the sample, θ, in case
(b).

$$\Lambda \ = \ \lambda/ \ 2 \ \sin \phi \tag{1}$$

The resulting spatially modulated carrier density forms a diffraction
grating; the angle of the diffracted light, γ, is given by,

$$\sin \gamma = \lambda \ / \ \Lambda. \tag{2}$$

The grating "washes out" because of carrier recombination and diffusion. Optical excitation provides equal numbers of electrons and holes so
that the diffusion in normal circumstances is "ambipolar", from regions of
high to low population density within the wells.

A counterpropagating (phase conjugate) geometry, Fig. 1b, produced a
short period grating perpendicular to the wells using two excite pulses
entering the sample from opposite directions at equal angles to the wells.
The probe was counterpropagating to one of the excite beams in this "phase
conjugate" configuration. The grating period, Λ, covers between 4 and 5
quantum wells. This grating is not normally observable in bulk semiconductors because it is rapidly washed out by carrier diffusion ($\sim 10^{-13}$s, in
bulk GaAs) over the modulation period of 115 nm. On the other hand, this
grating can be readily observed in semiconductor doped glasses because
carrier diffusion is inhibited[9]. Quantum wells provide an intermediate
case whereby diffusion of the carriers is restricted in one dimension,
therefore, the short period grating lifetime is significantly increased in
this geometry and dependent on cross well diffusion. However, note that

121

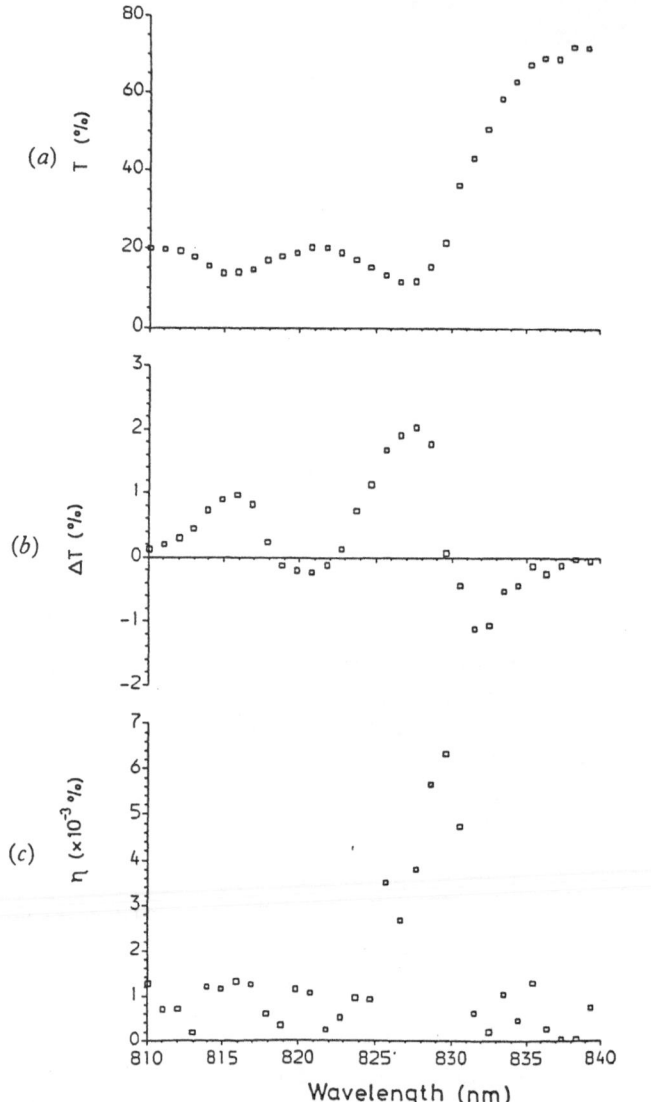

Fig. 2. The wavelength dependence of the low-power (a) transmission,
T, (b) differential transmission (ΔT) and (c) diffracted
probe efficiency, η, for sample KLB257 with a c.w. power of
~230 μW per excite arm.

the carriers are still free to move in the plane of the wells if the grating is angled with respect to the wells since a carrier concentration gradient can exist in this plane, Fig. 1c. In these measurements, the front excite beam was polarization modulated at 50 hKz for phase sensitive detection of the diffracted probe.

IV. FOUR WAVE MIXING

Fig. 2 shows (a) low power transmission, (b) differential probe transmission and (c) diffracted probe efficiency, η, as a function of

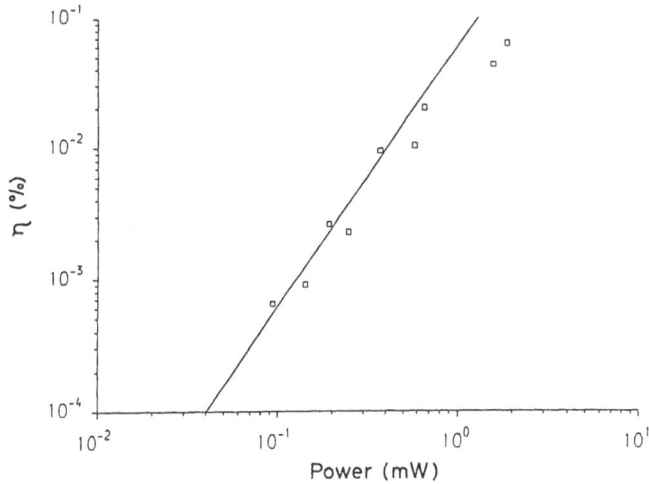

Fig. 3. Diffraction efficiency as a function of power in one excite
arm, taken at the wavelength for maximum diffraction
efficiency at 829.5 nm for sample KLB257.

wavelength for sample KLB257 using the geometry of Fig. 1a. The experiments were conducted with the probe delayed by 15 ps. Saturation and
broadening of the light and heavy hole excitons are clearly observed in
Fig. 2b. The peak diffraction efficiency, Fig. 2c, is on the long wavelength side of the 830 nm heavy hole exciton feature indicating a dispersive rather than absorptive grating. The diffraction efficiency, η, is
given by a first order Bessel function which for moderate excitation can
be approximated to,

$$\eta = (\frac{\pi\, n_{eh} N\, 1_\alpha}{\lambda})^2\, \exp(-\alpha 1) \qquad (3)$$

where n_{eh} is the nonlinear refraction cross section, α the averaged absorption coefficient, 1 is the sample thickness, 1_α is an effective sample
thickness which includes the effect of absorption , and N is the density
of carriers at the peak of a fringe. These low power results imply, n_{eh}
$\sim 0.4 \times 10^{-19}$ cm^3, slightly lower than that measured by Chemla et al[1].

The dependence of diffraction efficiency on excite beam power should
be quadratic at low input powers. This is confirmed, Fig. 3, up to 300 μW
(per excite beam) or a carrier density of 2×10^{17} cm^3, corresponding to
the limit of nonlinear refraction from excitonic saturation. This gives an
index change of about 0.01. At our maximum powers (~3 mW per arm), we
measured diffraction efficiencies as high as 0.1%, i.e. a refractive index
change of ~0.05 due to band filling.

V. TRANSIENT GRATINGS

In the configuration shown in Fig. 1a, the optically induced grating
spacing can be varied by altering the angle of incidence of the two excite
beams, equation 1. Fig. 4a shows the decay of diffraction efficiency with

123

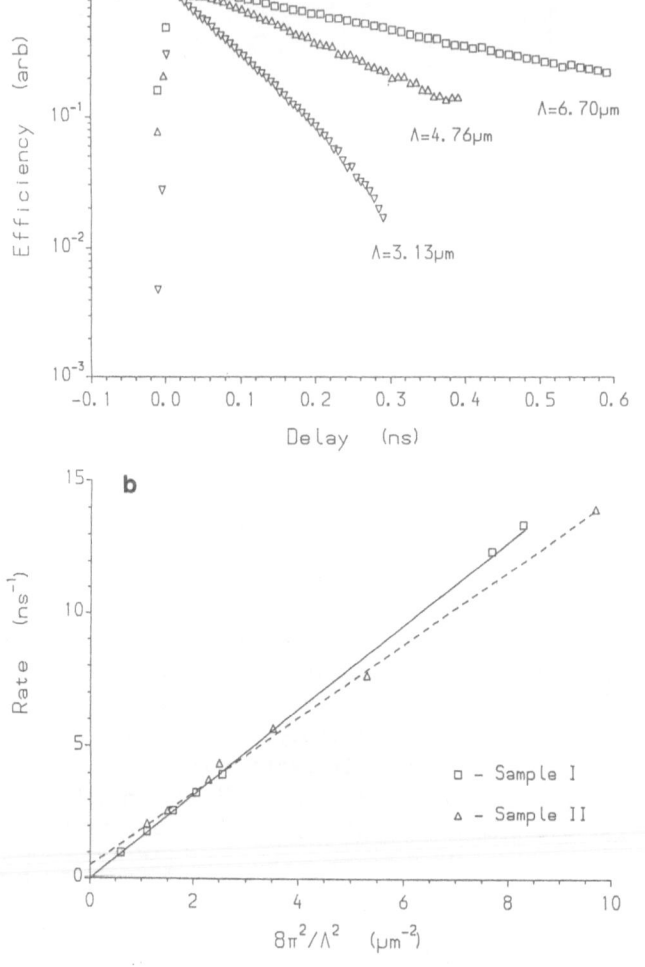

Fig. 4. (a) Diffracted signal decay rates measured in the forward
travelling configuration at different grating spacings, Λ, by
altering the angle of the excite beams, ϕ, sample KLB257;
(b) Rates against $8\pi^2/\Lambda$. (□ KLB257, △ KLB269).

time for three grating spacings (sample KLB257) measured at the wavelength
corresponding to maximum diffraction efficiency. The straight lines indicate
simple exponential decays; the different decays show the importance of carrier diffusion in washing out the grating. Efficiences are proportional to
the square of the modulation depth of the refractive index, thus, the
decay rate of the diffracted signal, r, is given by

$$r = \frac{8\pi^2 D_a}{\Lambda^2} + \frac{2}{\tau_R} \tag{4}$$

where D_a is the ambipolar diffusion coefficient and τ_R is the carrier recombination time. Fig. 4b shows the grating decay rates measured at different
angles, ϕ, for both samples; the gradients imply in-well ambipolar diffusion coefficients, D_a, of 13.8 (KLB257) and 16.2 cm^2/s (KLB269).

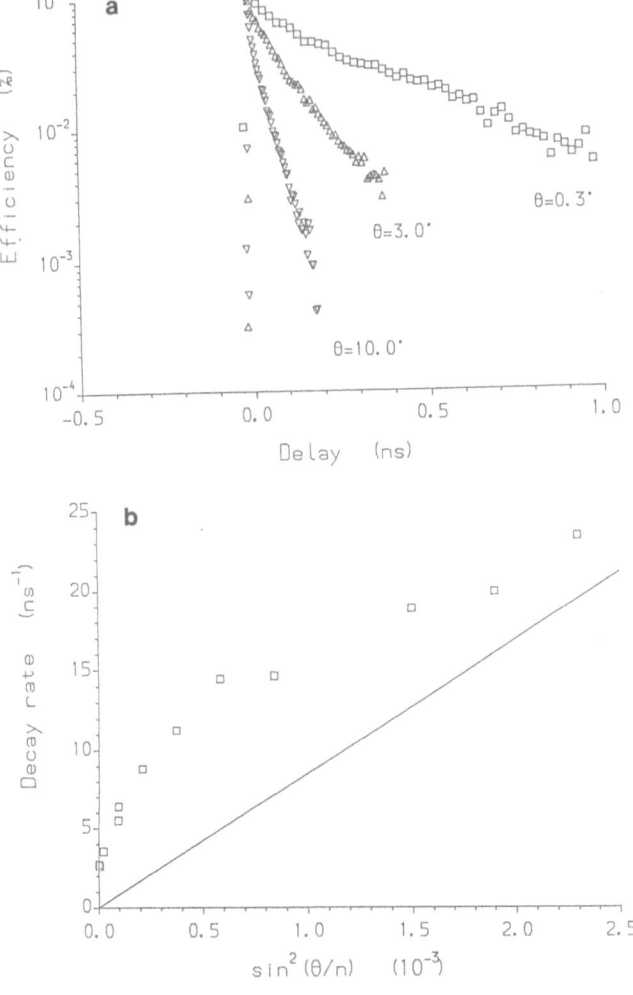

Fig. 5. (a) Diffracted signal decay rates measured in the counter-
propagating configuration for different rotations, θ, of
sample KLB269 from the normal; (b) Rates as a function of
angle. The solid line gives the predicted rate for purely
intra-well diffusion from the forward travelling measurements.

Fig. 5a shows the diffracted probe signal as a function of time for the
counter-propagating configuration of Fig. 1b. The angle, θ, is a measure
of the rotation from the normal position (grating perpendicular to the
wells) about a horizontal axis. Simple geometrical considerations of the
arrangement, Fig. 1c, predict that intra-well diffusion should contribute
to the observed decay rate when the grating is angled with respect to the
wells. Under these circumstances, the separation of carrier density maxima
along the wells is given by $L = n \Lambda / \sin\theta$, where n is the refractive
index, (Λ is essentially constant with sample rotation). If we assume that
the diffusion of carriers parallel and perpendicular to the wells is
independent, then equation 4 becomes,

$$r = \frac{8 \pi^2 D_a \sin^2 \theta}{n^2 \Lambda^2} + \frac{2}{\tau_\perp} + \frac{2}{\tau_R} \qquad (5)$$

where $\tau_\perp$ is the grating decay time due to cross-well diffusion. Fig. 5b plots the measured decay rates versus $\sin^2 \theta / n^2$ for sample KLB269. The intercept at zero angle implies a carrier grating decay time of ~1 ns. This timescale arises from a combination of the time taken to promote initially cold carriers out of the well by lattice heating and the time taken for these carriers to transverse 2 to 3 wells. Also plotted (solid line) is the anticipated relaxation rate due to intrawell diffusion given by Fig. 4b. In the simplest model, cross-well diffusion would be expected to give an angular independent contribution to the rate, but we also note an enhancement of the decay rate at small angles. This may be explained in terms of a differential emission rate over the barriers for electrons and holes (the holes have a lower barrier to cross). This leads to some degree of charge separation which would be rapidly cancelled by carrier movement along the wells.

VI. SOME CONSEQUENCES FOR OPTICAL BISTABILITY

One of the applications of band gap resonant optical nonlinearities in MQW structures is optical bistability and associated switching devices. We now consider some of the implications of the effects described above for Fabry-Perot bistable etalons.

Firstly, to achieve faster switching times, one technique employed for bulk semiconductors is to use surface recombination (by for instance leaving off the capping layers in MBE grown GaAs etalons). The restriction of carrier diffusion transverse to the layers means that rapid surface recombination could only be used in pixellated samples utilizing the fast diffusion direction along the wells. The possibility of density dependent recombination or diffusion times should also be taken into account, however no such dependences have been reported in MQWs at the carrier densities employed.

It is important to consider the fact that for MQWs there may only be a relatively small amount of active material, (summed thickness, D_{act}), for a given total thickness of etalon, D_{tot}. The full optimization criteria for minimizing the switching power needs to be considered. The optimization theories which have been developed, for "filled" absorbing etalons[10,11], assume an average cavity intensity, I_{cav}, so that the etalon transmission for a given incident intensity, I_i, is given by the solution of two equations linking I_i, I_{cav}, and the transmitted intensity, I_t, i.e.,

$$I_t \propto I_{cav} \qquad (6)$$

$$I_i \propto I_{cav} \left[1 + F \sin^2 \phi \right] \qquad (7)$$

where F is the coefficient of finesse, and ϕ is the round trip cavity phase. In the MQW case,

$$\phi = 2 \pi (n_0^{av} D_{tot} + n_2 D_{act} I_{cav}) / \lambda \qquad (8)$$

where n_0^{av} is an average linear index, n_2 is the nonlinear refraction of the active region. Wherrett et al[11] have derived an expression for the optimun thickness, D, for an etalon filled with nonlinear material in the high finesse limit (local nonlinearities),

$$D = (2 - R_F - R_B) / 4 \alpha \qquad (9)$$

where α is the absorption coefficient, R_F and R_B are front and back mirrors reflectivities. The switching criteria can be readily modified for etalons containing passive layers in the absence of spatial effects. We note that for bistability, an etalon containing nonabsorbing passive layers is equivalent in terms of switching conditions to one constructed without the passive layers (i.e. $D_{act} = D_{tot}$), so that given the same value of nonlinearity, n_2, they would both have the same switching intensity. Therefore, from equation 9 we see that if $D_{act} \ll D_{tot}$, optimization may require either a very thick total structure (many quantum wells) to increase the total active thickness of very high reflectivity mirrors. There is a difference between bulk and quantum well cases as mirror reflectivities approach 100%; in the bulk case, a $\lambda/2$ n thickness of material is the limit, whereas, in the partially filled etalon the optical thickness of the etalon may be the same, but there is still some flexibility in the choice of the number of wells. However, in more practical situations (reasonable reflectivities) the penalty to be paid for using MQWs is the extra thickness of material to grow optimized structures because of the inactive barrier regions.

The advantage of low dimensional structures may be thought to lie in enhanced nonlinearities. However, so far we have neglected the possibility of a limited amount of refractive index swing as is appropriate for the excitonic contribution. It then becomes important to design cavities which minimize both the required input power and the phase change required from each quantum well. Garmire[12] has recently considered this for bistable etalons and distributed Bragg reflectors. The conclusion was that for a Fabry-Perot structure, a figure of merit involving the saturated refractive index swing, Δn_{sat}, the absorption in the quantum well layer, and the wavelength, could be defined. From equations 6 and 7, we have derived the minimum index swing required to switch for values of R_F and R_B between 0 and 1, Fig. 6a, with the corresponding D_{act} (or number of wells) plotted in Fig. 6b. The absolute minimum index swing, when both R_F and R_B approach 100% can be solved analytically and is given by,

$$\frac{\Delta n_{sat}}{\alpha \lambda} > \frac{\sqrt{3}}{6\pi} \qquad (10)$$

which is slightly smaller than the Garmire result. By making comparison with the measurements of maximum refractive index change due to the exciton in GaAs quantum wells described in section 4, equation 10 confirms

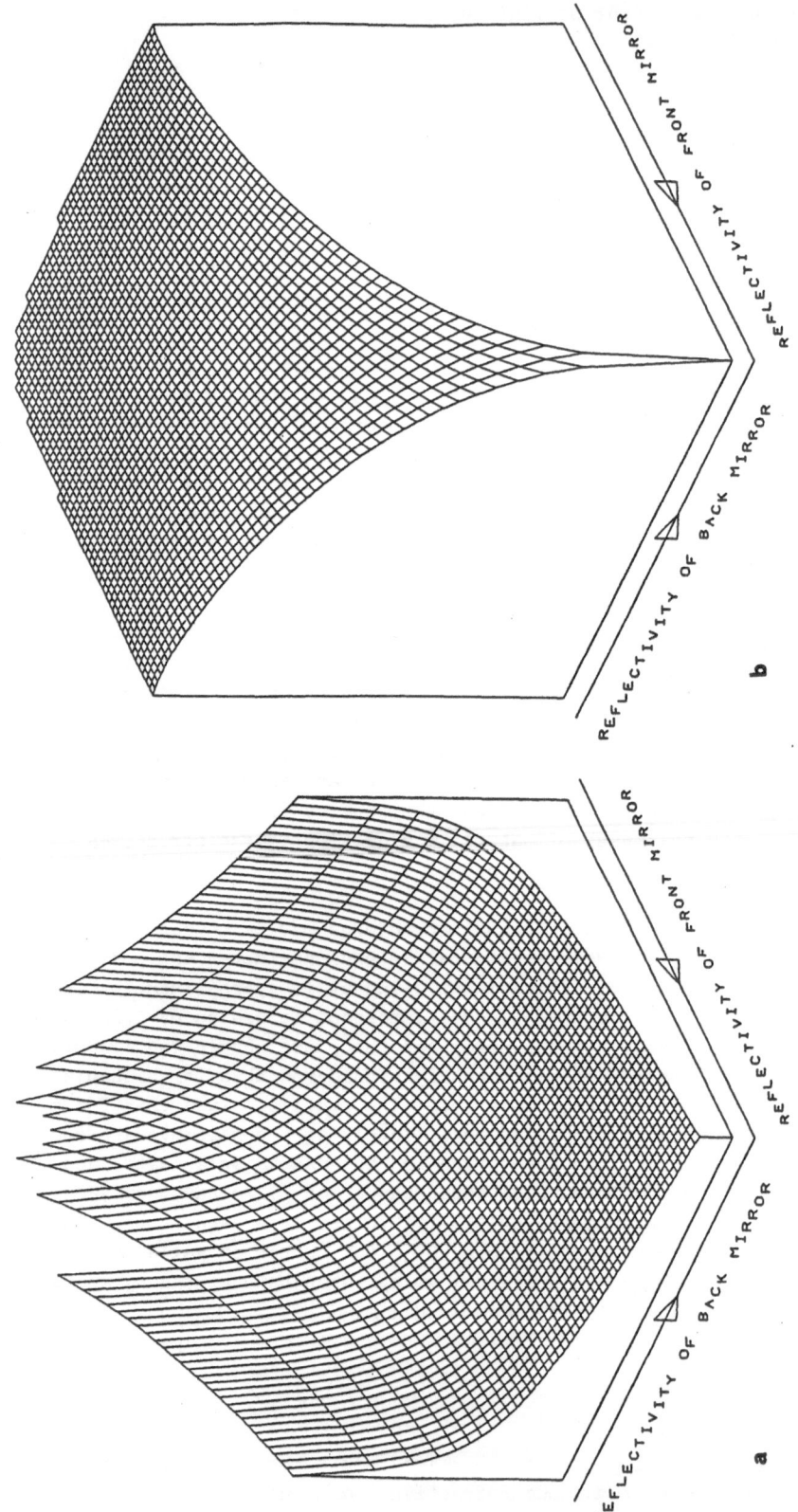

Fig. 6. (a) Minimized index swing versus front and back mirror reflectivities and (b) the correspnding dependence of D_{act}.

that it is not possible to achieve bistability at room temperature from this mechanism alone. The absorption is too high to utilize the large value of nonlinear refraction close to the exciton feature (the absorption does not completely saturate), but away from resonance, (smaller absorption) the excitonic nonlinear refraction is too small (see Fig. 2). On the other hand, the refractive index change due to band filling can be larger and falls off much less quickly with wavelength away from the band gap and thus satisfies this requirement.

VII. CONCLUSIONS

We have directly measured the saturated refractive index change due to the exciton in room temperature GaAs/AlGaAs MQWs as 0.01. Refractive index changes were measured up to 0.05 due to band filling at higher power levels. Transient grating measurements with sub-picosecond time resolution gave values of 13.8 and 16.2 cm^2/s for in-well diffusion for two different samples and revealed the consequencies of highly anisotropic carrier diffusion on band gap resonant four wave mixing in multiple quantum wells in the phase conjugate geometry.

The time constant associated with the short period grating in GaAs MQWs is shown to be enhanced by over three orders of magnitude compared to bulk material. A unique condition is created such that the temporal response can be controlled simply by sample rotation. These results may have implications on image processing using phase conjugation whereby the variable time constant means that the spatial resolution may be controlled. A variable time constant may also be useful within laser cavities where the operation of mode-lockers or phase conjugate mirrors which utilize counter-propagating beams could be influenced.

A theoretical analysis of cavity optimization for optical bistability for saturating nonlinearities shows that the high level of residual absorption at the wavelength of high refractive index change due to excitonic saturation in GaAs/AlGaAs means that bistability is not possible from this mechanism alone. Optical bistability in GaAs MQWs is therefore due to band filling which gives a larger index swing at lower absorption.

ACKNOWLEDGEMENTS

We thank K. Woodbridge of Phillips Research Laboratories for growing the samples used in these studies. We are grateful to B. S. Wherrett, D. Hutchings, A. Vickers and D. Herbert for useful discussions and D. Crust and K. Powell for experimental assistance.

REFERENCES

1. D. S. Chemla, D. A. B. Miller, P. W. Smith, A. C. Gossard, and W. Wiegmann, IEEE J. Quantum Electron. QE-20:265 (1984)
2. H. M. Gibbs, S. S. Tarng, J. L. Jewell, D. A. Weinberger, K. Tai, A.

       C. Gossard, S. L. McCall, A. Passner and W. Wiegmann, <u>Appl. Phys. Lett.</u> 41:221 (1982)

3. Y. Silberberg, P. W. Smith, D. J. Eilenberger, D. A. B. Miller, A. C. Gossard and W. Wiegmann, <u>Opt. Lett.</u> 9:507 (1984)

4. Y. H. Lee, A. Chivez-Pirson, B. K. Rhee, H. M. Gibbs, A. C. Gossard and W. Wiegmann, <u>Appl. Phys. Lett.</u> 49:1505 (1986)

5. D. A. B. Miller, D. S. Chemla, D. J. Eilenberger, P. W. Smith, A. C. Gossard and W. Wiegmann, <u>Appl. Phys. Lett.</u> 42:925 (1983)

6. S. Schmitt-Rink, D. S. Chemla and D. A. B. Miller, <u>Phys. Rev. B</u> 32:6601 (1985)

7. W. H. Knox, C. Hirlimann, D. A. B. Miller, J. Shah, D. S. Chemla and C. V. Shank, <u>Phys. Rev. Lett.</u> 56:1191 (1986)

8. R. J. Manning, D. W. Crust, D. W. Craig, A. Miller and K. Woodbridge, <u>J. Mod. Opt.</u> 35:541 (1988)

9. R. K. Jain and R. C. Lind, <u>J. Opt. Soc. Am.</u> 73:647 (1983)

10. A. B. Miller, <u>IEEE J. Quantum Electron.</u> QE-17:306 (1984)

11. B. S. Wherrett, D. Hutchings and D. Russell, <u>J. Opt. Soc.Am. B</u> 3:351 (1986)

12. E. Garmire, Criteria for optical bistability in a lossy saturating cavity in: "Optical Bistability IV", Les Editions de Physique, Paris (1988)

NONLINEAR OPTICS OF A SINGLE SLIGHTLY-RELATIVISTIC

CYCLOTRON ELECTRON

A.E. Kaplan and Y.J. Ding

Department of Electrical and Computer Engineering
The Johns Hopkins University, Baltimore, MD 21218

INTRODUCTION

The interaction of microwave and optical radiation with a tiny rela-tivistic single electron can result in strong nonlinear-optical effects[1-7] based on the most fundamental mechanism of nonlinear interaction of light with matter[3]. Even a slight relativistic change of mass of a single free electron may result in large nonlinear effects such as hysteresis and bistability in cyclotron resonance of the electron precessing in a dc magnetic field under the action of an EM wave[1]. The relativistic mass-effect consists in the increase of the effective mass of electron, m, as its speed v, or energy W, or momentum p, increase:

$$m/m_0 = \gamma = (1-v^2/c^2)^{-1/2} = (1+p^2/m^2c^2)^{1/2},$$

where m_0 is the electron rest mass, $\gamma = W/m_0c^2$ is the dimensionless electron energy, and c is the speed of light. Because of very low energy losses (which are due to synchrotron radiation), the relativistic change of mass to which the hysteretic resonance is attributed may be as strik-ingly small as 10^{-10} - 10^{-6}. Consistent with predictions[1], the hysteretic (bistable) cyclotron resonance of a free electron has been subsequently observed in an experiment[2] in which a single electron has been trapped in a Penning trap for a period of time as long as 10 months.

The hysteretic cyclotron resonance[1-3] occurs at the main frequency, i.e., in the situation when the driving frequency, Ω, is in the close vicinity of the unperturbed cyclotron frequency, Ω_0, see Fig. 1a (for presently available dc magnetic fields the maximum cyclotron frequency has a wavelength in the millimeter or submillimeter range). The strongly non-linear cyclotron resonance can also be excited by optical pumping with a driving laser frequency (or frequencies) much higher than the cyclotron frequency[4-7]. All these optical excitations are, in fact, multiphoton processes that can be described as relativistic nonlinear optics of a single electron. In particular, a strong cyclotron excitation can be obtained by biharmonic laser pumping when two laser frequencies, ω_1 and ω_2

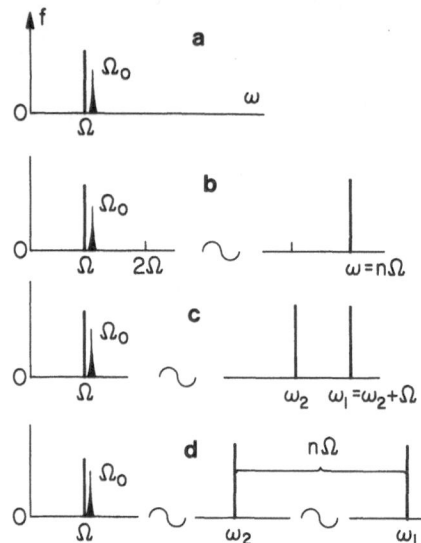

Fig. 1. Spectral arrangements of driving frequencies with respect to
the unperturbed cyclotron frequency Ω_0 for: (a) the main
resonance at the frequency $\Omega \cong \Omega_0$; (b) the optical subharmonic resonance at the frequency $\omega/n \cong \Omega_0$; (c) the first-
order cyclo-Raman resonance at the driving frequency $\omega_1 - \omega_2 \cong \Omega_0$
(d) the arbitrary order cyclo-Raman resonance with the driving
frequency $\omega_1 - \omega_2 \cong n\Omega_0$.

differ by either Ω (e.g. $\omega_1 - \omega_2 = \Omega$), see Fig. 1b, or 2Ω (i.e. $\omega_1 - \omega_2 = 2\Omega$), or even $n\omega$ (i.e. $\omega_1 - \omega_2 = n\omega$), where n is arbitrary integer[7], see
Fig 1d. These effects may be regarded as a three-photon, four-photon, and
n+2-photon interactions respectively or, more specifically, as stimulated
cyclo-Raman scattering of first, second and or n-th orders; for some
particular propagation configurations they can exhibit the so called
isolas[6,7]. It was also shown recently[5] that a single electron can exhibit
high-order cyclotron subharmonics in which the ratio of driving (single)
laser frequency to the frequency of cyclotron excitation is an arbitrary
integer, n (this corresponds to an n-photon process), see Fig. 1c. The
unified theory[4] of all these effects is based on the decomposition of
electron motion into a purely cyclotron component and noncyclotron components, the latter ones including higher-order oscillations with all
possible frequency combinations. This theory allows one to obtain general
results which are valid for an arbitrary energy of electron excitation,
and for arbitrary order of interaction.

Since all of the characteristics of the nonlinear interactions of EM
radiation with a single electron are directly related to fundamental constants, they can be used for measurement of these constants with greatly
enhanced precision. The hysteretic excitation of a single electron can be
viewed as the ultimate multistable interaction of light with matter, since
it relies on the bistable interaction of an EM wave with the single
simplest microscopic physical object and is based on a fundamental effect,
the relativistic change of its mass. This effect also suggests optical
bistability[8] based on the intrinsic property of a microscopic object rather

than on its macroscopic property in a nonlinear medium. In the future, it could be important to study hysteresis in the vicinity of its threshold, which will provide a unique opportunity to explore the quantum limit of bistable oscillators.

In the multiphoton interaction of an _optical_ laser with an electron, cyclo-Raman resonances offer a new method of excitation of the cyclotron motion which may prove more advantageous than conventional methods utilizing either mw or rf oscillators. Indeed, since the optical frequencies, ω_1 and ω_2, can be provided by two modes of the same laser, optical excitation allows for easily tunable control over the difference frequency ($\omega_1 - \omega_2$). The power of laser light required to obtain cyclotron excitation is sufficiently low to allow for the use of lasers in a cw or quasi-cw regime.

The proposed effect may also be used for particle acceleration[4,6]; even in a simple Penning trap, kinetic energy of an excited electron as high as a few Mev may be obtained. For the nth order cyclo-Raman resonance as well for n[th] order subharmonics, the excited electron can have n possible phases of cyclotron excitation (which differ by π/n); the phase which is excited depends on the initial conditions. This is a manifestation of a new type of optical bistability which may be regarded as phase multistability[4,6,9] (i.e., based on multistability of the phase of oscillation rather than on multi-stability of its amplitude). The high-order subharmonics[5,10] may provide coherent links between laser and rf or microwave frequency standards. For example, to divide the frequency of a CO_2 laser ($\lambda \cong 10\mu m$) by a factor of 100 down to $\lambda \cong 1mm$ in one step, cw laser as low as $10^{-6}W$ is sufficient[5]. It is worth noting that most nonlinear effects discussed above are also feasible[11,12] in some narrow-gap semiconductors, making use of the pseudo-relativistic behavior of the effective mass of their conduction electrons[13,17].

I. LORENTZ EQUATION AND RELATIVISTIC NONLINEARITIES

Consider a single electron in a homogeneous magnetic field H_0 with the unperturbed (nonrelativistic) cyclotron frequency $\Omega_0 = e\, H_0/m_0 c$. The EM wave impigent on the electron may, in general, consist of any number of plane traveling waves $\vec{E}_j(\omega_j t - \vec{k}_j \cdot \vec{r})$ where ω_j and $\vec{k}_j$ are, respectively, the frequency and wave vector of the jth traveling wave. We treat the problem classically; the motion of an electron with an arbitrary momentum $\vec{p} = m_0\, \gamma\, \vec{v}$ is governed therefore by the relaxation-modified Lorentz equation[17-23]:

$$\frac{d\vec{p}}{dt} = e \sum_j \vec{E}_j + \frac{e}{\gamma m_0 c}\, \vec{p} \cdot \sum_j (\vec{k}_j \cdot \vec{E}_j/k_j) + \frac{e}{\gamma m_0 c}\, (\vec{p} \cdot \vec{H}_0) + e\vec{G}_{tr} + \vec{F}_1 \quad (1)$$

where e is the electron charge. The second term on the right-hand side of Eq. (1) is the Lorentz radiation force of the incident EM waves; it is attributed to the magnetic field of the EM wave, $\vec{H}_j = (\vec{k}_j \cdot \vec{E}_j)/k_j$. A trapping dc electric field $\vec{G}_{tr} = \vec{G}_{tr}(\vec{r})$ is provided by the Penning trap potential.

The term $\vec{F}_1$ in Eq. (1) represents energy loss of the electron. For a single electron it arises only from the so-called synchrotron radiation[18] of the revolving cyclotron electron. In the non-relativistic case ($|v| \ll c$), this term can be written as[19] $\vec{F}_1 = (2e^2/3c^3)d^2\vec{v}/dt^2$. If the electron

133

oscillates[19] at a single frequency Ω_0 (e.g., the cyclotron frequency), with all other oscillations being negligibly small, than $d^2\vec{v}/dt^2 \cong -\Omega_0^2\vec{v} \cong +$ $-\Omega_0^2\vec{p}/m_0$, and the damping term can be written as $\vec{F}_1 \cong -\Gamma\Omega_0\vec{p}$, where Γ is the dimensionless bandwidth, or damping parameter, of cyclotron resonance:

$$\Gamma = 2e^2\Omega_0/3m_0c^3 = 2r_ek_0/3 \lll 1 \tag{2}$$

where $r_e = e^2/m_0c^2 = 2.8\times10^{-13}$ cm is the so-called classical electron radius and $k_0 = \Omega_0/c$. The same parameter Γ determines the dimensional bandwidth of the cyclotron resonance, $\Gamma = \Delta\Omega/2\Omega_0$. For presently available dc magnetic fields, synchrotron radiation damping is extremely small; e.g., for $B_0 = 6$ Tesla^2, $\lambda_0 = 0.2$ cm, and $\Gamma \cong 0.7\times10^{-11}$. It is precisely this fact that gives rise to the strikingly low driving threshold for observation of the non-linear effects discussed here. For example, in order to observe a hysteretic cyclotron resonance at the main frequency (see Section III below), the relativistic shift of mass, $\Delta m/m_0 \cong \beta^2/2$ where $\beta = v/c$, must exceed roughly speaking, the half-width of the resonance, which corresponds to very low kinetic energy. In fact, the experimental conditions (with typical kinetic energy 0.05 - 0.5eV, i.e. $\beta^2/2 \cong 10^{-7}-10^{-6}$) correspond to the situation in which this threshold is exceeded by a few orders of magnitude. In a more general case in which the electron is excited to arbitrary energy, one obtains[4]:

$$\vec{F}_1 = -2e^2\Omega_0^2\vec{v}/3c^3 \cdot (1 - \beta^2)^2 = -\Gamma\Omega_0\vec{p}. \tag{3}$$

We now introduce dimensionless momentum $\vec{p}$, fields $\vec{f}_j$, unit vector of magnetic field $\vec{h}$ and unit wave vectors $\vec{q}_j$, and trapping field $\vec{g}$ as follows: $\vec{p} = \vec{p}/m_0c$; $\vec{f}_j = \vec{E}_j/H_0 = e\vec{E}_j/m_0c\Omega_0$; $\vec{h} = \vec{H}_0/H_0$; $\vec{q}_j = \vec{k}_j/k_j$; $\vec{g} = \vec{G}_{tr}/H_0 = e\vec{G}_{tr}/m_0c\Omega_0$, after which we rewrite Eq. (1) as

$$\Omega_0^{-1}\dot{\vec{p}} + \Gamma\gamma\vec{p} - \vec{g} = \Sigma_j\vec{f}_j + \gamma^{-1}\vec{p} \cdot \Sigma_j(\vec{q}_j \cdot \vec{f}_j) + \gamma^{-1}\vec{p} \cdot \vec{h} \tag{4}$$

For most of the problems discussed in this paper, in particular, for all calculations of steady-state electron excitation, the trapping potential, and therefore $\vec{g}$, can be excluded from consideration (see below).

In order to develop a procedure that will enable us to describe all the high-order nonlinear (including multiphoton) effects, we have to distinguish pure cyclotron electron motion from all the other motions of the electron at noncyclotron frequencies. Assuming that the dc magnetic field H_0 is sufficiently strong in order for pure cyclotron motion to be dominant, we can treat all the noncyclotron motions as small perturbations. Therefore, the total momentum $\vec{p}$ can be written in the form

$$\vec{p} = \vec{p}_c(t) + \vec{p}_{nc}^{(1)} + \vec{p}_{nc}^{(2)} + \ldots; \quad |\vec{p}_{nc}| \ll |\vec{p}_c|, \tag{5}$$

where p_c is a "cyclotron" component of momentum describing a pure precession of the electron around some fixed center ($\vec{r} = 0$) with the frequency $\Omega \cong \Omega_0/\gamma_{c(s)}$. $\vec{p}_c$ is orthogonal to $\vec{H}_0$. The various orders of "non-cyclotron" components $\vec{p}_{nc}^{(s)}$ include oscillations with all the other, non-resonant, frequencies and may have any orientation. The dynamics of the cyclotron component $\vec{p}_c$ are governed then by the equation[4]:

$$\Omega_0^{-1}(d\vec{\rho}_c/dt) - \gamma_c^{-1}(\vec{\rho}_c \times \vec{h}) + \Gamma\gamma_c\vec{\rho}_c = \vec{F}_c^{(1)}(t) + \vec{F}_c^{(2)}(t) + \dots \qquad (6)$$

where $\gamma_c = \sqrt{1 + \rho_{2(s)}^2}$ is cyclotron motion energy. In Eq. (6) we introduced nonlinear forces $\vec{F}^{(s)}$ of different orders "s" each of which is related to the respective noncyclotron component $\vec{\rho}_{nc}^{(s)}$; "c" in $\vec{F}_c^{(s)}$ labels the components that oscillate at the cyclotron frequency Ω and are normal to $\vec{H}_0$. Once the nonlinear force $\vec{F}_c^{(s)}$ is determined, the respective noncyclotron component $\vec{\rho}_{nc}^{(s)}$ can be found from the equation:

$$\Omega_0^{-1}(d\vec{\rho}_{nc}^{(s)}/dt) - \gamma_c^{-1}(\vec{\rho}_c^{(s)} \times \vec{h}) + \gamma_c^{-3}(\vec{\rho}_{nc}^{(s)} \cdot \vec{\rho}_c)(\vec{\rho} \times \vec{h}) = \vec{F}^{(s)} - \vec{F}_c^{(s)}. \qquad (7)$$

In particular, the force $\vec{F}^{(1)}$ responsible for the hysteretic resonance at the main frequency (see below, Section II), as well as for subharmonic oscillations (see below, Section III) is defined as

$$\vec{F}^{(1)} = \sum_j \vec{f}_j(\omega_j t - \vec{k}_j \cdot \vec{r}_c(t)) + \gamma_c^{-1}\vec{\rho}_c \times \sum_j(\vec{q}_j \times \vec{f}_j), \qquad (8)$$

where $\vec{r}_c = c\gamma_c^{-1} \times \vec{\rho}_c dt = -c(\Omega\gamma_c)^{-1}(\vec{\rho}_c \times \vec{h})$ is the "cyclotron" radius vector. All the higher order nonlinear forces $\vec{F}_c^{(s)}$ can be written as

$$\vec{F}^{(s)} = \vec{F}_D^{(s)} + \vec{F}_L^{(s)} + \vec{F}_R^{(s)}; \quad (s>1) \qquad (9)$$

where each of the $s\underline{\text{th}}$ order forces $\vec{F}_D^{(s)}$, $\vec{F}_L^{(s)}$, and $\vec{F}_R^{(s)}$ is defined as a sum of all terms of $s\underline{\text{th}}$ order in f_j originating, respectively, from the first, second, and third terms on the rhs of Eq. (4) respectively, in which all the lower order terms of $\vec{\rho}$ in Eq. (5) (the highest of which is $\vec{\rho}_{nc}^{(s-1)}$) are taken into account. In particular, the force $\vec{F}^{(2)}$ responsible for cyclo-Raman resonances (see below, Section IV) was found in Ref. (4).

In eq. (9) we distinguish three main mechanisms of nonlinear interaction, each of which is related to the respective first three terms on the rhs of Eq. (4). The spatial oscillations of the electron make it see the phases $\vec{k}_j \cdot \vec{r}$ of the incident fields $\vec{f}_j(\omega_j t - \vec{k}_j \cdot \vec{r})$ (the first term on the right hand side of Eq. (4) rapidly modulated since $\vec{r} = c\gamma^{-1}\int\vec{\rho}dt$. Essentially, this modulation is due to the Doppler effect; hence the designation "Doppler" nonlinear mechanism, with the nonlinear force $\vec{F}_D^{(s)}$ defined as $\vec{F}_D^{(s)} = (\sum_j\vec{f}_j(\omega_j t + - \vec{k}_j \cdot \vec{r}(t)))^{(s)}$. The Lorentz radiation force $\vec{F}_D$ originated by the magnetic field of incident waves (the second term in rhs of Eq. (4)) gives rise to the "Lorentz" nonlinear mechanism, with the nonlinear force $\vec{F}_L^{(s)} = (\gamma^{-1}\vec{\rho} \times \sum_j(\vec{q}_j \times \vec{f}_j))^{(s)}$. Finally, the slight modulation of relativistic mass $m(t) = m_0\gamma$ results from small noncyclotron modulation of the energy term γ^{-1} in the last, cyclotron, term on rhs of Eq. (4); hence, the designation "relativistic" nonlinear mechanism, with the force $\vec{F}_R^{(s)}$ defined as $\vec{F}_R^{(s)} = (\gamma^{-1}\vec{\rho} \times \vec{h})^{(s)}$. Contributions from all these three mechanisms can be of the same order of magnitude. In general, none of them can be neglected; however, for particular propagation and polarization configurations some of them may dominate. It is worth emphasizing, however, that once the cyclotron motion is excited, it is only the relativistic mass-effect (the term γ_c^{-1} on lhs side of Eq. (6) that acts to limit the excitation energy and to form a hysteretic resonance.

The dynamics of cyclotron momentum, ρ_c, are described assuming

135

$$\vec{\rho}_c = \rho_c(\hat{e}_x \sin(\Omega t + \phi) + \hat{e}_y \cos(\Omega t + \phi) \tag{10}$$

where ρ_c and ϕ are the (unknown) slowly varying amplitude and phase of the cyclotron momentum respectively, whose dynamics are determined by Eq. (6) as:

$$\Omega_0^{-1} d\rho_c/dt + \Gamma\gamma_c\rho_c = \Phi_\rho; \quad \Omega_0^{-1}(d /dt) - (\gamma_c^{-1} - \Omega/\Omega_0) = \rho_c^{-1} \Phi_\phi, \tag{11}$$

with parameters Φ_ρ and Φ_ϕ defined as

$$\Phi_\rho = (\vec{F}_c^{lowest} \cdot \vec{\rho}_c/\rho_c)_0 = 0 \ (\Gamma); \quad \Phi_\phi = (\vec{F}_c^{lowest} \cdot (\vec{\rho}_c \times \vec{h})/\rho_c$$

$$= O(\Gamma), \tag{12}$$

where "0" labels zero-frequency components and $\vec{F}_c^{lowest}$ is a non-zero cyclotron component of the lowest order. Since $\Gamma <<< 1$ and $\rho_2 >> \Gamma$ (which is immediately attained once the threshold of excitation is exceeded), in steady state one has

$$\gamma_0 \cong \Omega_0/\Omega + O(\Gamma), \quad \rho_c = \sqrt{(\Omega_0/\Omega)^2 - 1} + O(\Gamma). \tag{13}$$

I.a. Trapping Potential and Propagation Configurations

The first experimental observation of hysteretic cyclotron resonance[2] at the main frequency ($\Omega \cong \Omega_0$) predicted earlier in Ref. (1), was accomplished using an electron confined in the relatively weak quadrupole potential of a Penning trap to keep the electron from escaping. The Penning trap in Ref. (2) consists basically of a positive ring and two negative cap electrodes kept at liquid helium temperature in the core of a 60-kG superconducting magnet. The surface of the quadrupole system (the caps and the ring) is formed by a hyperboloid of revolution symmetric with respect to $\vec{H}_0$. In such a potential, the confined electron exhibits, in general, quite a complicated behavior. Its motion may consist of the (dominant) cyclotron motion, which is the fastest one ($\cong$164 GHz in Ref. (2)), much slower magnetron motion ($\cong$60 MHz), and even much more slower axial oscillations along $\vec{H}_0$ ($\cong$ 1.6 KHz). The latter oscillations were used[2] to measure a relativistic mass increase through the axial frequency shift. Since we concentrate on the steady states, both magnetron and axial oscillations can be neglected. (We note though[1] that both of them could form a significant component in the transient behavior of electron if either driving frequency or amplitude is swept too fast). In fact, all the steady-state hysteretic and nonlinear effects with a single electron can be obtained in a dc magnetic field without any electric static trapping potential. However, when the electron is subjected to the action of only one traveling wave, the Lorentz radiation force of this wave acting upon the electron can force it to move away from its initial position unless a trapping potential is applied. In order to avoid this complication as well as the necessity to use the trapping potential in our calculations, we will always use propagation configurations which involve only a standing wave (i.e. couples of counter-propagating traveling waves) for each driving frequency.

136

The type of nonlinear interaction of Em radiation with a synchrotron electron, as well as the nonlinear mechanisms involved, depends critically on the direction of propagation and polarization of driving waves with respect to the dc magnetic field $\vec{H}_0$. Although almost all of the interactions discussed here are feasible for quite a general propagation configuration, there are a few propagation arrangements each of which can be regarded as the most optimal geometry for some particular nonlinear interaction. We will use a two-letter classification of propagation configuration using the first letter to describe the wave propagation direction and the second one to describe the wave polarization with respect to $\vec{H}_0$. The "pure" excitation of hysteretic cyclotron resonance at the main frequency (with no other "side" nonlinear effects) corresponds to the "PC" propagation configuration in which driving waves propagate parallel ("P") to $\vec{H}_0$ (i.e. $\vec{q}_j$ = ±h) and are circularly ("C") polarized in such a way that the polarization vector precesses in the same direction as the electron does. The high-order subharmonics are most efficiently excited by the "NN" propagation configuration when driving waves propagate normally ("N") to $\vec{H}_0$ and are linearly polarized in the direction also normal to $\vec{H}_0$, whereas cyclo-Raman excitations are stimulated by driving waves in the "NP" propagation configuration (i.e. by linearly polarized wavs incident normally to $\vec{H}_0$).

II. RELATIVISTICALLY-HYSTERETIC CYCLOTRON RESONANCE

The hysteretic resonance at the main cyclotron frequency, see Fig. 1a (i.e. when $\Omega \cong \Omega_c$, where Ω is the driving frequency and Ω_c is the cyclotron frequency of <u>excited</u> electron) is perhaps the most simple and fundamental relativistic nonlinear effect based strictly on the relativistic change of electron mass[1-3]. This is a first-order effect; Eq. (8) shows that for the PC propagation configuration (see inset in Fig. 2) none of the other mechanisms are engaged. Indeed, assuming driving waves in the form $\vec{f}_j$ = (f/2) x x $(\hat{e}_x \sin(\Omega t + k\vec{q}_j \cdot \vec{r}) + \hat{e}_y \cos(\Omega t + k\vec{q}_j \cdot \vec{r}))$ where j = 1,2, $\vec{q}_j = (-1)^j h$, k = Ω/c, and f = $2eE\sqrt{2}/m_0 c\Omega_0$ (with $|E|^2$ being the intensity of each of the waves), and assuming that the center of the electron orbit is located at the origin, $\vec{r}$ = 0, we obtain $\vec{F}^{(1)} = \vec{F}^{(1)}(t) = \vec{f}_1 + \vec{f}_2 = F(\hat{e}_x \sin\Omega t +$ + $\hat{e}_y \cos\Omega t)$. The dynamics of ρ_c and ϕ are governed by Eqs. (11) with Φ = $f \cos\phi$, Φ_ϕ = -f sinϕ. The steady-state solution (d/dt = 0) is thus determined by

$$f^2 = \rho_c^2(\Gamma^2\gamma_c^2 + (\gamma_c^{-1} - \Omega/\Omega_0)^2; \quad \tan\phi = -(\gamma_c \Omega/\Omega_0 - 1)/\Gamma\gamma_c^2. \tag{14}$$

Since $\Gamma \lll 1$, the hysteretic (bistable) resonance can be achieved at very low excitation , i.e., $\rho_c \cong \beta_c \ll 1$, $\gamma_c \cong 1 + \beta_c^2/2$, in which case Eq. (14) reduces to

$$f^2 = \beta_c^2(\Gamma^2 + (\Delta + \beta_c^2/2^2), \tag{15}$$

where $\Delta = \Omega/\Omega_0 - 1 \ll 1$ is a dimensionless resonant detuning parameter, see Fig. 2.

Under threshold conditions:

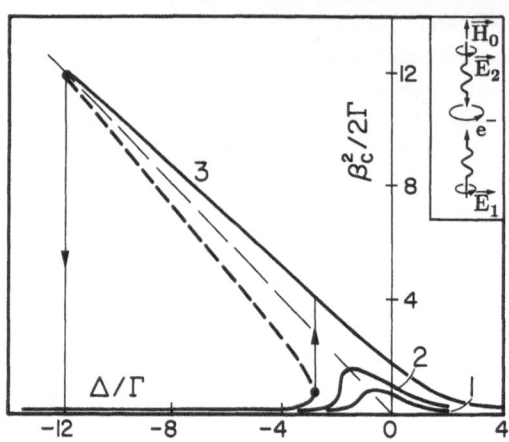

Fig. 2. The plot of normalized kinetic energy of the electron $\beta_c^2/2$
vs normalized resonant detuning Δ/Γ for various driving
amplitudes of incident EM field. Curves: (1) $f < f_{th}$, (2) $f =$
$= f_{th}$, (3) $f \gg f_{th}$. The inset depicts the propagation configuration of driving radiation.

$$f^2 > f_{th}^2 \equiv (16/3\sqrt{3})\Gamma^3, \quad \Delta < \Delta_{th} \equiv -\Gamma\sqrt{3}, \tag{16}$$

Eq. (15) yields a three-valued solution for β_c with two of the values stable
solution and one an unstable solution (Fig. 2). At the threshold point the
excitation is $\beta_{th}^2 = 2\Gamma\sqrt{3}$ (curve 2 in Fig. 2), and the orbit radius is r
$= \beta_{th}/k \ll \lambda$.

For the experimental conditions[2], $\Gamma \cong 0.66\times10^{-11}$, and Eq. (16) yields
$E_{th} \cong 2.65\times10^{-10}$ V/cm which corresponds to a classical threshold intensity
as strikingly low as $\cong 10^{-18}$ W/cm^2. The classical estimate (16) in electron
which is nearly α^{-1} times smaller than a quantum limit $\hbar\Omega$ ($\alpha = e^2/\hbar c = 1/137$
is the fine-structure constant). Therefore, in the close vicinity of the
threshold (16), only the quantum approach can give an adequate description
of the phenomenon, whereas, for a sufficiently strong driving field (E^2
$\gg E_{cr}^2$) the classical results (in particularly, hysteretic jumps) remain
valid. The actual excitation used in the experiment[2] was a few orders of
magnitude higher than this threshold, i.e., the experimental data obviously
represent the classical limit. The further improvement of the stability and
bandwidth of the source of the driving radiation may hopeful bring the experiment to the threshold level and, therefore, to the quantum limit of
the hysteresis effect.

In the case of the three-valued solution, the examination of Eqs. (11)
linearized in the close vicinity of the steady state solutions, Eqs. (14),
shows that only those states are stable which satisfy the energy criterion[1]
$d(\beta_c^2)/d(f^2) > 0$ (solid branches of the curves in Fig. 2); otherwise, they
are unstable (dashed branches in Fig. 2). The smallest of the hysteretic
jumps, the up-jump from the lower excitation branch to the higher one,
occurs in the close vicinity of the unperturbed resonance, whereas the much
stronger down-jump may occur considerably far from that point (curve 3 in
Fig. 2). When the driving radiation is sufficiently strong but the motion

is still low-relativistic (i.e., $f^2 \gg f_{th}^2$, but $\beta_c^2 \ll 1$) the kinetic energy of the excited electron $\gamma - 1 \cong \beta^2/2$ at the upper branch of hysteretic curve depends almost linearly on the detuning Δ:

$$\beta_c^2/2 \cong -\Delta. \tag{17}$$

The experimental data[2] are in good agreement with Eq. (17); according to them, $|2\Delta/\beta^2| \cong 0.92$ which is quite close to the value 1.0 suggested by Eq. (17). This confirms that the nonlinearity of the system is attribut-able to the relativistic mass-effect. In terms of nonlinear oscillation theory, the hysteresis in a nonlinear oscillator based on a low-relativ-istic electron resembles the so-called vibration hysteresis in driven anharmonic oscillators[21] (in particular, the so-called Duffing oscilla-tor, e.g., the pendulum) or in nonlinear parametric systems[22].

In principle, it is feasible to obtain electron with quite a rela-tivistic energy in a standing wave configuration even with modest driving power. For example, in order to excite an electron up to $\gamma_c = 188$ (i.e., to the energy $\cong$ 95MeV with orbit diameter $\cong$ 6cm) with $H_0 = 100$kG ($\lambda_0 \cong$ 1mm) one needs a driving source at $\lambda \cong 18.8$cm with an intensity still as low as 0.07W/cm^2.

III. HIGH-ORDER OPTICAL SUBHARMONICS

The high order subharmonics with $\omega = n\Omega$, see Fig. 1b, in nonlinear cyclotron resonance give a very interesting example of a multiphoton, highly nonlinear process which is still due to the first-order nonlinear force $\vec{F}^{(1)}$, Eq. (8). The most efficient generation of these subharmonics occurs with the NN propagation configuration, in which case the Doppler and Lorentz nonlinear mechanisms contribute to the nonlinear interaction. The electron is driven by two counter-propagating plane waves $\vec{f}_j = \hat{e}_y f \cos(\omega t + (-1)^j (kx - \psi))$; $j = 1,2$; $k = \omega/c$ (with the same frequency ω and amplitude f) both of them propagating along the axis x normal to $\vec{H}_0$, polarized along the axis y, also normal to $\vec{H}_0$ (see inset in Fig. 4), and forming a standing wave pattern. When the subharmonic of the nth order is excited, the momentum of electron is described by Eq. (10), where $\Omega = \omega/n$.

In our further calculations, we assume that the center of the orbit is located in the center of a trap and coincides with a maximum of the total field $\vec{f}_1 + \vec{f}_2$ (which corresponds to $\psi = 0$) when n is odd, and with a node of the total field (i.e. $\psi = \pi/2$) when n is even. Separating the cyclotron component $\vec{F}_c^{(1)}$ out of $\vec{F}^{(1)}$, and using Eqs. (12) one finds that in the steady state ($\vec{a}/dt = 0$) the energy of the excitation is $\gamma_c = n\Omega_0/\omega + \Delta\gamma_c$ with $\Delta\gamma_c = O(\Gamma) \lll 1$. The small detuning $\Delta\gamma_c$ for $n \geq 2$ is found to be

$$\frac{\Delta\gamma_c}{\Gamma} = \pm \frac{2J_n(n\beta_c)}{\beta_c^2\gamma_c} \sqrt{\mu^2 - \frac{\beta_c^2\gamma_c^4}{(J_{n-1}(\beta_c) - J_{n+1}(n\beta_c))^2}} \tag{18}$$

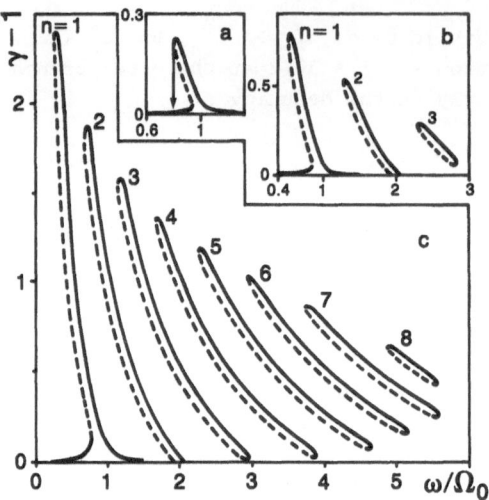

Fig. 3. Kinetic energy of the cyclotron motion $\gamma - 1$ vs. the dimensionless frequency of a driving field ω/Ω_0 for various
orders of subharmonics n (n = 1 is the main resonance, n = 2
is second suharmonic, etc.) and for the fixed amplitude
parameter $\mu = f/\Gamma = E/E_0$ of a driving field. (a) $\mu = 0.9$,
(b) $\mu = 3.5$, (c) $\mu = 16$. The thick solid branches correspond
to stable states, and thick broken branches to unstable
states. All the branches are stretched out along lines
determined by the formula $\gamma = n\Omega_0/\omega$ for each order n.

where J_ν is an ordinary Bessel function of the νth order and $\mu = f/\Gamma =$
$3Ec^2/2e\Omega_0^2$ is a driving parameter. The frequency of subharmonic $\Omega = \omega/n$
follows closely the effective cyclotron frequency $\Omega_c = \Omega_0/\gamma_c$ determined
by the relativistic mass-effect $m/m_0 = \gamma_c$, see Fig. 3.

A stability analysis using Eq. (11) shows that the positive upper
sign in Eq. (18) corresponds to stable states, and the negative sign to
unstable states. For each stable value of the energy γ_c, the cyclotron
motion can have n equally possible different equidistant states of the
phase ϕ^s, $\phi = \phi_c(n,f,\omega) + 2\pi s/n$, s being integral, $0 \leq s < n$. This
property is common for any subharmonic oscillations of the nth order
regardless of their origin[22,23].

For any order n > 2, there are both upper and lower limits on the
energy of excited subharmonics[5] for any fixed driving amplitude f (or
μ), see Figs. 3 and 4, which give rise to the formation of isolated
branches of excitation for each individual subharmonic with n > 2. The
initial "jump-start" required for the electron to reach the desirable
subharmonic, can be provided either by injecting electrons with proper
energy, or by triggering the system with the microwave oscillator having
its cyclotron frequency as in Section II (or some of its higher
harmonics) near the cyclotron frequency, or with optical biharmonic
pumping[3,6] by a laser having $\omega_1 - \omega_2 = \Omega$, see Section IV below. For any n $\geq$
2, there is a minimum (threshold) driving amplitude μ_{th} which is an
increasing function of n with $\mu_{th}(n = 2) = 1$. The threshold amplitude

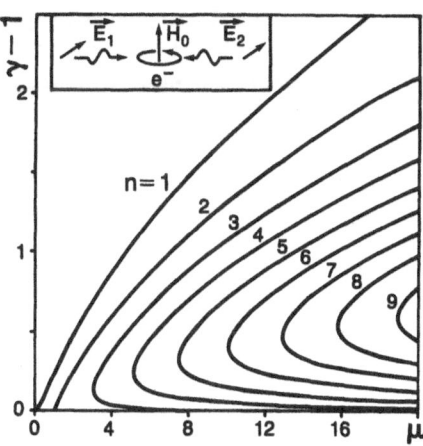

Fig. 4. Extremum kinetic energy $\gamma - 1$ of the cyclotron motion vs. driving parameter μ for various orders of subharmonics n. Inset: incident light configuration with respect to the dc magnetic field $\vec{H}_0$.

$\mu_{th}(n)$ corresponds to some certain energy of excitation which increases (although very slowly) as n increases.

The intensity of driving laser radiation required to excite even very high-order subharmonics, is very low. Indeed, the threshold amplitude for excitation 2-nd order subharmonic ($\mu_{th} = 1$) is

$$E = 2e\Omega_0^2/3c^2 = (2/3)(m_0c^2/e)k_0^2 r_e \qquad (19)$$

For $\lambda_c \cong 1$ mm, E_0 corresponds to as low intensity as $\cong 2\cdot10^{-10}$ W/cm^2. Consider now an example when the laser wavelength is $\lambda = 10$ μm (CO_2 laser), whereas $\lambda_c = 1$ mm, and therefore n = 100, i.e., the laser frequency is divided by a factor 100 in one step. The calculations[5] give $\mu_{th} \cong 560$ such that even if $\mu \cong 4\mu_{th}$, the resulting intensity is still very low, $\cong 10^{-2}$ W/cm^2. With the area of the beam $\lambda \times \lambda_c \cong 10^{-4}$ cm^2 this translates into a total driving power as low as 10^{-6} W.

High-order subharmonics have been observed in many resonant non-linear systems (both mechanical and electrical ones). For example, in a simple nonlinear circuit using a biased diode as a nonlinear capacitor, a one-step frequency division by a factor up to 500-100[0] in the ultra-high rf-range has been observed and studied[23]. The subharmonics in ref. (23) were attributable to the self-synchronization of parametric oscillations induced by a driving force. The same principle has later been proposed[24] to obtain low-order subharmonics in the optical range using an optical parametric oscillator. The subharmonic excitation of cyclotron motion by radio-frequency or microwave driving sources is well known; in fact, the synchrotron[25] and synchro-cyclotron[26] principles of particle acceleration are based on driving a particle beam at a frequency equal to the multiplied cycling frequency of accelerated particles. A similar principle was recently proposed[10] to obtain mw subharmonic radiation of electrons using a laser as a driving source.

In both mw accelerators and the laser schemes just described, the important condition is that the driving field must be highly inhomogeneous in space and act upon particles over a distance much shorter than the orbit circunference. On the other hand, the subharmonic described here can be obtained with homogeneous plane standing (or traveling) waves acting upon a particle along its entire cyclotron orbit. The laser power required for such a subharmonic excitation is so low that virtually any cw infrared laser can be used for this purpose. The quest for optical one-step multiple transformation of frequency (by either multiplication or division) stems from the need to cover a gap between optical and microwave time and frequency standards. The conventional techniques are based on frequency multipliers[27], complex frequency synthesis chains[28], frequency division based on locking both a laser and a rf source to a cavity[29], etc. The high-order cyclotron subharmonics discussed here have potential to provide a promising alternative method for obtaining a direct coherent link between lasers and microwave frequency standards.

IV. CYCLO-RAMAN OPTICAL RESONANCES

Electron excitation at the cyclotron frequency Ω_c by an optical biharmonic laser with two frequencies ω_1 and ω_2 ($\omega_1 > \omega_2$) such that each one of them is much higher than Ω and their difference equals to $\Omega \cong \Omega_c$, $\omega_1 - \omega_2 = n\Omega$, (see Fig. 1c) is essentially a three-photon process that can naturally be regarded as stimulated cyclo-Raman scattering (of the lowest order). More general stimulated cyclo-Raman scattering of arbitrary order n is feasible with $\omega_1 - \omega_2 = n\Omega$ (see Fig. 1d). The cyclo-Raman scattering with lowest orders (n = 1,2) was considered in Ref. (4) for the PC propagation configuration at low excitation energy, and in Ref. (6) for the NP propagation configuration and n = 1 at arbitrary excitation energy. For any order n > 2, the cyclo-Raman resonances in the NP configuration exhibit "prohibited" and "allowed" cyclotron orbits (which results in multiple isolated branches of solutions[6], the so-called isolas) n possible equidistant phase state (which results in phase multistability for n > 1), the optical Stark-shift (i.e., an intensity-dependent shift of the eigen-frequency), and multi-wave mixing effects.

We assume here that neither ratio $(\omega_1 + \omega_2)/\Omega$ nor $\omega_{1,2}/\Omega$ is integer (i.e. higher-order subharmonics, Section III, are excluded by proper frequency tuning). Because of such a choice, $\vec{F}_c^{(1)} = 0$ in Eq. (6). Therefore, cyclo-Raman scattering is attributed to the second-order nonlinear force $\vec{F}^{(2)}$. We also choose the NP configuration in which all the optical traveling waves $\vec{f}_j$ propagate in the plane normal to $\vec{H}_0$ with their polarization parallel to $\vec{H}_0$ (see inset in Fig. 5). For this configuration $\vec{F}^{(2)}$ is reduced to

$$\vec{F}^{(2)} = \gamma_c^{-1} \sum_j \vec{q}_j (\vec{f}_j \cdot \vec{\rho}_{nc}^{(1)}) - \frac{\vec{\rho}_c \times \vec{h}}{2\gamma_c^3} (\vec{\rho}_{nc}^{(1)})^2 \qquad (20)$$

where the first noncyclotron component of momentum, $\vec{p}_{nc}^{(1)}$, Eq. (7), is now[6,7]

$$\vec{p}_{nc}^{(1)} = \Omega_0 \int \vec{F}^{(1)} dt = \sum_j (\Omega_0/\omega_j)\, \vec{f}_j \, (\omega_j t - \vec{k}_j \cdot \vec{r}_c - \pi/2). \qquad (21)$$

If the driving radiation at both frequencies ω_1 and ω_2 forms standing wave patterns with all the waves propagating along the same axis x, i.e., $\vec{q}_{i\pm} = \mp \hat{e}_x$, the electric fields of these standing waves are expressed as $\vec{f}_{\omega i} = \vec{f}_{i+} + \vec{f}_{i-}$ (i = 1,2) with $\vec{f}_{i\pm} = f_i \hat{e}_z \sin(\omega_i t \pm k_i x + \psi_{i\pm})/2$; i = 1,2, where $f_{i\pm}$ are counter-propagating traveling waves with the same amplitude $f_i/2$ and frequency ω_i; their phases $\psi_{i\pm}$ in general could be different. Steady state cyclotron motion is achieved when the center of the cyclotron orbit coincides with a node of one of the standing waves and simultaneously with the maximum of the other one if n is odd, or when the center of the cyclotron orbit coincides with either nodes or maxima of both standing waves if n is even. At such a point, the average radiation forces of all the waves acting upon the electron cancel each other; one of the choices for the phases is then $\psi_{1+} = n\pi$, $\psi_{1-} = 0$, and $\psi_{2\pm} = 0$.

In the steady-state regime the small detuning $\gamma_c \Omega/\Omega_0 - 1$ is found to be[6,7]

$$\gamma_c \Omega/\Omega_0 - 1 = S \pm \rho_c^{-1}(Q_-(\alpha) - \beta_c^2 Q_+(\alpha)/n \sqrt{\mu^2 - \rho_c^2 \Gamma^2 \gamma_c^4/Q_+^2} \qquad (22)$$

where S is the so-called optical (i.e. intensity-dependent) Stark-shift of the relativistic cyclotron frequency Ω_c[6,7], and Q_+ are defined as $Q_\pm \equiv J_{n-1}(\alpha) \pm J_{n+1}(\alpha) + n\Omega$ x $(J_{n-1}(n\beta_c) \pm J_{n+1}(n\beta_c))/(\omega_1 + \omega_2)$. In these equations, $J_\nu(z)$ is the $\nu^{\underline{th}}$ order Bessel function of first kind, $\alpha_{1,2} \equiv \omega_{1,2}\rho_c/\Omega_0$, $\alpha \equiv \alpha_1 + \alpha_2$, $\beta_c = v/c = \rho_c/\gamma_c$, and μ is a driving parameter defined as $\mu = f_1 f_2(\Omega_0/\omega_1 + \Omega_0/\omega_2)/8$. Stability analysis of the dynamic equations[6,7] shows that the upper sign in Eq. (22) corresponds to stable states, and the lower to unstable ones. In Fig. 6, stable branches are shown by solid lines and unstable ones by broken lines. It is interesting to note that the $n^{\underline{th}}$ order cyclo-Raman resonances (with n > 1) exhibit phase multistability essentially similar to phase multistability in the $n^{\underline{th}}$ order subharmonics (Section III).

Consider first the lowest order resonance (n = 1) with sufficiently low excitation energy (i.e. $\alpha_{1,2} \ll 1$ and therefore $\rho_c \cong \beta_c \ll \Omega_0/\omega_{1,2}$, and $r_c \ll \lambda_{1,2}$). In such a case, Eq. (22) is reduced to Eq. (15), where instead of amplitude f of the resonant driving wave, one has to substitute now an "effective" amplitude of a three-photon interaction μ. Therefore, in order to obtain hysteresis, curve 1 in Fig. 6, one has to have μ exceed a threshold $\mu_{cr} \equiv 1.75\Gamma^{3/2}$. In the example with λ_0 = 2 mm and $\lambda_{1,2}$ = $2\pi k_{1,2} \cong 0.69$ μm (He-Ne laser), with $f_1 = f_2$, we obtain a critical amplitude, $\hat{E}_{cr} \cong 6V/cm$, which corresponds to a 48 mW/cm^2 intensity. If the beam is focused to a spot of $\cong$2 μm diameter, this amounts to a total power of only 1.5 x 10^{-6}W.

When the driving intensity and, therefore, the excitation electron energy increases, a new feature appears, which consists of the formation

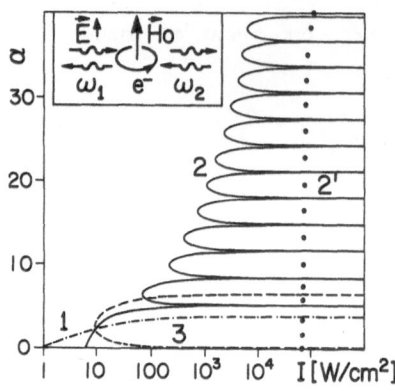

Fig. 5. The maximal and minimal values of excitation characteristic
α vs. the intensity of driving waves I. Curve 1 (broken line
with dots) and the lower branch of curves 2 (solid lines)
correspond to the maximum of the main hysteresis for the
cyclo-Raman scattering with n = 1 and n = 2 respectively;
next above the lower branch of curves 2 corresponds to the
first isola, etc. Curve 3 (broken line) corresponds to the
maximal and minimal excitation of the first isola for n =
3. Areas surrounded by each curve correspond to allowed
excitation; areas between them, to prohibited excitation.
Curve 2' (dots) corresponds to the maximal excitation for a
different configuration of the cyclo-Raman scattering with n
= 2.[4] The inset depicts the propagation configuration.

of isolated branches of excitation (the so-called "isolas"[6] known also in
other areas of nonlinear physics[30]). This feature is peculiar to the
configuration in which the waves propagate normally to the dc magnetic
field and therefore form a spatially oscillating pattern in the plane of
cyclotron motion. The isolas in consideration can be obtained even at low
excitation, $\rho_c^2 \ll 1$ (but with sufficiently large parameter α). When μ
exceeds some level, there are ranges of momentum ρ_c (such that $\rho_c <$
$(\rho_c)_{max}$), in which the steady-state excitation does not exist[6,7], i.e,
some orbits are "prohibited". For sufficiently large α, one obtains as
the radii of prohibited orbits, $r_{proh} \cong (2l + 1)\bar{\lambda}/8$, where l is an
integer and $\bar{\lambda} = 4\pi c/(\omega_1 + \omega_2)$, see Fig. 5. Prohibited orbits correspond
to the destructive interaction of both of the waves with respect to the
electron, as opposed to the constructive interaction leading to "allowed"
orbits. As the intensity of driving waves increases, the first isola is
formed, then the second, and so on. The formation of isolas in the case
n = 1 is illustrated in Fig. 6, in which the excitation characteristic α
is depicted vs. the frequency detuning parameter $\xi = sign(\Delta) \cdot \sqrt{2 \mid \Delta \mid}$ ·
$(\omega_1 + \omega_2)/\Omega_0 (\Delta = \Omega/\Omega_0 - 1$, and "sign" is the sign function) for various
driving amplitudes μ. The first isola (see curve 3 in Fig. 6) appears[6] at
$\alpha \cong 5$; CO_2 laser intensity as low as $\cong 77 W/cm^2$ ($\lambda_0 \cong 2mm$, $\lambda_{1,2} \cong 10~\mu m$,
$f_1 = f_2$) is required to observe it. Fig. 6 shows also a curious feature
of these regimes, the self-crossing of steady state amplitude that occurs
both in the isolas and in the main "mother" curve. It is unlikely,
though, that this feature can be seen experimentally, since one of the

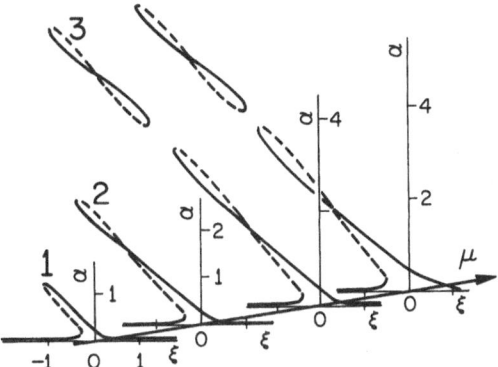

Fig. 6. The first-order cyclo-Raman (n = 1) excitation characteristic α vs. its frequency detuning parameter ξ (see
explanation in the text) for the fixed driving parameter μ.
Curves: 1 - $\mu < \mu_{s.c.}$ ($\mu_{s.c.}$ is a critical magnitude of μ
for the first self-crossing to occur); 2 - $\mu \cong 7\mu_{s.c.}$ (the
first self-crossing appears in the main "mother" curve); 3 -
$\mu \cong 2.6\mu_{cr}$ (μ_{cr} - a critical magnitude of μ for the first
isola formation); 4 - $\mu \cong 4\mu_{cr}$, the formation of the second
isola. The solid branches in curves 1-4 correspond to stable
states, the broken ones to unstable states.

self-crossing branches is unstable. Resonances with $n \geq 3$[7] do not have a
hysteretic branch; only isolas excitation is possible, although the
critical laser intensity remains relatively low (e.g., an intensity $\cong$
$9.7 W/cm^2$ is needed to observe the first isola for n = 3).

The total power P_c of the synchrotron radiation at the cyclotron
frequency Ω_c in the low-relativistic case ($\rho_c^2 \ll 1$) can be written as[19]
$P_c = \Gamma \rho^2 \Omega_0^2 m_0 c^2$. One can see from Eq. (21) that due to Doppler phase modulation, the noncyclotron momentum $\vec{\rho}_{nc}^{(1)}$ (parallel to $\vec{H}_0$) oscillates at
combination frequencies $\omega_{1,2} \pm l\Omega$, where l is integer, and therefore,
gives rise to stimulated dipole radiation at these frequencies with its
polarization parallel to $\vec{H}_0$. In nonlinear optics, such a process is sometimes referred to as multi-wave mixing whereby several waves with
different frequencies and wave-vectors are coupled via the nonlinear
interaction. At low excitation the power P_2 absorbed from the higher
laser frequency, ω_1, and the power P_1 radiated at the lower frequency,
ω_2, obey the Manley-Rowe relationships, $P_1 = - P_c \omega_1/n\Omega$, $P_2 = P_c \omega_2/n\Omega$,
which reflect a quantum balance of optical emission and absorption in the
system and suggest a stimulated emission at the frequency ω_2 analogous to
stimulated Raman scattering. The radiation at multi-wave optical
frequencies $\omega_{1,2} \pm l\Omega$ and at the mw cyclotron frequency Ω may provide an
experimental method for the observation of cyclo-Raman scattering of any
order.

CONCLUSION

   We have discussed the nonlinear interaction of EM (in particular,
optical) radiation with a single cyclotron electron, and demonstrated the
feasibility of many strongly nonlinear effects. They include hysteresis and
bistability at the main cyclotron resonance and various multi-photon processes (in particular, subharmonics and stimulated cyclo-Raman scattering
of arbitrary order) which exhibit isolas, prohibited and allowed orbits,
phase multistability and Stark shift. All of these effects are based on the
most fundamental relativistic properties of both electron and radiation
such as excitation-dependent mass-effect, the Doppler effect, and the
Lorentz radiation force. Most of these effects can be obtained using the cw
regime of any conventional laser. Some of these nonlinear effects have the
potential for applications (e.g. cyclotron mw excitation of electrons by
lasers via cyclo-Raman resonance, coherent links between the mw, millimeter, and optical ranges via high-order subharmonics, etc.). Further
research should investigate the "non-ensemble" quantum theory of the phenomenon.

REFERENCES

1. A. E. Kaplan, Phys. Rev. Lett., 48, 138 (1982)
2. G. Gabrielse, H. Dehmelt, and W. Kells, Phys. Rev. Lett., 54, 537
    (1985)
3. A. E. Kaplan, Nature, 317, 476 (1985);
    A. E. Kaplan, IEEE J. Quant. Electr., QE-21, 1544 (1985)
4. A. E. Kaplan, Phys. Rev. Lett., 56, 456 (1986)
5. A. E. Kaplan, Opt. Lett., 12, 489 (1987)
6. Y. J. Ding and A. E. Kaplan, Opt. Lett., 12, 699 (1987)
7. Y. J. Ding and A. E. Kaplan, Phys. Rev. A, Brief Reports (September
    1988)
8. H. M. Gibbs, "Optical Bistability", Academic Press, N.Y., 1985
9. A. E. Kaplan, "Optical Bistability III", Eds. H. M. Gibbs, P. Mandel,
    N. Peyghambarian, and S. D. Smith, Springer, N. Y., pp. 240-243
10. D. J. Wineland, J. Appl. Phys., 50, 2528 (1979);
    J. C. Bergquist and D. J. Wineland, Proc. 33rd An. Symp. Frequency
    Control, 1979, pp. 494-497
11. A. E. Kaplan, and A. Elci, Phys. Rev. B, 29, 820 (1984)
12. A. E. Kaplan, and Y. J. Ding, Opt. Lett., 12, 687 (1987)
13. E. O. Kane, J. Phys. Chem. Solids, 1, 249 (1957)
14. B. Lax, in Proc. Seventh Int. Conf. Phys. Semiconductors, Paris,
    France, 1964 p. 253;
    A. G. Aronov, Sov. Phys. Solid State, 5. 402, (1963);
    H. C. Praddaude, Phys. Rev., 140, A1292 (1965);
    W. Zawadzki and B. Lax, Phys. Rev. Lett., 16, 1001 (1966)
15. P. A. Wolff, J. Phys. Chem. Solids, 25, 1057 (1964)
16. E. Yablonovitch, N. Bloembergen, and J. J. Wynne, Phys. Rev. B,
    3, 2060 (1971);
    J. J. Wynne, Phys. Rev. B, 6, 534 (1972)
17. W. Zawadzki, S. Klahn, and U. Merkt, Phys. Rev. Lett., 55, 983
    (1985)

18. H. A. Lorentz, "The Theory of Electrons", Teubner, Leipzig, 1909, p. 49, 253

19. L. D. Landau and E. M. Lifshitz, "The Classical Theory of Fields", Cambridge, MA, Addison-Wesley, 1971

20. N. Bloembergen, "Nonlinear Optics", Benjiamin, N.Y., 1965

21. J. J. Stoker, "Nonlinear Vibrations in Mechanical and Electrical Systems", New York: Interscience, 1950, Sec. 4.4; see also N. Minorsky, "Nonlinear Oscillations", Princeton, NJ: Van Nostrad (1962); see also L. D. Landau and E. M. Lifshitz, "Mechanics", New York: Pergamon, 1976

22. A. E. Kaplan, Yu. A. Kravtsov, and V. A. Rylov, "Parametric Oscillators and Frequency Dividers", Soviet Radio, Moscow, USSR, 1966, in Russian

23. A. E. Kaplan, Radio Engineering and Electronic Physics, 8, 1340 (1963); A. E. Kaplan, ibid, 9, 1424 (1964); A. E. Kaplan, ibid, 11, 1214 (1966); A. E. Kaplan, ibid, 11, 1354 (1966)

24. A. E. Kaplan, Radiophys. Quant. Electr., 11, 900 (1968)

25. E. M. MacMillan, Phys. Rev., vol. 68, pp. 143-144, Sept. 1945; see also V. Veksler, J. Phys. USSR, vol. 9, p. 153 (1945)

26. D. Bohm and L. L. Foldy, Phys. Rev., 72, 649 (1947)

27. K. M. Evenson, D. A. Jennings, F. R. Peterson, and J. S. Wells, in: "Proc. Third Int. Conf. on Laser Spectroscopy", J. L. Hall and J. L. Carlsten, Eds., Heidelberg: Springer-Verlag, 1977, vol 7, p. 56; D. J. E. Knight and P. T. Woods, J. Phys. E, 9, 898 (1976)

28. D. A. Jennings, C. R. Pollack, F. R. Peterson, R. E. Drullinger, K. M. Evenson, and J. S. Wells, Opt. Lett., 8, 136 (1983)

29. R. G. DeVoe and R. G. Brewer, Phys. Rev. A, 30, 2827 (1984)

30. See for example, M. Kubicek, I. Stuchl, and M. Marek, J. Comp. Phys., 48, 106 (1982); T. Erneux and E. Reiss, SIAM J. Appl. Math., 43, 1240 (1983); J. C. Englund and W. C. Schieve, J. Opt. Soc. Amer., 2, 81 (1985); A. E. Kaplan and C. T. Law, IEEE J. Quant. Electr., QE-21, 1529 (1985)

OPTICAL COMPUTING, NEURAL NETWORKS, AND INTERCONNECTS

PRINCIPLES OF OPTICAL COMPUTING

A.W. Lohmann

University of Erlangen
Erwin-Rommel-Str. 1, 8520 Erlangen, FRG

INTRODUCTION

The optical computing represents a very stimulating challenge for optics that is better suited than electronics for highly parallel computing[1,2]. The reason relies on the fact that by a very large number of small optical processors, tied together by a powerful connecting network, the optical signals travel as light rays in free space. Many independent rays may traverse the same portion of the 3D-space, without "crosstalk". In contrast, electronic connections are confined to material guides, usually in planar topology.

We believe we will succeed if nonlinear optical effects and devices are improved, if our architectures utilize the benefits of optics, if enough users are dissatisfied with all-electronic computers.

I. THE REASONS FOR AN OPTICAL COMPUTER

Why do we want to build a digital optical computer (DOC)? Because it is fun. Having fun is probably the most effective inspiration in science and elsewhere. Hence, we are not ashamed to mention "fun" as a primary driving force.

Digital electronic computers are certainly powerful: no wonder, after hundred billion dollars have been spent for R&D on such computers. Nevertheless, we hear people crying everywhere: super computers are too slow. What can be expected of electronic super computers in the foreseeable future.

Replacing Silicon by Gallium Arsenide might reduce the switching time from one nanosecond to a tenth of a nanosecond. An increase in speed by another factor of a thousand would be desirable, but no suitable electronic material is visible on the horizon.

Nonlinear Optics and Optical Computing
Edited by S. Martellucci and A. N. Chester
Plenum Press, New York, 1990

The obvious way out is to couple several computers together in order to get an even bigger computer. One direction is to couple two or four of the biggest existing machines of the CRAY class together. Progress by one order of magnitude is likely to be achieved, but beyond that ...?

Another direction is to combine a large number of relatively small processors into a massively-parallel processor. In the earliest attempts of this variety the individual processors were laid out in a cartesian grid. Each processor was connected with his four or eight nearest neighbours. The time for transporting data across such a grid is many cycle periods long. Hence, such a parallel processor is good only for computing algorithms with local interactions. The important Fourier transform does not belong to the class of local algorithms, since every output value depends on all input values across the data field. The execution of the Fourier transform requires a global algorithm, which in turn requires a parallel computer with a global interconnection network, which can shuffle data over long distances at high speed.

Computer architects have tried to build parallel processors with fast global interconnections. THe best known representative of this class are based on the hypercube network. A K-dimensional cube has $2^k = N$ corners. A processor sites on each corner. One needs only K = log N steps to get to the farthest processor, which is situated at the opposite end of the principal diagonal of the cube. The hyper cube idea sounds fantastic. But it has its limitations, since the number of simultaneous dialogues among processor pairs will be close to the theoretical maximum of N/2 only very rarely. More typical are roughly K non-conflicting dialogues. The number of lines NK/2 is simply not large enough. One would need (N-1)!/2 lines if every processor had permanent lines to every other processor. For N = 1.000.000 the factorial number has about a million digits.

Such a network is completely out of the question electronically, even for modest values of N (the number of processors). Optical connections, as light rays in free space, could support a somewhat larger number of processors. Nevertheless, a permanent network with (N-1)!/2 is out of the question, unless the number of processors is very small. We must aim at interconnection networks which can be re-configured according to changing dialogue demands.

So far we have emphasized the significance of parallelism. To implement massive parallelism one needs a powerful network, which will be the topic of another section called "Optical Networks". Yet another section is devoted to the architecture of parallelism at the gate level. That section is entitled "Optical Parallel Logic". Before closing this section let us compare electrons and photons as carriers of information (Fig. 1).

Two things occur in a computer essentially: a) LOGIC INTERACTIONS, and, b) TRANSPORT COMMUNICATION.

Two electrons interact noticeably. They repell each other due to the Coulomb force and because they are Fermions. Interaction is necessary for performing logic. But interaction may disturb the communication as cross-

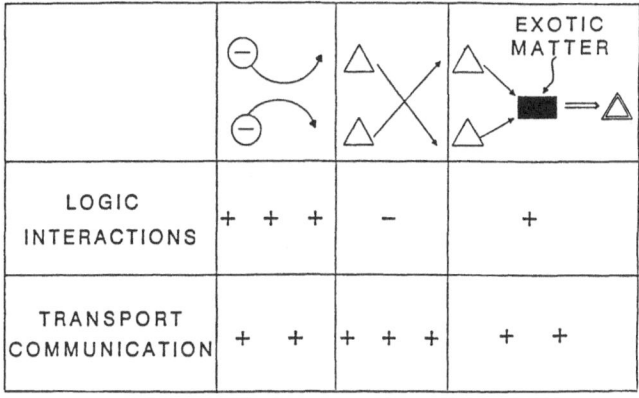

ELECTRONS OR PHOTONS ?

Fig. 1. Electrons or photons? The advantages (+) and handicaps (-)
of electrons and photons and the role of exotic matter for
enabling photons to perform logic.

talk. Photons in free space do not interact. Hence, today's computers are
all-electronic and tomorrow's computers might be hybrid (Fig. 2). Eventually, suitable exotic materials and components will be available. They
should be localized where interaction is wanted. Hence, with photonics we
may get the best of two worlds. The "exotic materials and components" con-
stitute the weakest leg of the platform, upon which the digital optical
computer will grow. However, recent results in nonlinear optics are en-
couraging.

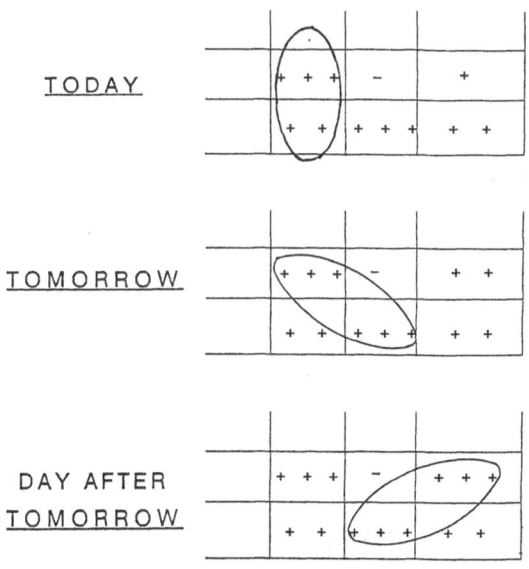

Fig. 2. From today's all-electronic computer towards an all-optical
computer via an intermediate hybrid system.

II. OPTICAL PARALLEL LOGIC

In the previous section I we emphasized the advantage of photons as information carriers in a parallel processor. Parallelism can occur at different levels. One may couple two or four complete CRAY processors, or one may couple thousands of primitive one-bit processors. The more processors there are, the more lines for communication are needed. That is the domain where optics can contribute significantly. Hence, we concentrate on "low-level parallelism".

A lens connects millions of object points simultaneously with millions of associated image points. Hence, image formation is a highly parallel communications process. One may modify an image forming system in the spirit of Ernst Abbe's experiments, which today are called "spatial filtering". Such an experiment, which performs logic operations on close to a million points simultaneously is shown in Fig. 3, 4 and 5. The local binary states of the two inputs objects are indicated by the orientation of pieces of a diffraction grating (Fig. 3). The grating orientation is called "theta", and the method "theta modulation"[3].

Light which has sensed the local binary state of object 1 will appear at orthogonal locations at the first Fourier plane (Fig. 4). There the filter 1 may eliminate the light which represents one of the two binary states. The filtered image of object 1 falls upon object 2, where again grating diffraction takes place. A proper choice of the two filters will produce as output various logic results (Fig. 5).

Another approach to parallel logic is called "symbolic substitution"[4]. Certain input patterns are replaced in two steps by associated output patterns. The first step can be executed by well-known optical pattern recognition methods. The second step is like an act of willful blurring.

Theta modulation and symbolic substitution do not compete. They may be combined, but do not have to.

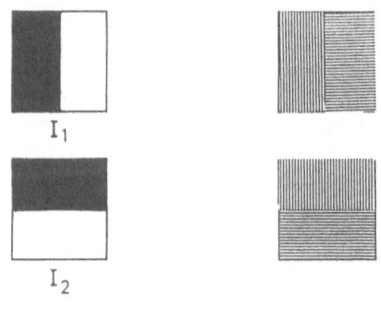

I_1

I_2

ORIGINAL OBJECTS   ENCODED MASKS

Fig. 3. Encoding of two binary objects $I_1(x,y)$ and $I_2(x,y)$ by
assigning pieces of a grating with associated orientations
to the logic states: <u>Yes</u>, bright, horizontal; <u>No</u>, dark,
vertical.

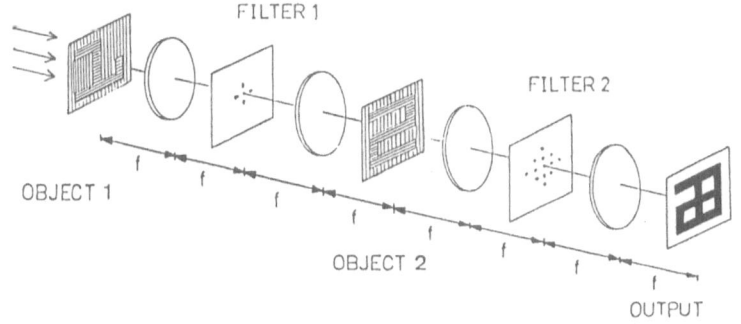

Fig. 4. Optical system for parallel logic based on theta modulation.
Two spatial filtering setups are cascaded.

III. OPTICAL NETWORKS

The digital optical computer is especially promising if it consists of
a large number of simple processors. These processors must exchange data
and control signals in order to cooperate efficiently. We believe that
optics is especially well suited to perform the communication of data for
reasons explained in the first paragraph. The job of the communication
network is in many ways similar to that in a central telephone station,
which might sit at the center point of a star. The job of the central
station is performed by a special purpose computer, also called "switching
machine".

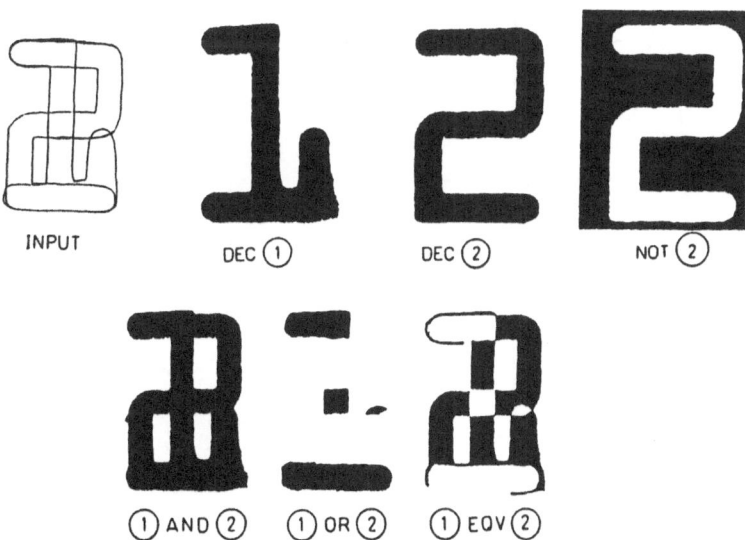

Fig. 5. A few results of theta logic. Shown are the shapes of the
two inputs, detection of input 1, negation of input 2, AND,
OR and EQV = NXOR.

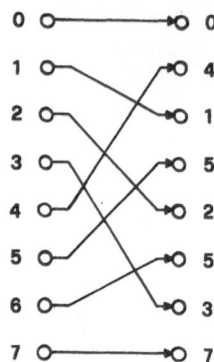

Fig. 6. The structure of the permutation of one layer of the perfect shuffle network. It consists usually of at least 2 log N such layers.

A switching machine may have for example 16 entrances and 16 exits. Each call station has its own permanent entrance and each receiving station has its own permanent exit of the switching machine. (A genuine station has an entrance as well as an exit, of course). The job of the switching machine is to "permute" the 16 entrance channels into the desired configuration of 16 exits. A so-called crossbar network with 16-square switches can do the job easily, but at the expense of N^2 switches (here N = 16).

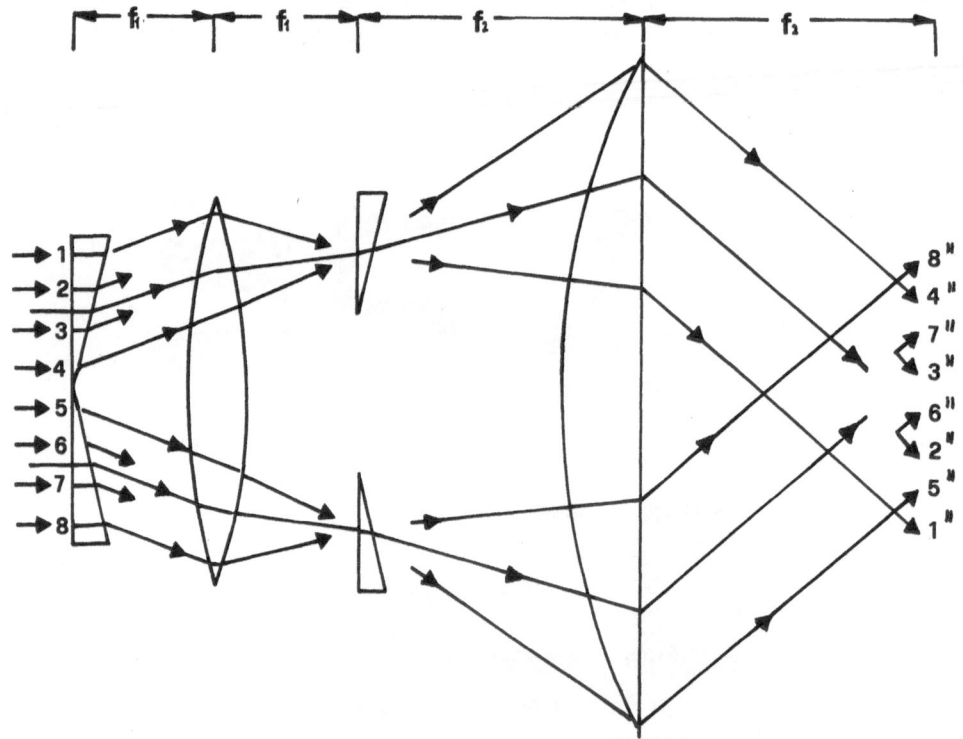

Fig. 7. The optical setup for implementing one layer of the perfect shuffle network.

The next link of our argument in favour of optics will be based on a common piece of knowledge about the fast Fourier transform (FFT). The traditional Fourier transform requires N^2 operations (one multiplication and one addition). The fast Fourier transformation requires only on the order of N log N operations. The gain is computational speed and is based on a smart utilization of some symmetry features of the transform matrix. If the Fourier job can be done N-fold parallel we need only on the order of log N cycles.

Almost exactly the same symmetry features as in the FFT are exploited in the so-called perfect shuffle network, which is the centerpiece of some advanced switching machines. The concept of the perfect shuffle is illustrated in Fig. 6. The concept can be implemented by standard optical components such as lenses and prisms (see Fig. 7, and also ref. 5). Also, arrays of micro-holograms are used as parts of the switching machine. The job of the hologram array is to compress and to separate data channels to avoid pictorial crosstalk[6].

REFERENCES

1. A. Huang, Proc. IEEE, 72, 780 (1984)
2. T. Bell, IEEE Spectrum, 23, 8, 34 (1986)
3. J. Weigelt, Opt. Eng., 26, 29 (1987)
4. K. H. Brenner, A. Huang and N. Streibl, Appl.Opt., 25, 3054 (1986)
5. A. W. Lohman, W. Stork and G. Stucke, Appl. Opt., 25, 1530 (1986)
6. A. W. Lohman and F. Sauer, Appl. Opt., 27 (1988)

OPTICAL ASSOCIATIVE MEMORY

K. Hsu and D. Psaltis

California Institute of Technology, Department of
Electrical Engineering
Pasadena, California 91125

INTRODUCTION

A neural computer is characterized by three properties: it consists
of a large number of simple processing units (the "neurons"), each processing element is connected to many others, typically several hundred or
several thousand, and the network is programmed to respond appropriately
to inputs by adjusting the weights of the connections between neurons during
a learning phase. Interest in this type of computer has arisen largely
because the above characteristics are also part of the models that attempt
to describe the operation of biological nervous systems. It is hoped that by
building a computer that shares some of the characteristics of the biological systems, we will be able to address problems such as image understanding
which animals do exceedingly well but current machines do not.

There has been considerable progress on the theoretical side to justify
optimism about future applications and this has focused attention on the
hardware realization of neural architectures. The computational power of
neural computers arises from matching the computer architecture and the
physical properties of the devices used in the implementation, to the requirements of the problem. In other words, a neural computer is highly
specialized and it is therefore very difficult to derive its potential
advantages from a general purpose computer. This provides a strong impetus
for advancing the technologies for the physical realization of neural computers in parallel and interactively with the development of theoretical
neural network models.

Electronics (analog, digital or hybrid) and optics are the two approaches under consideration for the hardware realization of neural networks. There are two basic components that need to be implemented: neurons
and connections. The neurons are typically simple thresholding elements that
can be implemented by a single switching device (i.e. transistor). The
switching speed and the accuracy required for the neurons are not beyond
the capabilities of current electronic technology. A practical neural computer may require millions of neurons operating in parallel. This require-

ment by itself is also achievable in electronics. However, each of the neurons must be connected to several thousand other neurons and these connections must be modifiable so that learning can take place. While this massive connectivity is relatively difficult to achieve electronically, optics is particularly well suited for the realization for interconnections.

In the basic architecture of an optical neural computer the neurons are arranged as 2-D arrays in planes that are separated from each other by an optical system that specifies the interconnection between each neuron and other neurons in the same or adjacent neural planes. A variety of optical architectures for the realization of optical neural computers have been proposed[1-9] and most fit this basic architectural design. Spatial light modulators are used to simulate the two dimensional array of neurons at the neural planes whereas holograms, transparencies and other passive optical elements comprise the optical interconnecting system.

The feature of the optical implementation that gives it an advantage when compared with electronic implementations is the fact that it is constructed in three dimensions. This allows the active devices at the neural planes to be populated by processing elements only, since the interconnections are external to the plane of the neurons. The third dimension is used to store the information that is required to specify the connections between the neurons. It is important to keep in mind that in a large network (e.g. 10^6 neurons) that is densely connected (e.g. 10^3 connections per neuron) the weights represent a large database. The ability to store this information in 3-D in the form of holographic interconnections provides the optical system with particularly high storage density[10,11]. A second advantageous feature of the optical implementation is the relative ease with which learning can be accomplished as dynamic holography in photorefractive crystals[12,13]. In this chapter we describe, as an example of this type of architecture, an optical associative memory that we have demonstrated experimentally[13,14]. This system is a single layer neural network architecture simulating a 2-D array of approximately 10^5 neurons on which images can be represented. Each neuron corresponds to a resolvable element of a 2-D spatial light modulator, in this case a Hughes liquid crystal light valve (LCLV). The 2-D array of neurons is fully interconnected via holograms and the system is organized as an auto-associative memory with feedback. A set of images is stored in a hologram and when an external image is projected into the system, one of the stored images emerges as a stable state of the system. In what follows we describe the performance of the system and the optimal choice of the system parameters. These include the input-output dynamics, the ability of the system to recognize distorted versions (e.g. rotated, shifted, or scaled) of a stored image, and the trade-off between distortion invariance and discrimination capability.

DESCRIPTION OF THE SYSTEM

A schematic diagram of the optical associative memory loop is shown in Fig. 1 and a photograph of the experimental apparatus is shown in Fig. 2. It is comprised of two cascaded Vander Lugt correlators[14] of which the first is used for calculating the degree of similarity between the external

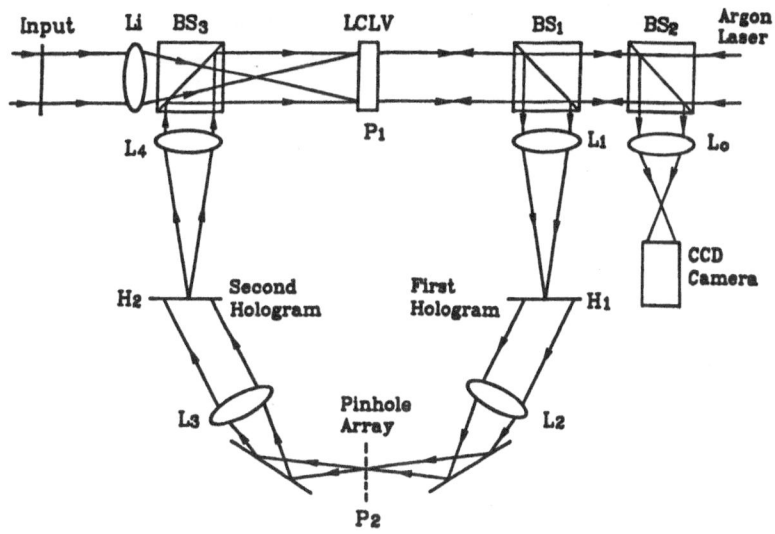

Fig. 1. Optical auto-associative loop.

input image and the images stored in the hologram. The second correlator uses the output from the first correlator to reconstruct the same images that are also stored in the second hologram to provide the feedback signal for the loop. In the system of Fig. 1, the LCLV at plane P_1, the beam-splitter cube BS_1, the lenses L_1, L_2, and the hologram H_1 form the first correlator, and the section of P_2, L_3, H_2, L_4, BS_3, and P_1 form the second correlator. The input pattern is imaged onto the LCLV by lens L_i and through beam splitter BS_3. A collimated argon laser beam illuminates the read-out side of the LCLV through beam splitters BS_2 and BS_1. A portion of the re-flected light from the LCLV that propagates straight through BS_1, is diverted by BS_2, and it is imaged by lens L_0 onto a CCD television camera. This provides real time monitoring of the activity of the system. The portion of the light that is reflected by BS_1 into the loop is Fourier transformed by lens L_1 and illuminates the hologram H_1. The correlation between the input

Fig. 2. Experimental setup of the optical auto-associative loop.

image and each of the stored images is displayed at plane P_2. The pinhole array at P_2 has center spacings corresponding to the spatial separations of the stored images. The remainder of the optical system from P_2 back to the neural plane P_1 is essentially a replica of the first half, with the hologram H_2 storing the same set of images at H_1.

The operation of this associative loop can be explained with the aid of the block diagram shown in Fig. 3a. In this example four images are spatially separated and stored in the Fourier transform holograms H_1 and H_2 as shown in Fig. 3b. When the input pattern A is presented as an input to the system, the first correlator produces the auto-correlation pattern along with three cross-correlations at plane P_2. The pinhole array at P_2 samples these correlation patterns at the middle of each pattern where the inner products between the input and each of the stored images form. Each of the four beams that pass through the pinholes act as delta functions, reconstructing from the second correlator the four images stored in hologram H_2. These reconstructed images are spatially translated in proportional to the position of each pinhole and superimposed at plane P_1. At the center of the output plane of the second correlator we obtain the superposition of the four stored images. The stored image which is most similar to the input pattern gives the strongest correlation signal hence the brightest recon- structed image. The weak read-out from the cross-correlation can be elimi- nated by thresholding in the LCLV. The output of the LCLV becomes the new input image for the loop and thus iterations take place. The stable pattern that forms as a recirculating image in the loop is the stored image that is most similar to the original input. Fig. 4 shows an example of an experiment performed with this loop. Fig. 4a is the external input, in this case a par-

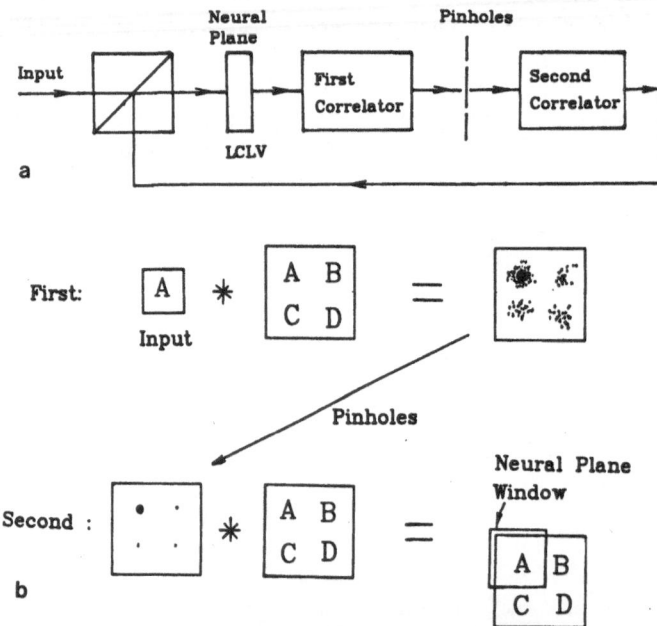

Fig. 3. Operating principles of the optical associative loop. (a) Block diagram; (b) example of recalling one of the stored images from an input.

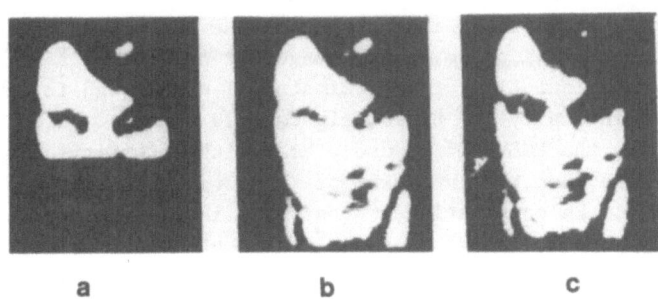

a   b   c

Fig. 4. Experimental demonstration. (a) Half-face image input into the loop; (b) overlap of the input image with the recalled complete image; (c) stable image in the loop after the input was cut off.

tially obstructed image of one of the stored patterns. Fig. 4b shows the response of the system with the external input still present, and Fig. 4c shows the stable state of the loop after the external input is removed.

The key elements in this optical loop are the holograms, the pinhole array, and the threshold device. The holograms in this system are thermo-plastic plates. They have a resolution of 800 lines per millimeter over an area of one square ich. If we put a mirror[10] or a phase conjugate mirror[4,6] at the pinhole plane P_2 to reflect the correlation signal bact through the system then we only need one hologram to form a closed loop. The use of two holograms, however, improves system performance. We make the hologram at H_1 with a high pass characteristics so that the input section of the loop has high spectral discrimination. On the other hand, we want the feedback images to have high fidelity with respect to the original images. Thus the hologram at plane H_2 must have broadband characteristics. We achieve this by using a diffuser when making H_2.

The pinhole array at plane P_2 samples the correlation signal between the image input from the LCLV and the images stored in the hologram H_1. The pinhole diameters used in these experiments are from 45μm to 700μm, depending on the images to be stored and the desired system performance. If the pinhole size is too small, then the light passing through it to reconstruct the feedback image is too weak to be detected by the LCLV and no iterations can occur. On the other hand, large pinholes introduce excessive blurring and cross-talk in the feedback and make the reconstructed images unrecognizable. The pinhole size also affects the shift invariance property of the loop. In order to be recognized, the auto-correlation peak from an external image should stay within the pinhole. Larger pinholes allow more shift in the input image. The system performance under different selections of pinhole diameter will be described in the following section.

The threshold device in this system has two purposes. The first is to provide a thresholding operation to the feedback signal so that the cross-correlation is reduced in successive iterations. The second is to provide gain to the feedback signal. The optical signal is attenuated in the loop due to the diffraction efficiencies of the Fourier transform holograms and the losses from pinholes as well as lenses and beam splitters. Therefore we need to have optical gain to compensate this loss. In our system this is achieved by adding an image intensifier at the photoconductor side of the LCLV. The microchannel plate of the image intensifier is sensitive to a minimum incident intensity of approximately 1 nW/cm^2 and it reproduces the input with an intensity 10^4 times brighter (10 μW/cm^2). This is bright enough to drive the LCLV. If we use a beam with intensity equal to 10 mW/cm^2 to read the LCLV then the intensity of the output light is approximately 1 mW/cm^2. Thus, the combination of the image intensifier and the LCLV provide optical gain up to 10^6. Fig. 5a shows the input-output characteristics of the optical thresholding element which is similar to a sigmoid function. The optical gain can be adjusted by changing the bias voltage of the image intensifier. Fig. 5b shows the relationship between the bias voltage applied to the image intensifier and the gain. In the following section we will see that the setting of the gain is the key parameter that mediates the trade-off between distortion invariance and the discrimination capability of the loop.

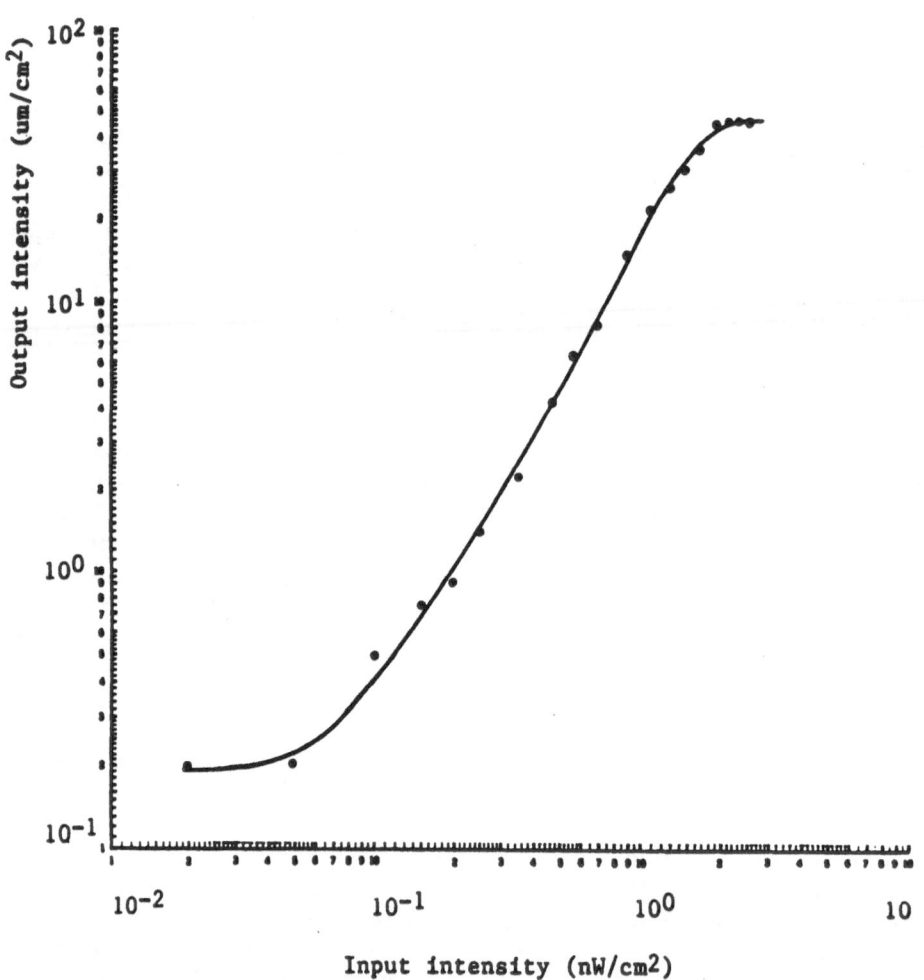

Fig. 5a. Characteristics of the optical neurons. Input-output relationship.

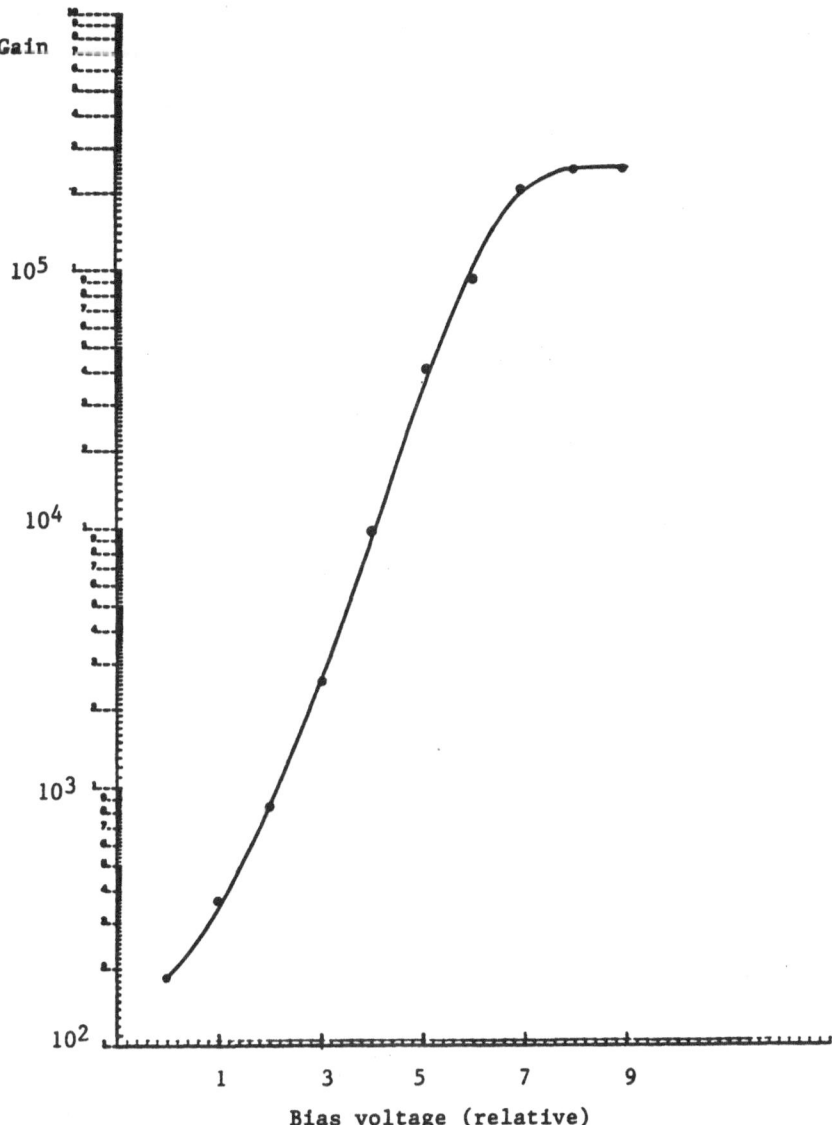

Fig. 5b. Neuron gain as a function of the bias voltage on the image
intensifier. Read-beam = 500 W/cm^2.

EXPERIMENTAL RESULTS

The time for the loop to reach a stable state depends critically on the
initial conditions. Fig. 6a shows the temporal response of the loop to an
input pattern. When the signal in the lower trace becomes high, it indicates
that the external input is ON. The upper trace shows the corresponding response of the loop. The initial rise is due to the presentation of the input
whereas the second rise is initiated by closing the feedback path. It is seen
that it takes about two seconds for the loop to reach a stable state, whereas
the rise time of the LCLV is approximately one second in the mode we operated
it. When the external input is turned off, the loop remains latched to a
stable state which is one of the stored images. Fig. 6b shows the same ex-

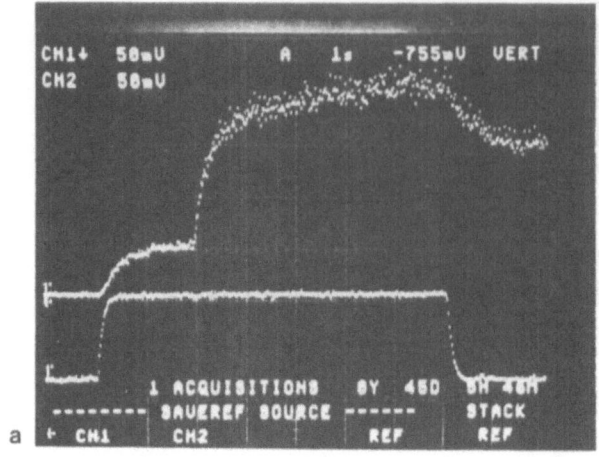

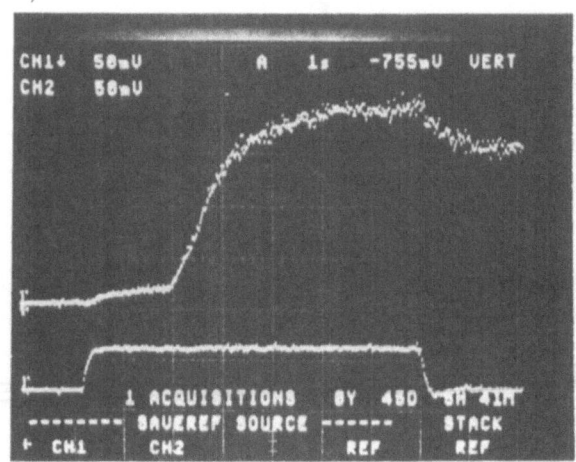

Fig. 6. Temporal response of the optical loop. (a) Loop response with
a strong initial condition; (b) loop response with a weak
initial condition.

periment but with input intensity reduced to one third of the first input.
The second rise of the upper trace shows that it takes approximately four
seconds for the loop to reach its stable state. However, after the input
is returned off the loop gives the same output intensity. This example shows
that initial conditions affect the dynamics of the loop but do not affect
its final state.

The dynamics of the recall process can be described by using an iteration map formed by the gain and loss curves as shown in Fig. 7. In the
figure the slope of the straight line is proportional to loop loss due to the
holograms and the pinholes, and on it is superimposed the input-output response of the neurons. The intersection points of the straight line with
the neuron gain curve at point Q_1 determines the threshold level and Q_2
represents a stable point. If the initial condition of the neuron is above
the threshold point θ_1, the signal grows in each iteration until it arrives

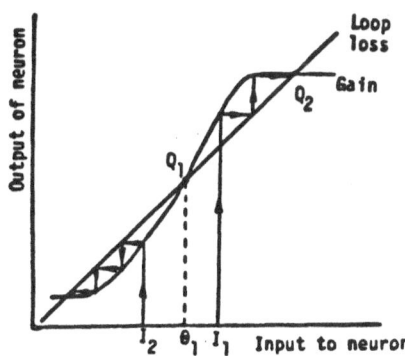

Fig. 7. Dynamics of the optical loop. θ_1 = threshold value of the loop, I_1, I_2 = initial conditions of the loop.

and latches at Q_2. On the other hand, if the initial condition is below θ_1 the signal will decay to zero. The number of iterations depends on the distance of the initial condition from the threshold. This explains the dynamics of Fig. 6.

Since the external input does not affect the shape of the final state, but rather it selects which state is produced, we can build a degree of invariance in the system since a shifted, rotated or scaled version of a stored image can cause the stored image to be recalled. The effect of such distortions of the input image are to decrease the level of the initial condition. Therefore, by raising the neuron gain, no matter how much we change the initial condition by rotating, shifting, and scaling the input image, the loop can always be made to produce an image as a stable state.

However, the ability to correctly recognize a stored image from a distorted input and the discrimination capability, i.e., the ability to distinguish images from one another, are two things that compete with each other. If there is too much gain than just shining a flash light at the input of the system causes it to lock on to one of its stable states. If the gain is set too low then even an input that is a slightly distorted version of one of the stored images is not recognized. There are two parameters under our control that can afect the gain in the loop: the gain of the neurons and the size of the pinhole.

We will use Fig. 3b as an example. Let $f_i(x,y), i = 1, 2, 3, 4$, represent the images of the letters A, B, C, D, respectively and let the pinhole size be W. Then the reconstructed images in the window at P_1 can be shown to be

$$\sum_{i=1}^{4} (g_{1i}(x,y)\text{rect}(\tfrac{x}{W})\text{rect}(\tfrac{y}{W})) * f_i(x,y) \tag{1}$$

where * represents the convolution operation, g_{11} is the auto-correlation of A and $g_{1i}, i \neq 1$, are the cross-correlations of A with B, C, D, respectively. We see that the images are blurred by the finite dimension of the pinholes. Decreasing W gives better image quality but we need to increase the gain of the neurons to compensate for the loss due to the small pinholes. At the other limit, if the pinhole size is increased we do not need very high

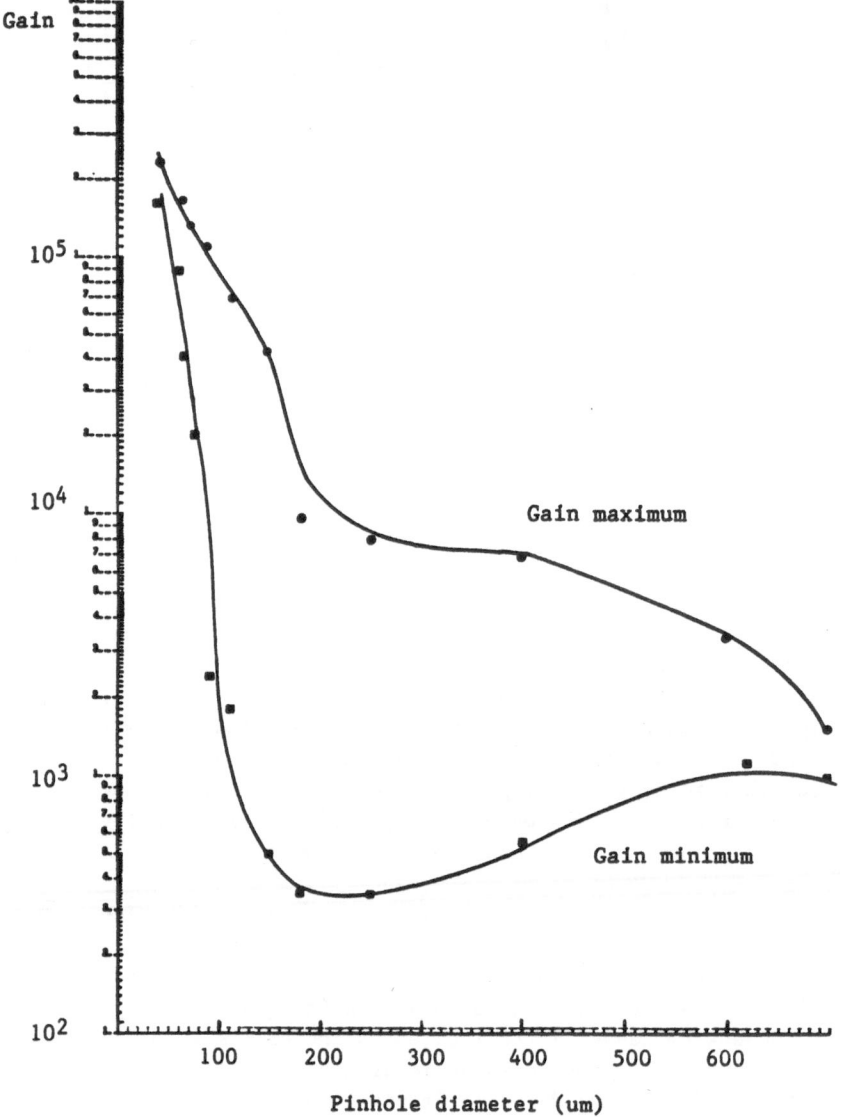

Fig. 8. The allowable gain region of the optical loop versus pinhole
size.

gain neurons but the image quality deteriorates. In the limit when W becomes
infinitely large, the reconstructed image in the window at P_1 becomes a
superposition of all the stored images, each equally strong, and severely
blurred. Thus there is an optimum pinhole size and an optimum neuron gain.

Fig. 8 shows the minimum gain required and maximum gain allowable for
the loop to sustain a stable memory as a function of pinhole size. Below
the minimum gain the loop can not recognize any image, in the sense that
once the external input is cut off the loop activity decays to zero. Above
the maximum gain the loop loses discrimination capability, meaning that any
input image, even a flash light, will induce the loop into a stable state.
This behaviour is consistent with our previous predictions. Note that the

minimum gain increase when the pinhole size is increased to more than 250 μm. This is because the reconstructed images are blurred so much that the correlation peaks are weakened and the losses in the loop are increased. Fig. 8 shows that the optimum pinhole size in this system is in the range of 70 μm to 150 μm. We choose 90 μm for the rest of the experiments.

Three kinds of invariances were studied: shift, rotation and scaling. The images stored in the holograms were four faces. The invariance capability was measured by presenting to the network one of the stored images rotated and/or shifted and/or scaled by varying amounts and monitoring the response of the system under various gain conditions. From Fig. 8, the minimum gain for the pinhole used is 2.8×10^3 and the maximum gain is 1.2×10^5. We made measurements under low gain ($\approx 3 \times 10^4$) and high gain ($\approx 10^5$) conditions. The results of the shift experiment are shown in Fig. 9. Fig. 9a shows that as the input image is shifted away from the memory position, the loop response time becomes longer because the correlation signal is shifted away from the center of the pinhole. This makes the initial condition of the loop weaker, thus it takes more iterations to reach a stable state. If the input is shifted too much then the correlation peak misses the pinhole completely and thus the input is not recognizable. However, the final output intensity is invariant so long as the shift is small enough such that the loop recognizes the input. Fig. 9b shows that the tolerance to shift can be increased by increasing the neuron gain. However, in this high gain region the loop has poor discrimination capability and it also incorrectly recognizes a similar face as one of the stored images.

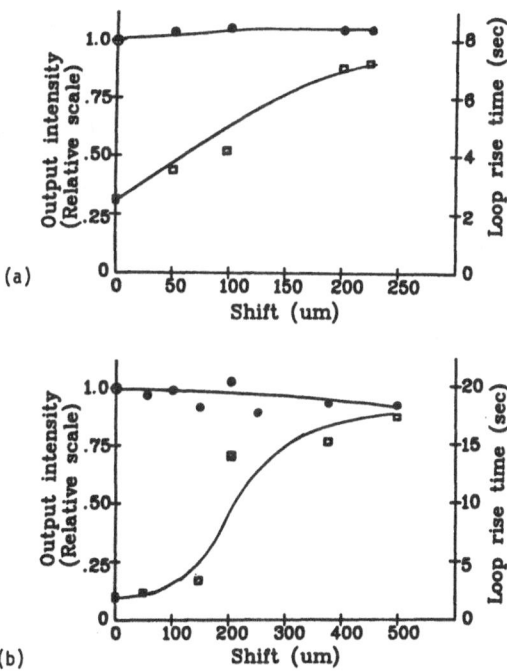

Fig. 9. Shift invariance of the optical associative loop. (a) Neuron gain = 3×10^4; (b) neuron gain = 1×10^5. The upper curve is loop output intensity and the lower curve is the loop rise time.

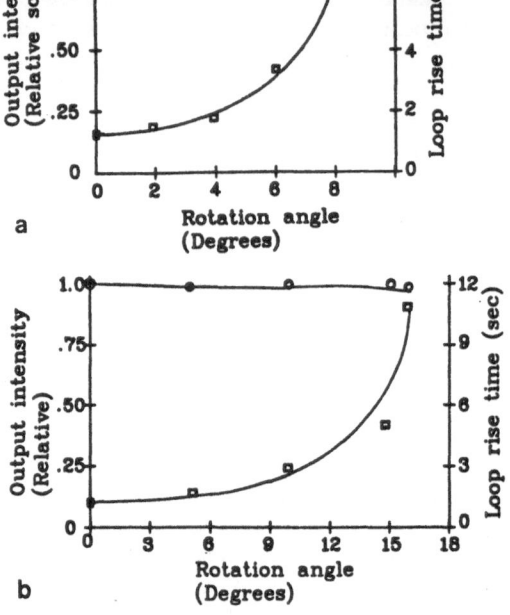

Fig. 10. Rotation invariance of the optical associative loop. (a)
Neuron gain = 3 x 10^4; (b) neuron gain = 1 x 10^5. The upper
curve is the loop output intensity and the lower curve is
the loop rise time.

The dynamics and invariance properties under rotation of the input were
also measured by using the same pinhole diameter and optical gain. The
results are shown in Fig. 10. It is seen that by increasing the optical gain
from 10^4 to 10^5 the allowable rotation angle for the input is increased from
8 degrees to 16 degrees. Again, the dynamics and rotation invariance are
consistent with our predictions. The experiments were repeated for scaling
th einput images and they showed similar invariance properties. When the
input image was scaled down to 85% size of the original stored images, the
time for the loop to converge to a stable state was increased by a factor of
3. When the input image was further scaled down to 70% size we had to increase the neuron gain from 10^4 to 10^5 for the loop to recognize the image.
However, this high gain result in such low discrimination that an input
image that is not stored is also incorrectly recognized.

ACKNOWLEDGEMENTS

Part of this work will be presented at the IEEE International Conference on Neural Networks, San Diego, July 1988. This research is supported by the Defense Advanced Research Projects and the Army Research
Office. The authors thank Dr. Eung Gi Paek who contributed greatly at the
early stage of this work and also David Brady and Claire Gu for many
helpful discussions on various aspects of this work.

REFERENCES

1. D. Psaltis and N. H. Farhat, Opt. Lett., 10, 2, 98 (1985)
2. Y. S. Abu-Mostafa and D. Psaltis, Scientific American, 256, 3, 88 (1987)
3. N. H. Farhat, Appl. Opt., 26, 10, 5093 (1987)
4. Y. Owechko, G. J. Dunning, E. Marom and B. H. Soffer, Appl. Opt., 26, 10, 1900 (1987)
5. D. Z. Anderson, Opt. Lett., 11, 45 (1986)
6. A. Yariv and S. K. Kwong, Opt. Lett., 11, 482 (1986)
7. A. D. Fisher, R. C. Fukuda and J. N. Lee, Proc. SPIE 625, 196 (1986)
8. C. Guest and R. Te Kolste, Appl. Opt., 26, 10, 5055 (1987)
9. R. S. Athale, H. H. Szu and C. B. Friedlander, Opt. Lett., 11, 7, 482 (1986)
10. D. Psaltis, J. Yu, X. G. Gu and H. Lee, Second Topical Meeting on Optical Computing, Incline Village, Nevada, March 16-18, 1987
11. D. Psaltis, D. Brady and K. Wagner, Appl. Opt.,,27, 9, 1752 (1988)
12. E. G. Paek and D. Psaltis, Opt. Eng., 26, 5, pp. 428-433 (1987)
13. K. Hsu, D. Brady and D. Psaltis, "Proceedings of the IEEE Conference on Neural Information Processing Systems", Denver, November 1987
14. A. B. Vander Lugt, IEEE IT-10, 2, pp. 139-145 (1964)

OPTICAL COMPUTING USING PHASE CONJUGATION

G.J. Dunning and C.R. Giuliano

Hughes Research Laboratories
3011 Malibu Canyon Road, Malibu, California, USA

I. PHASE CONJUGATION

One of the most interesting areas of research to emerge since the birth of laser is that of nonlinear optics. Many nonlinear phenomena such as second harmonic generation, parametric conversion, and stimulated Raman scattering have been extensively pursued as means of generating coherent optical radiation throughout the spectrum. One area of nonlinear optics, optical phase conjugation (OPC), has been receiving a great deal of attention over the past few years, not because of the production of new wavelengths, but because it offers attractive solutions to several problem area arising in a variety of applications. Such applications include adaptive optics, active imaging, high brightness lasers, laser energy and power scaling, and optical communications. Several works exist in the literature[1-3] dealing with the physics and applications of OPC; this work will focus on its potential application to optical data processing and optical computing.

The term "phase conjugation" is derived from the description of electromagnetic radiation in which the electric field is written as the real part of a complex expression. In order to understand the properties of conjugate fields consider an optical wave of frequency moving in the +z direction,

$$E = Re \left[\psi (x,y,z) \exp (i\omega t) \right] \tag{1}$$

where $\psi(x,y,z) = A (x,y) \exp i(-kz + \phi(x,y))$, A real. The phase conjugate of the wave E is defined as:

$$E = Re \left[\psi*(x,y,z) \exp(i\omega t) \right]$$

$$= Re \left[A (x,y)) \exp i(kz - \phi(x,y)) \exp (i\omega t) \right] \tag{2}$$

Note that to get the phase conjugate wave we take the complex conjugate of only the spatial part of E leaving the temporal part unchanged. The

Nonlinear Optics and Optical Computing
Edited by S. Martellucci and A. N. Chester
Plenum Press, New York, 1990

conjugate wave corresponds to a wave moving in -z direction with the phase $\phi(x,y)$ reversed relative to the incident wave. This is equivalent to leaving the spatial part of E unchanged and reversing the sign of t, hence the term "time reversal".

An intuitive appreciation of phase conjugation can be realized by comparing reflection from an ordinary mirror to reflection from a conjugate mirror. This is illustrated in Fig. 1. Here we show a diverging spherical wave striking an ordinary mirror at an angle θ and leaving the mirror at an angle $-\theta$ while continuing to diverge. In contrast the same wave striking the conjugate mirror is converted to a <u>converging</u> wave that retraces the path of the incident wave.

To elaborate a bit further, suppose the wave incident on the conjugator were aberrated. An example is a uniform plane wave passing through a distorting medium such a piece of bottle glass or a turbulent atmosphere or a severely strained optical element. Such a wave incident on a conjugator would result in an output wave that would be as severely aberrated as the input wave. However, when the output wave passes back through the aberrator it will emerge completely free of distortion. By comparison, if the conjugate mirror were replaced with an ordinary mirror, the distortion would be doubled. In fact it is this type of demonstration-passing a clean wave through an aberrator, into a conjugator, back through the aberrator to recover the clean wave again that is the subject of the first published observations in 1971 and 1972 of optical phase conjugation by researchers in the Soviet Union.

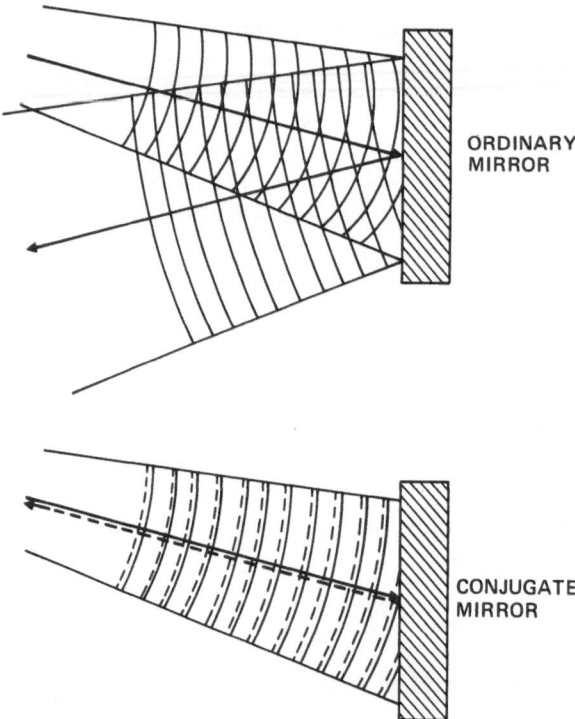

ORDINARY MIRROR

CONJUGATE MIRROR

Fig. 1. Comparison between an ordinary mirror and a phase conjugate mirror.

Zel'dovich and coworkers[4] and Nosach and coworkers[5] observed wave front
conjugation resulting from the nonlinear optical phenomenon, stimulated
Brillouin scattering, SBS. Stepanov and coworkers[6] observed conjugation via
real-time holography, a phenomenon that is similar to degenerate four-wave
mixing, DFWM, a nonlinear phenomenon recognized by Hellwarth[7] as a conjugation process and subsequently demonstrated by Bloom and Bjorklund[8]. Other
nonlinear optical processes that give rise to conjugate wave generation are
backward stimulated Raman scattering, SRS[9], and three-wave down conversion[10,11].

In the following two sections we will discuss briefly two of the most
promising nonlinear phenomena that yield conjugate waves and attempt to give
some intuitive physical understanding of how they work.

Stimulated Brillouin scattering, SBS, involves the generation of a
coherent acoustic wave through the interaction of an intense optical wave
with a nonlinear medium. The mechanism generally acts through electro-striction, that is the tendency of the medium to increase its density where the
electric field is large relative to where it is small. Without going into
details of the SBS mechanism it is observed that when an intense light wave

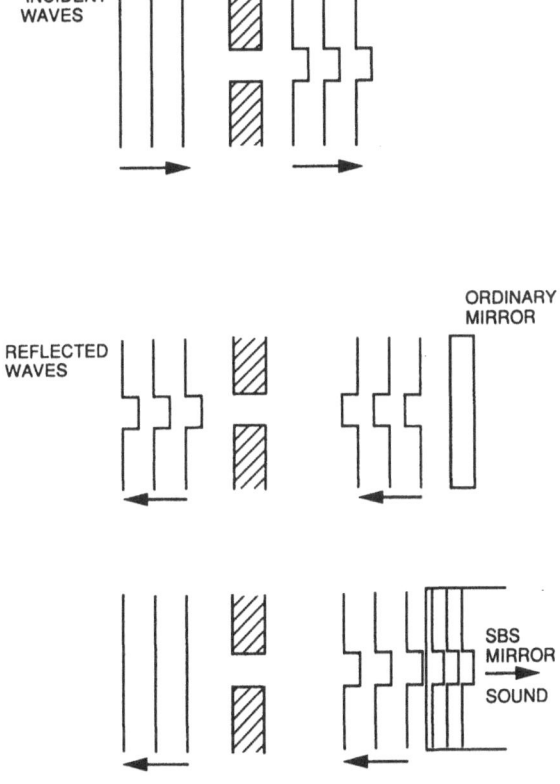

Fig. 2. Phase conjugation via SBS. The aberrated optical wave generates
an aberrated acoustic wave whose surfaces of constant phase
exactly match those of the light.

is allowed to propagate in a nonlinear medium, there results above a certain threshold intensity an intense back-scattered optical wave that is frequency down-shifted by an amount equal to the acoustic frequency. The collinear process, that is, the one that corresponds to counterpropagating incident and scattered waves has the highest gain of all the possible scattering processes and is the only one observed in SBS. In this process the acoustic wave propagates in the same direction as the incident wave and can be thought of as a moving mirror or dielectric stack from which the incident wave reflects to give a Doppler-shifted scattered wave. In fact, the wavelength of the SBS-generated acoustic wave is 1/2 that of the optical wave in the medium and as such is a moving half-wave dielectric stack.

When thinking of SBS in terms of electrostriction we note that the two counterpropagating optical waves interfere to form a moving interference pattern whose speed $V_{fringe\ pattern} = (\omega_1 - \omega_2)/(k_1 + k_2)$ is equal to the speed of sound in the medium. In fact it is this condition that allows for the buildup of the acoustic wave and the scattered optical wave at the expense of the incident optical wave.

Analysis by Zel'dovich et al.[4] and by Hellwarth[12] has indicated that under certain conditions the process for which the Brillouin gain is greatest is the one in which the scattered wave is the conjugate of the incident wave. An intuitive appreciation of the process can be realized when one considers that an aberrated input wave generates, through SBS, an equally aberrated acoustic wave with an exactly matching phase surface (see Fig. 2). The nonlinear process can be thought of as creating in the medium of a deformable mirror whose surface is just right to reverse the phase of the reflected wave over that of the incident wave so that the reflected wave, when it retraces the original path, has removed from it whatever phase errors were accumulated in the first pass. The idea of time reversal comes from the notion that if one were able to take a moving picture of the incident wave, the complete behavior of the conjugate wave would be obtained by running the movie backwards.

SBS can be made to occur in highly controlled manner and with efficiencies approaching unity especially under the conditions where it works best as a conjugator, i.e. in multimode optical wavelengths and under both cw and pulsed conditions.

Four-wave mixing, another way of generating the phase conjugate wave, is a nonlinear process in which three input waves mix to yield a fourth output wave. The geometry for degenerate four wave mixing (DFWM) is shown in Fig. 3. The three input waves consist of two counterpropagating pump waves, labeled E_f and E_b (f for forward and b for backward), taken to be plane waves and a probe wave E_p entering at an arbitrary angle to the pumps. Under these conditions the waves couple through the third-order susceptibility, $\chi^{(3)}$, to yield a fourth wave, E_s, which is proportional to the spatial complex conjugate of E_p. The third order polarization that yields the conjugate wave, $P_{NL} = \chi^{(3)} E_f E_b E_p$, is proportional to the product of the three input wave amplitudes. More specifically, for isotropic media the nonlinear polarization that yields the conjugate of E_p can be shown to arise from the contribution of three separate terms

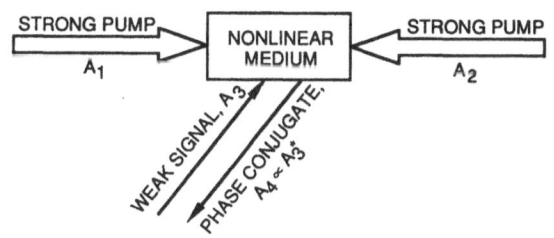

GENERAL FOUR-WAVE MIXING

$$\omega_4 = \omega_1 + \omega_2 - \omega_3$$
$$\vec{k}_4 = \vec{k}_1 + \vec{k}_2 - \vec{k}_3$$

DEGENERATE FOUR-WAVE MIXING

$$\omega_i = \omega$$
$$\vec{k}_1 + \vec{k}_2 = 0 = \vec{k}_3 + \vec{k}_4$$

Fig. 3. Geometry for degenerate four wave mixing.

$$P_{NL} = A(\theta)(\vec{E}_f \cdot \vec{E}_p^*)\vec{E}_b + A(\pi-\theta)(\vec{E}_b \cdot \vec{E}_p^*)\vec{E}_f + B\ (\vec{E}_f \cdot \vec{E}_b)\vec{E}_p^* \qquad (3)$$

The first two terms in equation (3) are responsible for the analogy
between DFWM and holography. Each contains a scalar product in parentheses
corresponding to the interference between one of the pump waves and the
probe wave which is then multiplied by the field of the other pump wave.
Thus each term corresponds to the creation of a hologram using one of the
pumps and the probe while simultaneously reading it out with the other
pump. This is illustrated in a simple way in Fig. 4, where we show the
holographic (or dual grating) picture of DFWM. In the figure the formation
process and readout process are shown separated for simplicity whereas in
fact they take place at the same time. The formation process is shown as
the generation of two overlapping grating structures (also shown separately
for simplicity) each one consisting of a series of planes whose normals are
in the direction $\vec{k}_f - \vec{k}_p$ and $\vec{k}_b - \vec{k}_p$ respectively and whose separations are
given by $D = \lambda/(2/\sin(\theta/2))$. We refer to the pattern arising from the interference between forward pump and probe as the large-spaced grating and the
one between backward pump and probe as the small spaced grating. The read-
out process occurs when the backward pump scatters from the large-spaced
grating and the forward pump scatters from the small-spaced grating yielding the conjugate wave. Thus one can view the phenomenon described by the
first two terms in equation (3) as one in which the refractive index of the
nonlinear material is spatially modulated by the interference between pump
and probe followed by scattering by the other pump.

The third term in equation (3), $B(\vec{E}_f \cdot \vec{E}_b)\vec{E}_p^*$, has no holographic
analogy. Here the scalar product between $\vec{E}_f$ and $\vec{E}_b$, corresponds to a nonlinear index which has no spatial modulation but which oscillates at a
frequency 2ω. The probe wave interacting with this driven coherent excitation at 2ω creates a polarization that results in conjugate wave generation.

The relative sizes of the coefficients A and B in (3) depend strongly
on the properties of the nonlinear medium chosen for the four-wave interac-

tion. In particular, if the nonlinear medium has an optical resonance for a single quantum transition at a frequency near ω, the frequency of the experiment, large enhancements[15] of the four-wave mixing signals arising from the first two terms in (3) are possible over that obtainable from a nonresonant system. For example, if the nonlinear medium consists of an ensemble of 2-level atoms, the near resonant contribution to the nonlinear index will manifest itself as a spatial modulation of the population of the lower state relative to the upper state. The gratings formed by interference between pumps and probe would be "population gratings" i.e., if one were to walk along the direction $\vec{k}_f - \vec{k}_p$ in the nonlinear medium one would notice that the population of atoms in the excited state relative to the ground state would vary sinusoidally with a period $D = \lambda/(2 \sin(\theta/2))$. Alternatively, if the medium possesses a pair of energy levels of the same parity such that they may be coupled coherently through an interaction involving 2 quanta, then the third term in (3) may be dominant in contributing to the four-wave signal.

Unlike SBS where the conjugate wave intensity cannot exceed the input

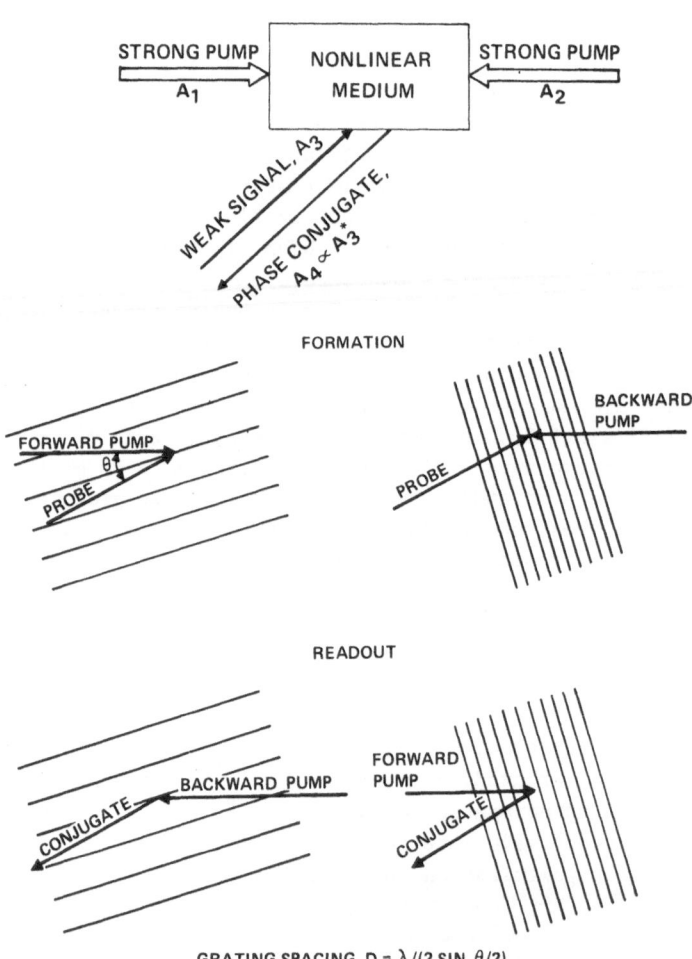

Fig. 4. The dual grating picture for DFWM.

intensity, DFWM allows for conjugate reflectivities (i.e., I_{signal}/I_{probe}) in excess of unity and in fact many such examples have been observed experimentally. This fact has practical implications for DFWM. Another characteristic of DFWM that can have practical impact involves its behavior for waves of different polarization. In fact, increased fundamental understanding of DFWM has been obtained through experiments in which the polarizations of pumps and probe waves have been manipulated to achieve the desired result[14].

Note that each of the terms in equation (3) involves the scalar product between two fields multiplied by a third field. We see that a given term will contribute to the nonlinear polarization only if the scalar product term is non-zero, that is only if the fields in the scalar product have polarization components along a common direction. This fact can be exploited to explore the fundamental properties of DFWM. For example by performing an experiment in which E_f and E_p are linearly copolarized while E_b is cross polarized one is examining the contribution of only the first of the three terms in (3) i.e., the large spaced grating. Hence, by appropriate selection of co- and cross-polarized combinations one can examine separately the contributions of the different terms in (3) for various nonlinear materials.

An interesting and potentially valuable application for phase conjugate optics in the use of a phase conjugator as an optical resonator. Fig. 5 shows schematically the essential features of a phase conjugate resonator (PCR), an optical resonator in which one (or both) of the conventional mirrors is replaced by a conjugate mirror. Several analytical papers[15,16] have been written predicting the properties of these devices and experimental demonstrations have taken place[17,18].

There are several unique properties that phase conjugate resonators are expected to exhibit. One of these is the curious property that such a resonator will not possess longitudinal modes that depend on the cavity length.

An ordinary optical resonator possesses longitudinal modes separated in frequency by $c/2L$, where L is the cavity length. This is a consequence of the boundary condition that after one round trip the wave that corresponds to a mode must constructively interfere with itself. Stated another way, the net accumulated phase after one round trip must be an integral

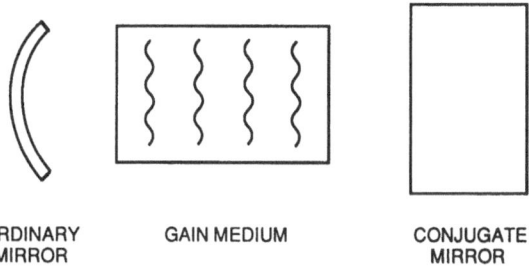

ORDINARY   GAIN MEDIUM   CONJUGATE
 MIRROR   MIRROR

Fig. 5. Phase conjugate optical resonator.

times 2π, i.e., the only waves that "fit" in the resonator are those for which $n\lambda = 2L$, n an integer. In contrast, in a phase conjugate resonator the phase that is accumulated as the wave propagates from the ordinary mirror to the conjugate mirror is subtracted by the same amount on the wave back to the ordinary mirror so that in one round trip the net accumulated phase is always zero. Consequently, a phase conjugate resonator of length L can support any wavelength consistent with the bandwidth of the gain medium and the conjugator itself. Moreover, a conjugate resonator oscillating at a particular wavelength will continue to oscillate at that wavelength independent of whether or not the cavity length is varied. This in contrast to an ordinary resonator whose spectral output will exhibit "mode hopping" and frequency drifting if the cavity length is allowed to vary.

Another property of a phase conjugate resonator, one which is highly attractive for application to high power oscillators, is that of compensation for intracavity optical distortion. It can be shown that if one extracts the light from the "ordinary" mirror end of the resonator the transverse phase of the wave will only depend on the figure of the output mirror itself and not on any other sources of distortion within the body of the resonator[15]. This feature has been qualitatively demonstrated in laboratory experiments[18].

II. OPTICAL NEURAL NETWORKS

Implementing neural networks is an area of research which can benefit from optical computing techniques. There is a significant overlap between the requirements of a neural network processor and the strengths of optical computing. Presently, there are several network models which have been proposed[19,20]. The common feature of these models is a network of nodes which performs relatively simple computations, summations or integrations and then a nonlinear thresholding operation. The computational power of a neural network comes from the high degree of connectivity between these nodes and the parallel processing of the information.

There have been two basic approaches to implement optical analogs of neural networks. The first approach uses vector matrix multiplication type optical processors using spatial light modulators and a combination of source and detector arrays[21-24]. The second approach uses processors which include holograms either in nonlinear media with saturable gain or within resonator cavity[25-33].

In order to illustrate the power of phase conjugation, we will present a detailed discussion of an all-optical associative memory. An associative memory is a fault tolerant subclass of neural networks, which can reconstruct an entire image when prompted by a partial or distorted version of the stored image. In this device, phase conjugation provides regenerative feedback, amplification and thresholding. A hologram is used as the memory element and is located in a phase conjugate resonator bounded by two phase conjugate mirrors. Holograms have been chosen because they are capable of storing a large number of three-dimensional objects and the stored information is processed in parallel. The information is stored globally throughout the volume of the hologram providing a large degree of redundancy.

Further, there is an inherent association between the object and the reference beams created by the recording process.

The recording of a hologram involves the exposure of a light-sensitive medium by two coherent waves. The amplitudes of the object a and b are given by $A(u,v)$ and $B(u,v)$. When the hologram is irradiated by a distorted or incomplete version of $A(u,v)$, $\hat{A}(u,v)$, the amplitude diffracted by the hologram is proportional to

$$\hat{A}|A + B|^2 = \hat{A}(|A|^2 + |B|^2) + \hat{A}AB^* + \hat{A}A^*B, \tag{4}$$

where the asterisk indicates the complex conjugate.

As can be seen, there is no term proportional to the complete stored image $A(u,v)$. The last term of this expression can be considered the convolution of the reference b with the correlation of $\hat{a}$ and a^*. For most natural objects there is sufficient phase variation so that if $\hat{a}$ is identical, or close, to a, their correlation provides a sharp peak and b is faithfully reconstructed. Most associative memories are designed so that given $\hat{a}_i$, one is interested in reconstructing a_i, the stored object rather than its reference b_i. Therefore by re-addressing the hologram with the last term in Eq. (4) one obtains

$$\hat{A}A^*B|A + B|^2 = \hat{A}A^*B(|A|^2 + |B|^2) + \hat{A}A^*BAB^* + \hat{A}A^*BA^*B \tag{5}$$

$$= \hat{A}A^*B(|A|^2 + |B|^2) + \underline{(\hat{A}A^*)}|B|^2A + \hat{A}A^{*2}B^2 \tag{6}$$

For most objects a and b, $|A|^2$ and $|B|^2$ result in uniform intensity distributions at the hologram. These terms will only slightly alter the transmitted amplitude, leaving its phase almost unaffected. For the case $\hat{A} \equiv A$, the phase is perfectly regenerated. Therefore, the underlined term represents a close reconstruction of the field distribution A, which reconstructs the stored object a. The previous discussion treats only a single image recorded on a hologram and is analyzed with a linear approximation model. The various terms in equation (6) can be physically separated by where they occur in space. However, if multiple images are recorded, analysis would show that there would be an overlap between the desired image and the cross terms. The addition of a nonlinear element that could provide thresholding and feedback is necessary to improve the discrimination.

An architecture for an associative memory which can improve the discrimination is shown in Fig. 6, and uses a single hologram, and two phase conjugate mirrors (PCM). Angular multiplexing of the reference waves is used in the recording of the hologram. The memory consists of a Fourier transform hologram in which the ith stored object, a_i, is recorded by interfering it with the ith reference wave, b_i. The two legs of device consist of a reference leg, to the right of the hologram, and an object leg, to the left. A partial or distorted input object, $\hat{a}_i$, addresses the hologram from the left and generates distorted reference beams, $\hat{b}_i$. These distorted references can be interpreted as the correlation of the input wavefront with the stored wavefronts and the result convolved with the corresponding undistorted references. This output, $\hat{b}_{i'}$, is focused by a lens into the phase conjugate mirror, PCM 1. The PCM by virtue of a nonlinear interaction selects the reference with the strongest correlation. The complex conjugate, $\hat{b}_i^*$, thus generated propagates back toward the hologram. This restored complex conjugate of the reference then illuminates the hologram and generates a more complete version of the stored object, a_i. If the PCM gain is greater than the losses in the system, the output will be a real image of the complete stored object a_i.

Several different objects can be stored simultaneously by recording a multiple exposure hologram, using angular multiplexing of the reference waves. This technique improves both the signal to noise ratio and the

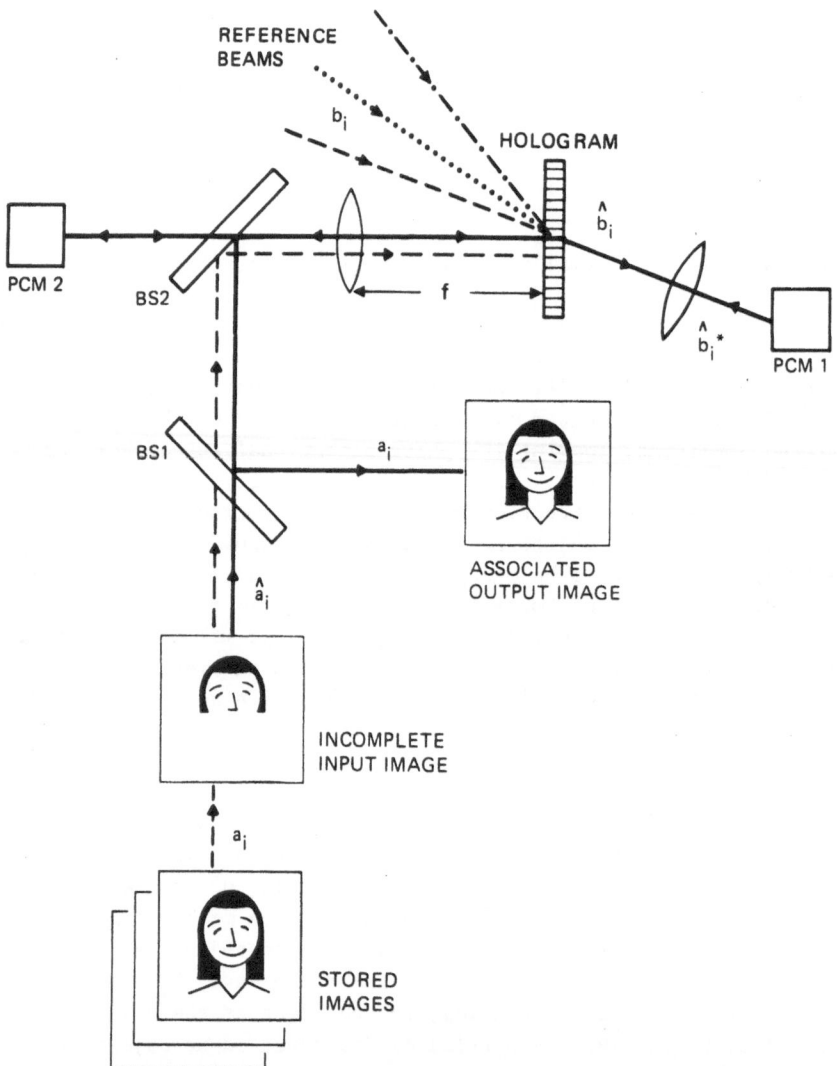

Fig. 6. Schematic diagram of an all-optical associative memory incorporating holography and phase conjugation.

storage capacity of the device by spatially separating a significant portion of the correlation noise from the signal. The reference leg of the device is self-aligning with respect to the hologram because of the retro-reflection nature of phase conjugation. However, there is an alignent requirement between the partial input image and the stored hologram. The system can be made partially shift invariant if a thin fourier transform hologram is used. The input can be translated perpendicular to the optic axis if the undesired objects which are reconstructed off-axis do not enter the field of view at the output plane.

Hetero-associations can be obtained by using spatially modulated reference beams in the recording of the hologram. In one case, a second complex object could serve as the reference for a given object. Therefore, for a partial input, the output would be an associated object and not a complete version of the stored object. In the second case, several objects could be recorded with the same reference. In this case addressing the memory with a partial version of any one of the recorded objects would recall the entire set of complete objects.

Experimentally we have demonstrated many aspects of an all-optical associative memory and processor. First, a complete two-dimensional gray-scale image with high spatial frequencies has been reconstructed with high fidelity from an incomplete input image. Second, the system selectivity between multiple objects has been proved by storing a pair of completely superimposed two-dimensional images and reconstructing either complete image from its associated partial input. Third, we have shown the system's invariance to the translational position of the input. For the following experiments we did not implement the full resonator design described above but instead used a single pass configuration. The full resonator design should have better discrimination between stored images because of the processing gain obtained through iteration. In our first experiment we demonstrated the complete reconstruction of a two-dimensional gray-scale image when the system was addressed by only a portion of the stored image. For this experiment, a single image hologram was recorded on a thin thermoplastic film. The reference beam was a converging spherical wave, f = 1 m, designed to come to a focus within the pumped volume of the phase conjugate mirror. The intensity ratio of the reference to the object beam was typically 10:1. Both beams were initially p-polarized; however, there was some depolarization of the object beam resulting from the diffuser plate placed in contact with the object. The diffuser was placed in contact with the image in order to sharpen the auto-correlation peak of the object relative to the sidelobes, and to spread the image information over the entire recording area of the hologram. The stored object was a photographic slide consisting of the gray-scale two-dimensional portrait shown in Fig. 7a.

The phase conjugate mirror was produced by degenerate four wave mixing (DFWM) in a photorefractive crystal of $BaTiO_3$. The crystal was externally pumped by two counterpropagating pump beams. In order to produce the required high phase conjugate reflectivities the pumps were p-polarized in order to access the large r_{42} electro-optic coefficient. Both the pump-probe and c-axis grating-$\vec{k}$ angle were optimized to produce sufficiently

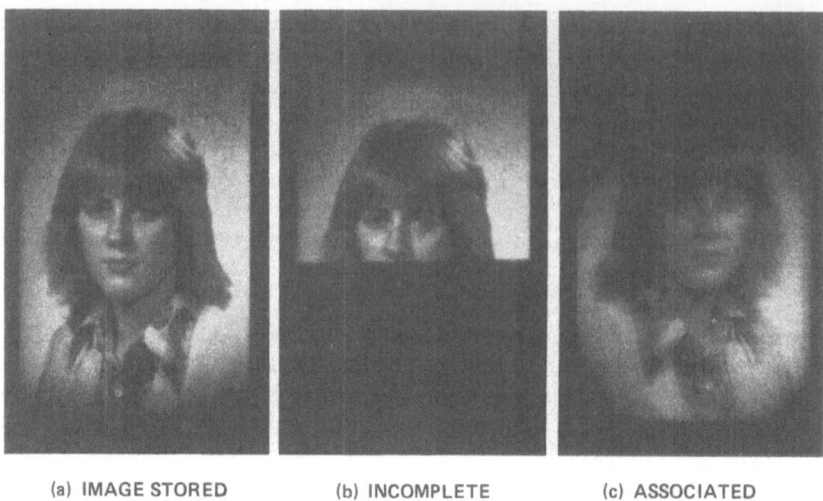

(a) IMAGE STORED
IN MEMORY

(b) INCOMPLETE
INPUT IMAGE

(c) ASSOCIATED
OUTPUT IMAGE

Fig. 7. Photographs of experimental results: (a) image stored in
memory; (b) incomplete input image; (c) associated output
image (inversion due to mirror reflection).

large values of the phase conjugate reflectivity, necessary to overcome the
system losses and to produce a good signal to noise ratio.

The input to the system was a partial version of the stored data and
is shown in Fig. 7b. When this partial input addressed the system, the
complete image of the entire face was regenerated, as shown in Fig. 7c;
figure reproduced courtesy of Opt. Lett., Vol. 12, May 1987, Dunning,
Marom, Owechko and Soffer. As can be seen, the reconstructed image is a
faithful, good fidelity reconstruction of the stored gray scale image. By
using an Air Force resolution chart as the object, we measured the resolution of the system to be 16 lines pair per millimeter. This does not,
however, represent a fundamental limit of the resolution of the system.
By using fast optics and minimizing diffractive losses the ultimate
resolution when using DFWM is on the order of the wavelength of light
interacting in the nonlinear medium.

In order to demonstrate the system's invariance to the position of
the input, we recorded a Fourier transform hologram. The object was
located at the front focal plane of a lens with the holographic recording
plane located at the back focal plane. After recording the hologram, we
could translate the input object perpendicularly to the optic axis of the
lens and still reconstruct the complete associated stored image.

As a way to show the system's selectivity between multiple stored
objects we recorded two separate images. The words "OPTIC" and "WAVES"
were recorded in superposition by double exposing the hologram. Each word
was in contact with its own individual diffuser. The two words, Fig. 8a,
are shown displaced so that each word can be easily read. In Fig. 8b, the
words are shown as they were recorded, superimposed completely overlapping. Each stored image was positioned at the same spatial location and

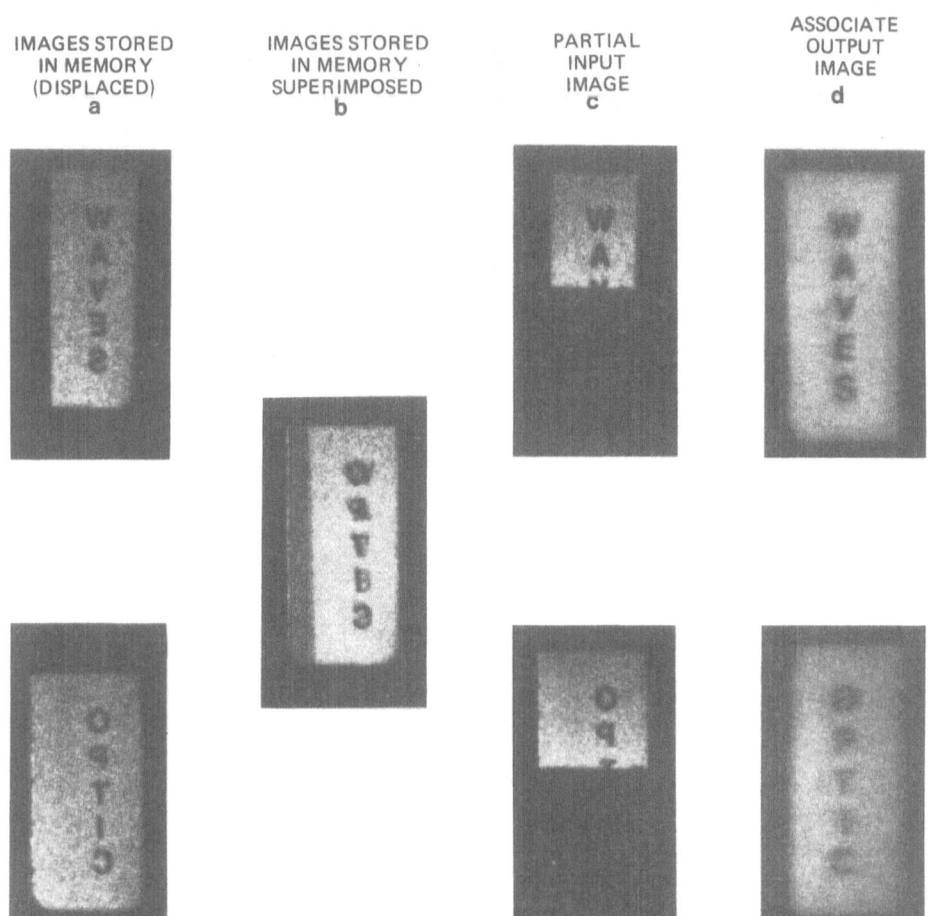

| IMAGES STORED IN MEMORY (DISPLACED) a | IMAGES STORED IN MEMORY SUPERIMPOSED b | PARTIAL INPUT IMAGE c | ASSOCIATE OUTPUT IMAGE d |

Fig. 8. Photographs of multiple image experimental results: (a) images stored in memory shown displaced; (b) superimposed images stored in memory as recorded; (c) partial input images; (d) corresponding associated output images (inversion due to mirror reflection).

was recorded with its own reference wave. When a portion of either word was input to the system (for example, the "WA"), the entire word "WAVES" would appear at the output, suppressing the word "OPTIC". Conversely, when a few letters of the word "OPTIC" was input, the complete word was produced at the output. The corresponding experimental results are shown in Figures 8c and 8d. The fidelity of the output demonstrates the high degree of cross-talk suppression. In further experiments we were able to fully reconstruct the entire word when the system was addressed with only one letter or less than 20% of the stored information.

We presently have ongoing experiments in which we are using the photo-refractive material LiNbO$_3$ as the holographic recording medium. Because the recordings are made in the bulk of the material there are several advantages over a thin recording material. First, higher diffraction efficiencies can be obtained and this has enabled us to operate the full resonator configuration. Second, more holograms can be stored in the same cross section of material due to the constraints imposed by the Bragg condition. This feature greatly improves the storage capacity of the device. Third, by using a photorefractive crystal the holograms can be either modified or re-recorded in "real-time", allowing the stored information to be updated.

In conclusion, we have discussed an all-optical associative memory which combines a holographic memory element and phase conjugation. We have been able to store multiple superimposed images and reconstruct a complete image when only a portion of the stored image was input to the system. We have also demonstrated the system's invariance to translation of the input images.

III. IMAGING THRESHOLD DETECTOR

As a second illustration of the use of phase conjugation for optical computing, we will discuss a high resolution imaging threshold detector[34]. This device is designed to determine which pixels in an input image are above or below a preset intensity threshold value. In an ideal device, an infinitesimal change in the input signal near the threshold value will cause a large change in the output signal.

We have chosen to use a photorefractive phase conjugate resonator as the basic element for the threshold detector because it has several desirable properties. First, there is a sharp threshold associated with optical resonators, determined when the gain exceeds the resonator loss. Near this threshold, a small change in either the gain or the loss causes a significant change in the output. The exact threshold value can be easily adjusted by varying either the gain or the loss of the system. A second advantage is derived from the phase conjugate nature of the mirrors. The modes of a PCR can support higher spatial frequencies over a larger cross section than those of conventional resonators. This feature enables the parallel processing of very high resolution images. Third, the output signal-to-noise ratio for an oscillator in the "on" state to the "off" state can be very high.

One architecture for the device is shown in Fig. 9 and consists of a phase conjugate resonator bounded by two phase conjugate mirrors, generated by degenerate four wave mixing in photorefractive crystals. The desired spatial information to be threshold, is impressed on an incoherent erase beam collinear with the optic axis of the resonator. The erase beam modifies the resonator gain by altering the photorefractive gratings formed during oscillation[35-37]. The input information is imaged into the read-in PCM at the interaction region, defined by where the resonator mode and counter-propagating pumps overlap. The intense portions of the input image erase the photorefractive gratings formed during the resonator oscillation and locally extinguish the resonator output. However, oscillation will still occur at locations in the image where the intensity is too weak to erase the grating. The resultant resonator pattern at the read-in PCM is a highly nonlinear, contrast inverted image of the input.

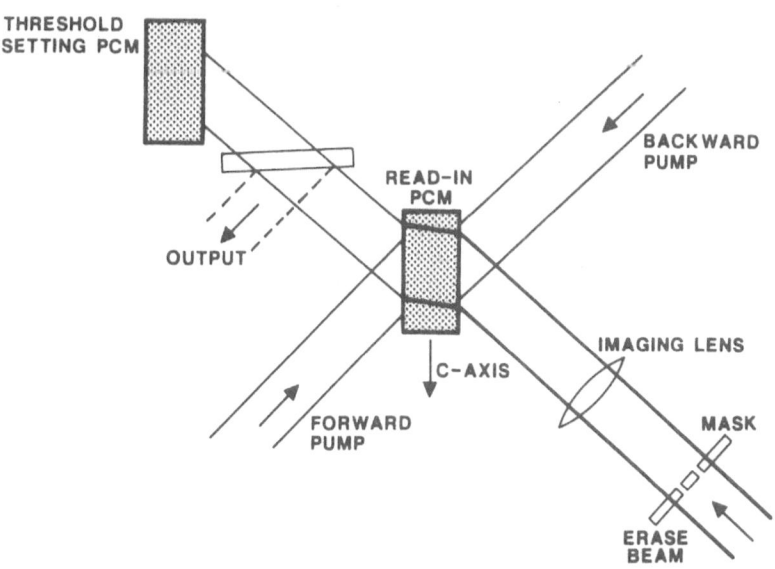

Fig. 9. Architecture for imaging threshold detector using a double
phase conjugate resonator and an incoherent erase beam.

In this device, each of the phase conjugate mirrors is generated by
using a single domain $BaTiO_3$ crystal pumped by two counterpropagating
beams, each with the same frequency and polarization. The erase beam propagates collinear with the resonator axis and must be noninterfering with
the pump beams. This avoids the formation of undesired extraneous gratings
which would degrade the device performance. The erase beam can be generated
by one of several different techniques; first, it could be a different
wavelength from the pumps; second, it could be the same wavelength as the
pumps but a different polarization; third, it could be the same wavelength
as the pump, but it could be made incoherent by increasing the path length
difference between the pumps and the erase beam to a value greater than the
coherence length of the laser; fourth, it could be generated by using an
incoherent light source.

In a photorefractive crystal if two interfering beams of amplitude E_1
and E_2 and an incoherent erase beam of amplitude E_3 are incident on the
crystal, the refractive index change Δn is given[37] by

$$\Delta n \sim E_1 E_2^* / (|E_1|^2 + |E_2|^2 + |E_3|^2). \tag{7}$$

Therefore the spatial intensity variation of the erase beam is transferred
to a local variation in refractive index change in the PCM. Since the
undepleted DFWM reflectivity is proportional to Δn, this spatial vairation
is consequently transferred to a spatial variation in resonator intensity,
with the contrast reversed and the resultant output threshold. The exact
threshold level can be controlled by adjusting the ratio of the pumps in
the second PCM.

The transfer curves for the DFWM signal versus the input erase beam
intensity is shown in the upper portion of Fig. 10. Note that for any

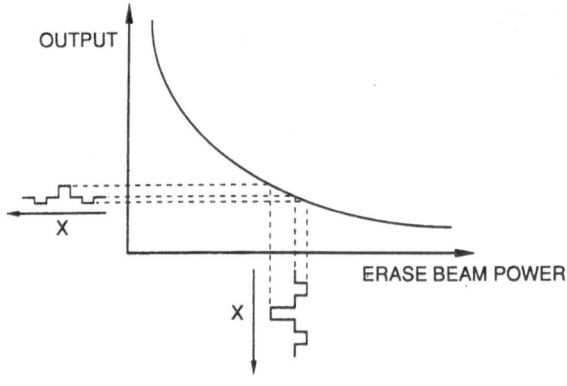

LINEAR SYSTEM

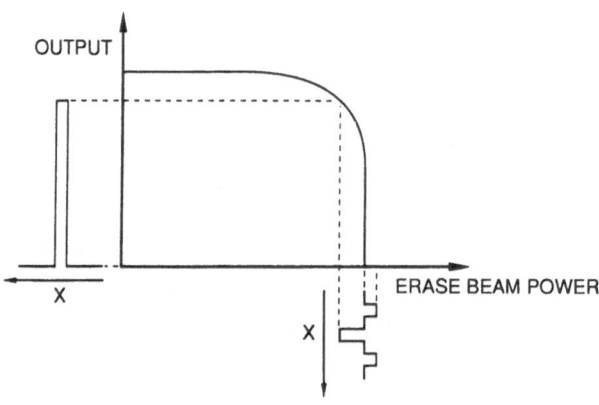

THRESHOLDING SYSTEM (PCR)

Fig. 10. Transfer curves for degenerate four wave mixing and phase conjugate resonator based devices utilizing an erase beam.

operating point the oputput signal is linearly related to the input. The lower curve corresponds to the output of the phase conjugate resonator based device versus the erase beam power. If the input operating point is chosen to be near the threshold level, notice the existence of a nonlinear threshold operation. Those portions of the erase beam higher in intensity than the threshold produce no output, while those portions below the level produce a signal. Another benefit of the PCR based system is that the output signal will oscillate at close to maximum output power level. This improves the device signal to noise ratio for pixels in the "on" state to those in the "off" state.

In the construction of the device it is important to maintain the highest resolution possible. With this goal in mind we have demonstrated thresholding using three different device configurations. One device consisted of a PCR resonator composed of a single PCM and a high reflector. The second device was constructed using a double PCM phase conjugate resonator. The third device, which is similar to the one shown in Fig. 9, used two PCMs with the addition of an intracavity lens which imaged one PCM onto the other. Assuming that the resolution at the read-in PCM is limited by diffraction in the resonator, then the three designs have different resolution capabilities. In the first design, the smallest resolvable picture element at the read-in PCM is determined by the acceptance angle of the pumped region after the mode propagates one round trip in the resonator. In the second design the resolution is determined by the pumped region in the second PCM. Because of the phase conjugate nature of the second mirror, the pixel information exactly retraces itself minimizing diffractive loss. An added benefit of a second PCM is that the over-all resonator threshold can be independently set by the pump ratio in the second or threshold-setting PCM. In the third case, the smallest resolvable picture element is determined by the f-number of the intracavity imaging lens. This design also benefits from the threshold-setting second PCM.

The resolution condition can be expressed analytically by the condition $F \cong 1$, where F is the Fresnel number, given by

$$F = aD/L\lambda. \tag{8}$$

In the above expression, a, is the diameter of the picture element at the read-in mirror, D, is the diameter of the diffraction-limiting element in the resonator, L, is the distance from the read-in mirror to the diffraction-limiting element, and λ is the wavelength. Therefore, for a given D and λ, in order to minimize the diffractive losses one would want to minimize L. Two different compact resonator configurations were constructed. The spacing between the PCM and either a high reflector or a second PCM was 1.3 cm. The diameter of the pumped regions in the crystals were approximately 200 μm. Using Eq. (8), we find that the diameter of the smallest resolvable picture element at the read-in mirror would be a $\cong 75$μm for the single PCM resonator and a $\cong 40$ μm for the double PCM resonator. To control cross talk and losses due to diffraction at higher spatial frequencies, it is desirable to further reduce the spacing of the PCR mirrors. However, a spacing reduction below 1.3 cm cannot be achieved due to the physical obstruction of the pump beams by the intracavity beam splitter used to couple out the signal. Therefore we investigated the use of an intra-cavity lens designed to image the read-in in PCM onto the second or threshold setting PCM.

As a way to characterize the resolution of the devices, a specially generated transmission mask was placed in the erase beam to impress spatial information into the PCR. The mask consists of four-dot pixel arrays on a clear substrate. The individual pixels varied in size from 10 μm to 100 μm in diameter and the centers were spaced two times the pixel diameter. By illuminating the array and imaging the pixels into the read-in PCM, we were able to obtain PCR oscillation replicating the four dots.

For diagnostic purposes we built a system which had the capability of storing and making quantitative measurements on gray-scale images. A highly linear video camera was used to detect the output image. The camera's externally controlled gain and black level allowed us to use it for photometric measurements. A video image analyzer was used to produce

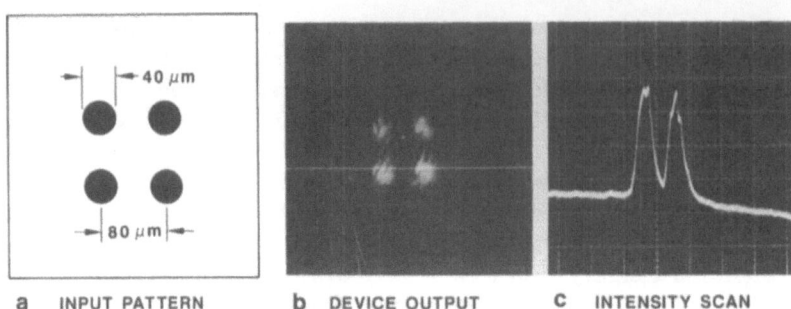

a   INPUT PATTERN   b   DEVICE OUTPUT   C   INTENSITY SCAN

Fig. 11. (a) Input array pattern of 40 μm pixels; (b) imaging
threshold detector output and (c) an intensity scan of
the output.

an isometric projection of the intensity versus X-Y position. The
analyzer also produced a pseudo-color image by assigning a different
color to each video signal voltage range. The results were then displayed
on a color video monitor. The video data could be recorded on a profes-
sional 3/4 inch tape video recorder so that we could access recorded data
for quantitative comparisons. In addition, we could simultaneously record
a sound track describing the experiment and the parameters on a real-time
basis. An individual video frame could be stored by a digital frame
grabber to be examined for extended periods of time and because of the
digital format of the output the unit could be interfaced to a computer
for further analysis.

In the experiments with the PCR utilizing an intracavity lens, the
lens had an f/number of f/1, and a focal length of 7.6 cm. By placing the
lens at a distance of twice the focal length from both PCMs, we imaged
one onto the other with a unity magnification. The pump beams were at
5145Å, p-polarized, and the erase beam was 4880Å. Using Eq. (8) (with L
= 15,2 cm and D = 7,6 cm), we calculate a ≅ 1 μm, indicating the higher
resolution of this cavity configuration. In order to characterize this
resonator, we illuminated the mask described above with an erase beam
operating at 4880Å. The four pixels shown in Fig. 11a of 40 μm diameters
on 80 μm centers are reproduced on the PCR output shown in Fig. 11b. Note
the four dot PCR mode pattern and the absence of oscillation outside the
dots. An intensity scan versus position is shown in Fig. 11c, and shows
comparable intensity for each of the pixels. We also illuminated the
other pixel arrays and obtained oscillation for pixels as small as 20 μm.
The limiting resolution was determined by the erase beam optics and not
the PCR.

In order to further study the importance of diffraction on the
spatial properties of the double-PCM PCR, we placed a variable iris near
the imaging lens on the side facing the threshold-setting PCM. We have
shown[38] the use of a variable aperture can control the spatial and
temporal properties of PCRs. By decreasing the size of the aperture, the
output from the PCR can be made to resemble the spatial distribution of a
conventional laser TEM_{00} mode. The size of the aperture did control the
physical scale of the structure in the PCR output. The results of these

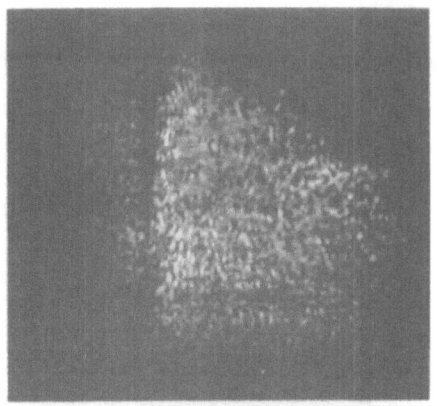

A. 12.2 mm DIAMETER APERTURE

B. 3.8 mm DIAMETER APERTURE

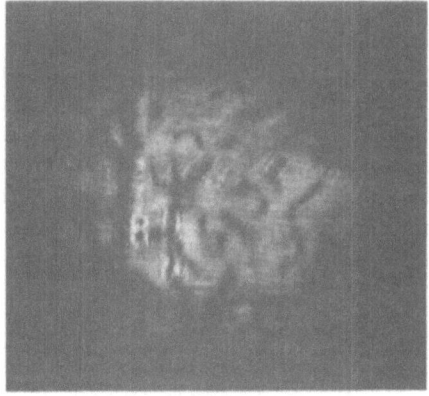

C. 0.95 mm DIAMETER APERTURE

Fig. 12. Interaction region of phase conjugate mirror used in a
phase conjugate resonator containing an intracavity lens
and various diameter intracavity apertures.

experiments are shown in Fig. 12. We have imaged the interaction plane of
the read-in PCM (i.e. that plane where the desired spatial information
would be imaged by the erase beam) onto a video camera. Fig. 12a shows the
PCR mode when the intracavity iris diameter was 12.2 mm. When the intracavity iris was stopped down to 3.8 mm, the mode structure was coarser, as

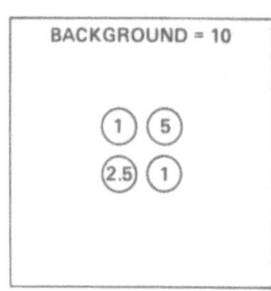

BACKGROUND = 10

(1) (5)
(2.5) (1)

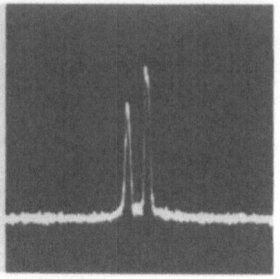

a   RELATIVE ERASE   b NEAR-FIELD DEVICE   c   INTENSITY SCAN
     INTENSITY   OUTPUT

Fig. 13. (a) Input pattern showing relative transmission background
         and pixels 75 μm in diameter; (b) near field device output;
         (c) intensity distribution along cursor shown in middle
         photograph.

can be seen in Fig. 12b. For a still smaller iris opening of 950 μm, we
observed an increase in the scale size of the spatial structure, along
with a reduction in complexity. Based on previous experience, we extra-
polate that the use of a still smaller iris size would produce a stable
mode resembling a TEM_{00} laser mode. However, having this size aperture
within the cavity would not permit oscillation of very high spatial patterns. For a given desired feature size (e.g. 75 μm), one can calculate a
limiting aperture diameter D by rearranging Equation (8). The use of this
aperture would allow the resolution of the required feature, without producing fine structure within the feature. Using a = 75 μm and L = 14 cm, the
required aperture diameter D ≅ 1 mm. Obviously, higher resolution would
require a larger aperture.

     There are two other factors which influence the resolution of this
device. First the interaction length d (typically defined by the crystal
thickness) cannot exceed the depth of focus for the smallest resolution
element in order to avoid spatial overlap of separate picture elements in
the crystal. For Gaussian beams, this corresponds to the condition that the
interaction length cannot exceed the confocal parameter b:

$$d < b = \pi n a^2 / 2\lambda \tag{9}$$

where n, is the refractive index and a, is the diameter of the focal spot.
A condition similar to Eq. (9) differing only by a numerical factor, was
derived by Marrakchi et al.[37] using geometric optics. For n ≅ 2.4 (for
$BaTiO_3$), λ = 500 nm, and a spot diameter of 10 μm (corresponding to a
spatial frequency of 50 lines/nm), we find d < 0.75 nm. The second factor
that influences resolution is the orientation of the erase-beam axis with
respect to the axis of the PCR. Following the analysis for the DFWM erase-
mode device[37], we find that the k-vector mismatch Δk is minimized when the
erase beam is aligned parallel to the probe beam. For this optimum alignment,
the condition d < π/Δk also yields an expression differing only by a numerical factor from equation (9).

     In order to demonstrate the thresholding properties of the device, we

used a different mask in the erase beam. This mask consisted of various
pixel arrays, composed of four 75 μm dots on a clear substrate, each pixel
having a different transmission. These experiments were performed using the
resonator constructed from a single PCM and high reflector. The output from
a cw argon-ion laser at 5145 Å was rotated to p-polarization and divided by
a beam splitter to form two counterpropagating pump beams in the $BaTiO_3$ PCM.
Both the pump-probe angle and grating-k c-axis angles were optimized to
produce large phase conjugate reflectivities, typically 25. The spacing
between the PCM and the mirror was 1.3 cm. The erase beam was obtained from
the same laser used to pump the PCM. However, the polarization of this beam
was rotated by 90° to s-polarization to avoid interference with the pump
beam. The mask was placed in this erase beam and imaged into the interaction
region of the PCM. The output from the PCR was taken using an intracavity
beam splitter, where a relay lens was used to image the interaction region
of the read-in PCM onto the video camera.

The mask actually consisted of several different arrays of partially
transmitting dots (each 75 μm in diameter) on a clear substrate. One of the
arrays is illustrated in Fig. 13a. The output from the device when this
arrays was input is shown in Fig. 13b. The ratio of erase beam intensity to
pump intensity was set so that the upper right element was just below
threshold. The other three elements were seen to oscillate, with the lower
left element only slightly reduced in intensity from the lower right element.
The intensity profile, Fig. 13c, shows that even though the input intensity
varies by a factor of 2.5, the output differs by only 15%. This data set
clearly demonstrates the thresholding capability of the device (figure
reproduced courtesy of Opt. Lett., Vol. II, Sept. 1986, Klein, Dunning,
Valley, Lind and O'Meara).

In summary, we have described an all optical threshold detector
that can determine if one or more pixels in an array of pixels are above
or below a preset intensity threshold. The threshold level can easily be
adjusted by varying the ratio of the pump beams. Further, we have demonstrated 20 μm resolution and optimization of the optics should produce
1 μm resolution.

IV. ACKNOWLEDGEMENTS

We would like to acknowledge our colleagues: Bernard Soffer, Yuri
Owechko, and Many Marom for their contributions to the work presented in
Section II "Optical neural networks"; and, Marvin Klein, George Valley,
Richard Lind and Thomas O'Meara for their contributions to the work
presented in Section III "Imaging threshold detector". This work was
supported in part by the U.S. Air Force Office of Scientific Research.

REFERENCES

1. C. R. Giuliano, Applications of Optical Phase Conjugation, Physics
    Today, 34, 27 (1981)
2. R. A. Fisher, ed. "Optical Phase Conjugation" (Academic Press,
    New York, 1983)

3. D. R. Pepper, Nonlinear Optical Phase Conjugation, "The Laser Handbook", Vol. 4 (1985), ed. M. Bass and M. Stitch.

4. B. Ya. Zel'dovich, V. I. Popovichev, V. V. Ragul'skii and F. S. Faizullov, Soviet Physics JETP, 15, 109 (1972)

5. O. Yu, Nosach, V. I. Popovichev, V. V. Ragul'skii and F. S. Faizullov, Soviet Physics JETP, 16, 435 (1972)

6. B. I. Stepanov, E. V. Ivakin and A. S. Rubanov, Soviet Physics Doklady, 16, 46 (1971)

7. R. W. Hellwarth, J. Opt. Soc. Amer., 67, 1 (1977)

8. D. M. Bloom and G. C. Bjiorklund, Appl. Phys. Lett., 31, 592 (1977)

9. B. Ya. Zeld'dovich, N. A. Mel'nikov, N. F. Pilipetskii and V. V. Ragul'skii, JETP 1ETT., 25, 36 (1977)

10. A. Yariv, J. Opt. Soc. Amer., 66, 301 (1976)

11. P. Avizonis, F. A. Hopf, W. D. Bomberger, S. F. Jacobs, A. Tomita and K. H. Womack, Appl. Phys. Lett., 31, 435 (1977)

12. R. W. Hellwarth, J. Opt. Soc. Amer., 68, 1050 (1978)

13. R. L. Abrams and R. C. Lind, Opt. Lett., 2, 94 (1978) and Opt. Lett., 3, 205 (1978)

14. D. G. Steel, R. C. Lind, J. F. Lam and C. R. Giuliano, Appl. Phys. Lett., 35, 376 (1979)

15. J. F. Lam and W. P. Brown, Opt. Lett., 5, 61 (1980)

16. J. M. Bel'dyugin, M. G. Galushkin and E. M. Zemskov, Sov. J. Quantum Electron., 9, 20 (1979);
J. AuYeung, D. Fakete, D. M. Pepper and A. Yariv, IEEE J. Quantum Electron., QE-15, 1180 (1979);
P. A. Belanger, A. Hardy and A. E. Siegman, Appl. Opt., 19, 602 (1980)

17. D. M. Pepper, D. Fekete and A. Yariv, Appl. Phys. Lett., 33, 41 (1978)

18. J. Feinberg and R. W. Hellwarth, Opt. Lett., 5, 519 (1980);
R. C. Lind and D. G. Steel, Opt. Lett., 6, 554 (1981)

19. T. Kohonen, "Self Organization and Associative Memory", Springer-Verlag, New York (1984)

20. J. Hopfield, Neural Networks and Physical Systems with Emergent Collective Computational Abilities, Proc. Nat. Acad. Sci., 79, Biophysics

21. R. Athale, H. Szu and C. Friedlander, Optics Lett., 11, 482 (1968)

22. A. Fisher, C. L. Giles, and J. Lee, J. Opt. Soc. Am., A, 1, 1337 (1984)

23. G. Eichmann and H. J. Caulfield, Appl. Opt., 24, 2051 (1985)

24. N. Farhat, D. Psaltis, A. Prata and E. Paek, Appl. Opt., 24, 1469 (1985)

25. D. Anderson, Opt. Lett., 11, 45 (1986)

26. H. J. Caulfield, Opt. Comm., 55, 80-82 (1985)

27. M. Cohen, Self organization, association, and categorization in a phase conjugating resonator, Proceedings of SPIE Optical Computing, 625, 214-219, January 1986

28. G. J. Dunning, E. Marom, Y. Owechko and B. H. Soffer, Optical holographic associative memory using a phase conjugate resonator, Proceedings ot SPIE Optical Computing, 625, 205-213, January 1986

29. B. H. Soffer, G. J. Dunning, Y. Owechko and E. Marom, Opt. Lett., 11, 118, (1986)

30. M. Kim and C. Guest, Adaptive 2D holographic associative processor, Proceedings of SPIE Optical Computing, 625, 174-178, January 1986

31. H. Mada, Appl. Opt., 24, 2063-2066, July 1985

32. A. Yariv and S. Kwong, Opt. Lett., 11, 186-188, (1986)

33. A. Yariv, S. Kwong and K. Kyuma, Appl. Phys. Lett., 48, 1114-1116, (1986)

34. M. Klein, G. Dunning, G. Valley, R. Lind and T. O'Meara, Optics Letters, 11, 575, (1986)

35. A. Kamshilin and M. Petrov, is'ma Zh. Tekh. Fiz., 6, 337 (1980) (Sov. Tech. Phys. Lett, 6, 144 (1980))

36. Y. Shi, D. Psaltis, A. Marrakchi and A. Tanguay, Appl. Opt., 22, 3665 (1983)

37. A. Marrakchi, A. Tanguay, J. Yu, and D. Psaltis, Proc. Soc. Photo-Opt. Instrum. Eng., 465, 82 (1984)

38. G. Valley and G. Dunning, Opt. Lett., 9, 513 (1984)

NONLINEAR PHOTOREFRACTIVE EFFECTS AND THEIR APPLICATION

IN DYNAMIC OPTICAL INTERCONNECTS AND IMAGE PROCESSING

L. Hesselink, J. Wilde and B. McRuer

Dept's of Aeronautics/Astronautics and Applied Physics
Stanford University, Stanford, Ca 94305-2186

I. INTRODUCTION

I.1. Dynamic Optical Interconnects

Future generations of computing machines may require an optical swiching arrangement which provides a high speed, large size reconfigurable interconnect. The fundamental component of such a switch is based on a dynamic medium which can route optical beams in accordance with some external programming stimulus. In particular, photorefractive crystals have been invetigated for their potential to solve the dynamic interconnect problem[1,2]. These crystals allow for real-time recording of holograms which diffract signal beams to the desired optical detectors. Although the benefits of an ideal holographic interconnect scheme are tantalizing, the realization of such an approach is no simple matter. Establishing a communication network between points on a plane requires a very specific set of multiplexed holograms. The fact that a large number of holograms (most likely more than 10) is required before this approach becomes competitive with current electronic techniques places an inherent constraint on the ultimate performance of the crystal, since the diffraction efficiency of each hologram decreases with increased multiplexing.

In this paper several approaches for establishing optical interconnects involving photorefractive crystals are discussed. First, an architecture incorporating an overhead nonlinear crystal in conjunction with a phase conjugator for establishing communication channels between several processors located in a plane is considered. The phase conjugator is located above the nonlinear crystal and provides the wavefront necessary for establishing the interconnect pattern. The wavefront is generated from a superposition of sources located close to the required processor receiver positions. The time reversed wavefront in conjuction with the the wave emanating from the source(s) then establishes the interconnect pattern in the storage crystal. This two step process requires that all processors must provide mutually coherent light waves in close proximity to the detector locations. It may be possible, however, to design a single element which fulfills the role of both a source and detector. An attractive possibility, though perhaps not immediately available, would involve an integrated array of injection-locked diode lasers.

A second alternative approach takes advantage of several degrees of freedom available to the designer for establishing a hologram in a photorefractive medium such as SBN (strontium borium niobate)[3]. For example, photorefractive sensitivity is strongly wavelength dependent, allowing a hologram formed with green beams to be reconstructed with a red beam without significant erasure. In thick photorefractive media, however, reconstruction efficiency is severely limited, unless the Bragg condition is substantially satisfied. A configuration which allows the wavelengths for recording and reconstruction to be considerably different while still maintaining the Bragg condition involves placing the sources and receivers on conical surfaces. The apex of each cone coincides with the interconnect element and the height scales as the ratio of the wavelengths. The base diameter for each cone is the same. The projection of recording and reconstruction wave vectors onto the base are therefore also the same, implying that the Bragg condition is satisfied, despite disparity in wavelength. The attractiveness of this design stems from the reduction in the number of writing beams required to establish the interconnect. In general, to arbitrarily diffract n signal inputs into m outputs, requires nxm gratings, or double that number of writing beams. For the configuration discussed here, it can be shown that n+m writing beams are sufficient to establish nxm interconnects. Alternatively, by using polarization effects in conjunction with judiciously choosing the recording time of the holograms in SBN a very asymmetric recording/reconstruction cycle can be established without varying the wavelength[3]. Additionally, fibers of SBN give the freedom to synthesize an interconnect medium with very high multiplexing capability and little cross talk[4]. In the third example, it is shown that a set of spatially multiplexed holographic recordings in a photorefractive crystal constitutes an optical crossbar switch that globally interconnects a two-layer neural network. A photorefractive transmission hologram can be considered an optical crossbar switch, in which a set of vectors can be stored as a linear combination of outer-product interconnection matrices. Using a simple amplitude-phase encoding scheme allows a set of normalized vectors to be represented optically. Stored vectors may be recalled in a content addressable fashion by illuminating the crystal with a related input vector and allowing the holographic crossbar to perform the required matrix-vector multiplication operation. The output vector is then determined from amplitude and phase measurements of the diffracted signal beams. Nonlinear feedback may be added to enhance convergence to the desired stored vector.

I.2. Optical Image Processing

Optical processors based on holographic recording in photorefractive crystals allow a variety of image processing tasks to be carried out. For example, by varying the ratio of the object to reference beam intensity, the strength of the diffracted beam can be made to be proportional to $\sqrt{R}$ for R<1 or $1/\sqrt{R}$ for R>1, where R denotes the ratio of the object beam intensity to that of the reference beam intensity. By illuminating the crystal with the Fourier transform of the object, small defects in periodic objects or those with smooth backgrounds may be enhanced. Enhancements is accomplished by tuning the reference beam intensity to that of the Fourier transform of the signal component corresponding to the defect signal. For example, in this fashion tiny cracks of 0.15 μm width have been enhanced in heads used in magnetic disk storage devices.

Finally, by exploiting the enhancements of diffraction efficiency of holograms written in photorefractive crystals using time varying intensity patterns, a whole new class of optical processors may be developed. As an example, a velocity filter may be synthesized which enhances moving objects in certain direction and with velocities in a given range. This filter is based on the strong grating velocity sensitivity of the non-linear photo-refractive response in BSO. Using a schlieren imaging system a real-time velocity filter may be synthesized. The range of velocities that are enhanced is tunable by varying the writing intensities of the illumination beams.

II. DYNAMIC HOLOGRAPHIC INTERCONNECTION USING PHOTOREFRACTIVES

Schematically the holographic interconnect is pictured in Fig. 1. It consists of two photorefractive crystals situated above an array of micro-processor chips. The top crystal functions as a phase conjugate mirror while the bottom crystal is used as a holographic storage medium. For pur-poses of this discussion a planar square array is chosen for the processor geometry. However, this part of the design is flexible and may even be three-dimensional provided each processor can "see" the non-linear crystal and phase conjugate mirror located above the array. All processors must have mutually coherent sources along with detectors in close proximity to those sources. It may even be possible to design a single element which can serve as both source and detector. The fact that all sources must be coherent requires the use of fiber distribution from a common laser. A more attractive approach, although not immediately possible, would involve an integrated array of injection-locked diode lasers.

The formation of an interconnect between an arbitrary source and detector is a two-step process which results in the writing of a reflection hologram in the overhead nonlinear crystal. This hologram will then diffract the incident source light to one or more detectors. Hence, a communication link is established which is essentially bandlimited (possibly in the low

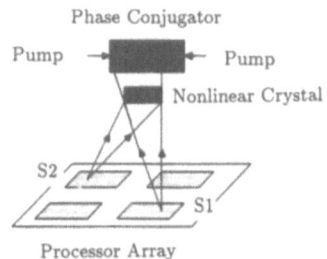

Fig. 1. Schematic of dynamic holographic interconnect.

GHz) only by the modulator and related electronics. Ideally, the source modulator should be able to deliver three light level intensities: high, low, and off. All three levels are used for writing the hologram but only low and off are used for communication.

Using the following process a holographic interconnect between two arbitrary points on the array may be established.

1. Step one. Source 1, in the low light level mode, illuminates the phase conjugate mirror after passing through the lower nonlinear crystal. The four-wave mixing phase conjugate mirror returns an amplified phase conjugate beam (now at high level) back towards the lower crystal. This return beam builds up in intensity with some time constant τ_1 determined by the crystal material and incident intensity. Note that amplification will compensate for absorption in the lower crystal.

2. Step two. After the phase conjugate beam is established, source 1 is turned off. The phase conjugate beam is no longer being stimulated by source 1 and will subsequently decay with the time constant τ_1 as the grating are read out. Source 2 is then turned on to a high light level and will interfere with the decaying phase conjugate beam. This interference pattern is recorded as a reflection hologram in the nonlinear crystal with a time constant τ_2, thus creating the required connection between the two points.

II.1. Proof of Concept

A single experiment has been carried out to demonstrate proof of concept. This experiment was not designed to follow the exact procedure outlined in the previous section, but only to provide proof of a concept. In addition, the preliminary investigation serves to analyze important issues such as:

- the effects of unwanted gratings due to spurious reflections;
- the relative recording sensitivity between the two elements;
- the spatial resolution of the interconnect.

The experiment is carried out using a single 1 cm^3 crystal of LiNbO$_3$. This one crystal serves as both the phase conjugate mirror and inter-connect recording medium. The asymmetry of the read-write cycle is strongly dependent on the amount of iron doping, which in our case is 0.015%, and the absorption coefficient is 0.39 cm^{-1} at 515 nm. Since the concentration of the traps greatly exceeds that of donors, the sensitivity for grating formation is quite low. Therefore, this material is not ideal for a system requiring fast time response, but the dramatic asymmetry in its write/read curve allows the one crystal to function as the two needed elements.

The crystal arrangement is shown in Fig. 2. In the first step, a grating is formed between source 1 ($I_{S1} \cong 3$ mW/cm^2) and reference beam 1 ($I_{R1} \cong 6$ mW/cm^2). The resulting reflection hologram is read out by ref-

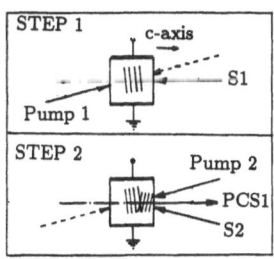

Fig. 2. Crystal arrangement for creation of a holographic interconnect in LiNBO$_3$.

erence beam 2 to produce the phase conjugate of source 1. Since the erasure of this grating is very slow, the intensity of reference beam 2 (I_{R2})
can be increased until the desired phase conjugate intensity is reached.
In our case I_{R2}= 23 mW/cm^2, which provides a phase conjugate intensity of
3 mW/cm^2, thereby emulating a phase conjugate mirror with unity reflectivity.
In the second step source 2 ($I_{S2} \cong$ 3 mW/cm^2) is turned on and interferes
with the phase conjugate of source 1. This interference pattern creates the
reflection hologram linking the two sources. For optimal recording of the
hologram the crystal c-axis must be oriented normal to the surface of incidence. The diffraction efficiency for this process is rather small, namely
0.1% necessitating approaches involving other materials. One such approach
involves polarization switching between recording and readout[3]. This proof
of concept, nevertheless, demonstrates the following interconnect features:

- reconfigurable or dynamic operation;
- direct data linking;
- easily-alignable and non-mechanical optical switching.

Since LiNbO$_3$ is relatively insensitive, this application requires
excessive energy to establish the interconnection. Materials with a faster
time response such as SBN must be investigated. Alternatively, new architectures may be devised for implementation of optical crossbar switches.

III. CONICAL INTERCONNECT CONFIGURATION

The optical crossbar is an attractive architecture for providing high
bandwidth interconnects, because once it is configured it will diffract,
refract or reflect optical inputs passively to their respective outputs.
When the interconnect pattern has served its purpose the crossbar can be
reconfigured for a new pattern. The major advantage of this device is that
while the reconfiguration time may be relatively slow (ms or μs), the
interconect has optical transmission bandwidth (GHz) and does not require
detection or regeneration of the signal as is needed for conventional
electronic switching.

The two-wavelength dynamic optical interconnect is a novel optical
crossbar architecture which reduces the order of switching complexity
from n^2 to n for a nxn switch, albeit at the expense of some restrictions
on the implementation of the interconnect pattern. This architecture inter-

connects a set of input channels to a set of output channels in some arbitrary configuration using a set of gratings written into a photore-fractive crystal. Transmission gratings are written or erased using planar writing beams of wavelength λ_w. The destructive readout problem of photo-refractive gratings is avoided by using signal readout beams with a longer wavelength λ_s.

In general, to arbitrarily diffract n signal inputs to m outputs, nm gratings are needed. This would require 2 nm writing beams. Using the geom-etry of Fig. 3, the number of writing beams is reduced by associating a writing beam with each of the n input and m output signal beams. The geom-etry ensures that a grating written by any pair of writing beams will satisfy the Bragg condition to interconnect the associated pair of signal paths. To record the grating K_{12} using writing beams $\vec{k}_{w1}$ and $\vec{k}_{w2}$ the fol-lowing k-vector condition is satisfied

$$\vec{k}_{w1} - \vec{k}_{w2} = \vec{K}_{12} \tag{1}$$

(note: All k-vectors discussed are inside the crystal). The Bragg condition for which signal input $\vec{k}_{s1}$ is diffracted into signal output $\vec{k}_{s2}$ by $\vec{K}_{12}$ is satisfied provided:

$$\vec{k}_{s1} - \vec{k}_{s2} = \vec{K}_{12}. \tag{2}$$

These relationships are satisfied in the conical geometry seen in Fig. 3. The writing k-vectors $\vec{k}_{wi}$ lie on the surface of a cone and emanate from the cone apex. The signal k-vectors $\vec{k}_{si}$ (may be either an output or an input) lie on a second cone which has the same base, but is lower in height. Each $\vec{k}_{wi}$ is associated with the $\vec{k}_{si}$ which has the same projection onto the cone base. The lengths from the apex to the base perimeter along the side of each cone are defined by $|\vec{k}_{wi}| = 2 \pi \phi / \lambda_w$ and $|\vec{k}_{si}| = 2 \pi \phi / \lambda_s$ where ϕ is the bulk index of refraction. The radius of the cone base r is the final degree of freedom

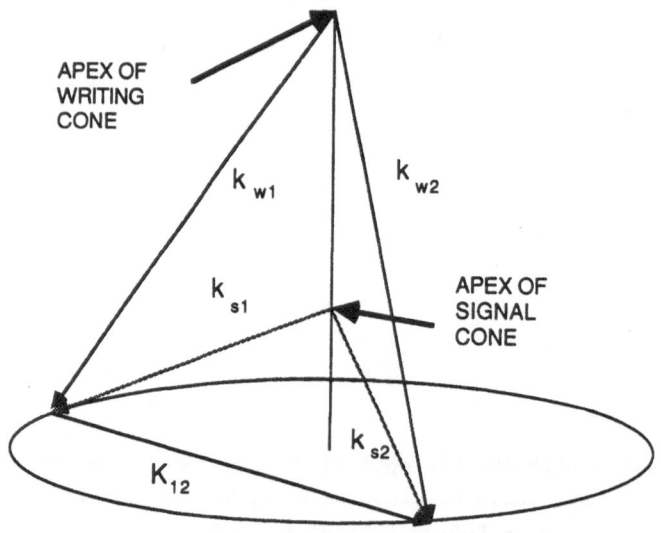

Fig. 3. Conical interconnect arrangement.

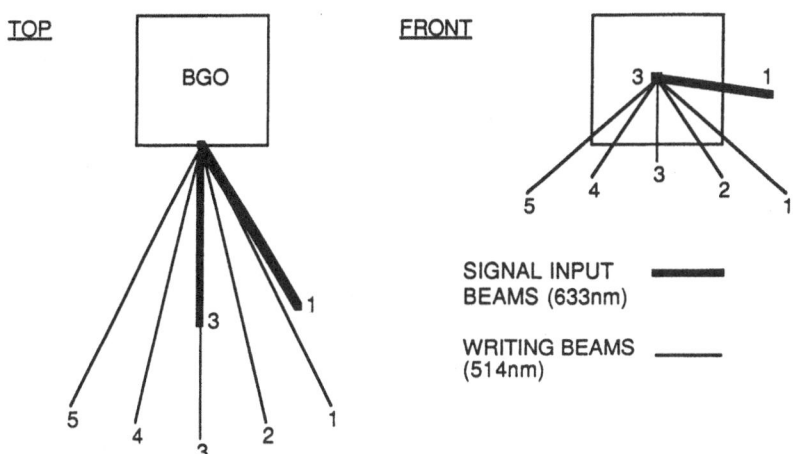

Fig. 4. Experimental implementation of a dynamic conical interconnect.

needed to define a unique cone geometry and is constrained by $0 < r < |\vec{k}_{si}|$. Input and output signals are assigned different $\vec{k}$-vectors because of the crosstalk problem of the undiffracted signals. This arrangement is attractive, because it allows nondestructive readout of multiple Bragg gratings.

III.1. Experimental Configuration

The experimental configuration for implementation of the conical interconnect architecture is shown in Fig. 4. A 2 x 3 optical crossbar switch is formed by using a 0.8 cm-sided cubic BGO crystal in the <110> orientation with an applied electric field of 5 kV/cm, perpendicular to the <110> face. Five writing beams with λ_w = 514 nm and two signal input beams with λ_s = 633 nm form the basis of a 2 input to 3 output switch. All beams are vertically polarized. The writing beams are labeled 1 through 5. All beams are aligned so that the $\vec{k}$-vectors intersect the cone base along a .21 radian arc of the base. Using only a small section of the cone ensures that a major component of the possible grating vectors is parallel to the applied field.

For example, to connect input 3 to output 2, writing beams 3 and 2 are turned on to write grating $\vec{K}_{32}$. To broadcast input 3 to outputs 2, 4 and 5 writing beams 2, 3, 4 and 5 are turned on simultaneously. Besides the required gratings $\vec{K}_{32}$, $\vec{K}_{34}$ and $\vec{K}_{35}$, cross gratings $\vec{K}_{24}$, $\vec{K}_{25}$ and $\vec{K}_{45}$ are also generated. These do not contribute to crosstalk since they interconnect outputs to outputs. More than one independent grating can be written in the crystal at once, by using a multiplexing method. Gratings $\vec{K}_{15}$ and $\vec{K}_{32}$ are written by alternately switching on each pair of writing beams at 300 Hz, much faster than the buildup or erasure times. As expected, the effective erasing of the $\vec{K}_{32}$ caused by the writing beams 1 and 5 significantly lowers the maximum steady-state diffraction efficiency below that of writing $\vec{K}_{32}$ alone. Crosstalk of signal input 1 into output 2 and that of signal input 3 into output 5 is below the measurement limit of -25 dB. Although the erasure sensitivity of BSO to HeNe light is relatively small, light from a diode laser at 780 nm may be effectively used for communication via the

optical interconnect pattern using the same recorded gratings by adjusting
the cone angle of the readout beams. Monitoring of the diffracted output
showed no apparent erasure, other than that caused by the dark current.

IV. OPTICAL ASSOCIATIVE MEMORIES BASED UPON PHOTOREFRACTIVE
    CROSSBAR SWITCHING

     Optical crossbar switches are not only useful for performing inter-
connect operations, but also allow fundamental mathematical operations
common to most digital associative memory models to be carried out optically.
Optical crossbar switches, as shown in the previous section, require that n
inputs are coupled to n outputs, and this arrangement is readily implemented
using transmission holography. In this section we show that a simple amplitude-phase encoding scheme allow a set of normalized vectors to be represented optically and stored as a linear combination of outer-product interconnection matrices using a photorefractive crossbar switch. A stored vector
is recalled in a content-addressable fashion by illuminating the crystal
with a related input vector and allowing the holographic crossbar to carry
out the required matrix-vector multiplication. An output vector is subsequently generated upon amplitude-phase detection of the diffracted signal
beams. If needed, nonlinear feedback can be added to the system so that
after a few iterations the output converges to the closest matched stored
vector.

     In a generalized crossbar switch (Fig. 5) n inputs are connected to n
outputs, with variable weights W_{ij}. In a holographic implementation, n inputs
diffract from a set of multiplexed gratings to produce n outputs. The
strength of the output depends on the grating from which it diffracted, and
each output beam is characterized by its amplitude and phase. In a volume
recording medium several interconnect patterns may be superimposed without
appreciable crosstalk by spatially multiplexing the gratings. The dynamic
nature of photorefractive crystals allows selective gratings to be modified
with the appropriate alteration of writing beam intensity and/or phase.

     An optical crossbar switch is well suited for the problem of auto-
associative memory[5]. In this model recall of a complete stored vector is
achieved by matrix vector multiplication of a storage matrix by the input

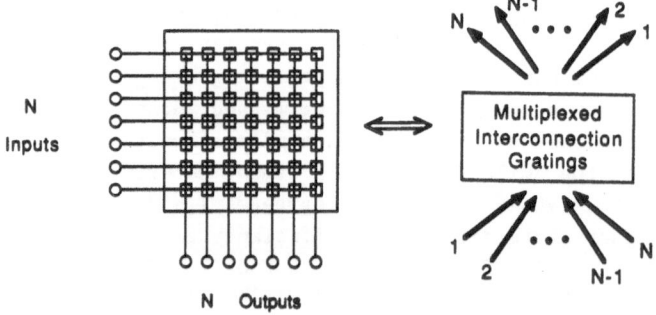

Fig. 5. Generalized crossbar switch.

vector. The input vector may represent partial or distorted information about the object. The storage matrix is symmetric and the diagonal elements are zero. For example, if the data set consists of M vectors $\vec{v}_i$ and each vector has n polar bits (+1, -1), then the interconnect matrix is

$$T = \sum_{i=1}^{M} \vec{v}_i \vec{v}_i^T. \tag{3}$$

In this model the recall procedure then represents a matrix vector multiplication:

$$\vec{v_0}ut = T\vec{v_i}n \tag{4}$$

and the output is returned to the input to provide a looped system from which a complete output vector can be extracted using only partial input data.

As an example of an optical implementation of the Hopfield model[5] we consider the autoassociative case, so that each data vector we wish to store is simply a discrete set of N real numbers. In an optical implementation each real number x is represented by a monochromatic plane wave with an electric field magnitude equal to the absolute value of x and a "phase factor" of 0 for positive x or π for negative x. The general expression for a plane wave is $Ae^{i(\vec{K} \cdot \vec{r} - \omega t + \phi)}$, but here "phase factor" refers only to the arbitrary additive phase ϕ. Hence, an N-bit data vector is given by

$$v = \begin{pmatrix} v_1 \\ v_2 \\ \vdots \\ v_N \end{pmatrix} = \begin{pmatrix} V_1 e^{i\phi_1} \\ V_2 e^{i\phi_2} \\ \vdots \\ V_N e^{i\phi_N} \end{pmatrix} \tag{5}$$

in this expression the time dependence and direction of the wave have been omitted, since all phase information is retained in the ϕ_i's.

Storage of the information is achieved by performing an outer product operation. The symmetry interconnect matrix for the autoassociative case is then $\vec{W} = \vec{v}\vec{v}^T$. This interconnect matrix can be realized optically by superimposing plane waves and using a binary phase encoding scheme. As an example, let us consider a single element $W_{ij} \equiv v_i v_j$. For the case of small modulation depth in the crystal, the response of the material is linear, and the general expression for the spatially varying index of refraction can be written

$$\Delta n_{ij} = const \cdot \frac{V_i V_j}{I_0} e^{i(\phi_i + \phi_j + \theta_0)}. \tag{6}$$

Comparison of equation 6 with the desired result, $v_i v_j = V_i V_j e^{i(\phi_i + \phi_j)}$, demonstrates that the product of two numbers is generated by the intensity pattern resulting from the interference of two optical plane waves that represent the individual numbers. The absolute value of the product is contained in the modulation depth while the sign is encoded in the phase of the grating.

An extension of this analysis shows that a complete outer-product interconnection matrix ($\vec{W} = \vec{v}\vec{v}^T$) is formed when all N beams representing an N-bit vector simultaneously illuminate the crystal. In component notation,

$$W_{ij} = \begin{cases} v_i v_j = \delta n_{ij} & i \neq j \\ 0 & i = j \end{cases} \qquad (7)$$

and the average intensity $I_0 = \sum_{i=1}^{N} v_i^2$. Here we have assumed that the crystal spatial frequency response is flat over the range of grating spacings involved. Thus holography is quite suited for establishing the interconnection matrix associated with a single vector. In a full implementation of the static Hopfield model or the adaptive learning models of Widrow and Hoff, a set of vectors must be stored. A complete memory matrix is stored as a linear combination of intermediate matrices, $\vec{W}^{(k)}$. The exact structure of these intermediate matrices depends on the particular model used, but in most cases the $\vec{W}^{(k)}$ are derived from outer-product operations. In mathematical terms we have,

$$\vec{W}^{net} = \sum_{k=1}^{M} \eta_k \vec{W}^{(k)}. \qquad (8)$$

where η_k is a weighting coefficient for each of the M matrices in the sum. Photorefractive crystals are ideally suited for performing the summation. Since the material has a variable time response dependent on the average intensity, time-division multiplexing of the intermediate matrices allows the crystal to generate the average interconnection matrix which is simply a normalized summation. The exposure time τ_k of any one intermediate matrix $\vec{W}^{(k)}$ determines its associated weighting coefficient η_k. The relationship between τ and η is one of direct proportionally when the average illumination intensity during each time slot $I_0^{(k)}$ is a constant for all k. Repetitive illumination of an M-step cycle then produces a net response given by

$$\vec{W}_{ij}^{net} = |\vec{W}_{ij}^{net}| e^{i(\phi_{ij} + \theta_0)} = \frac{1}{T_0} \sum_{k=1}^{M} \tau_k \vec{W}_{ij}^{(k)} \qquad (9)$$

So $\vec{W}_{ij}^{net}$, later referred to as Δn_{ij}^{net}, represents a steady-state grating that is simply the weighted average of M individual gratings.

Once the network has been programmed, memory recall is invoked through a multiplication of the storage matrix by an input vector. Denoting the input vector $\vec{a}$ and the output vector $\vec{b}$, the matematical result is simply $\vec{b} = \overleftrightarrow{W}^{net}\vec{a}$. This operation can also be implemented using volume holographic diffraction gratings. For example, consider multiple beam diffraction in which an input wave represents an input vector component $a_j = A_j e^{i\theta_j}$. Each output wave generated by the set of multiplexed interconnection gratings constitutes an output vector component $b_i = B_i e^{i\theta_i}$. In fact, b_i is actually a superposition of N-1 co-propagating plane waves, each wave having an amplitude and phase dependent on its corresponding input wave (a_j) as well as the grating from which its input diffracts (Δn_{ij}^{net}). Using standard Kogelnik analysis it can now be shown that apart from an overall complex constant we obtain the desired result $b_i = \Sigma_j \Sigma_k v_i^{(k)} v_j^{(k)} a_j$. The overall constant is the same for all components. It merely scales the amplitude and shifts the bias, or zero, phase point. To obtain the numerical value of an output component, we must measure both the amplitude and phase of its corresponding wave. The amplitude is simply the square root of the intensity. The phase is found by interferometrically combining the signal with a reference wave of known fixed phase.

V. OPTICAL IMAGE PROCESSING

In this section we discuss a few selected examples of optical image processing using photorefractive crystals[6,7]. The first example concerns the detection of defects in a periodic lattice, or enhancement of cracks in otherwise smooth or slightly rough surfaces such as the heads used in magnetic storage devices.

V.1. Defect Enhancement

Photorefractive crystals may be effectively used[8,9] to enhance defects in periodic masks or cracks in relatively smooth objects. In either case a photorefractive crystal is placed in the Fourier plane of the object to be inspected, as shown in Fig. 6. A hologram is formed by interfering the trans-form of the object with a plane reference beam. The hologram is recorded nonlinearly so that the readout results in an enhancement of the periodic structure.

Following standard Kogelnik analysis, the diffraction efficiency of a volume phase hologram is proportional to the square of Δn, the refractive index modulation. Through the electro-optic effect

$$|\Delta n_i| = \left[n_i^3 r_{ij} | |E_j| \right]/2 \tag{10}$$

where n_i is the unperturbed refractive index, r_{ij} is the electro-optic coefficient appropriate for the geometry, and E_j is the generated space--charge field within the crystal. The space-charge field then is the factor which determines how the diffraction efficiency will depend on the writing beams.

Under conditions usually encountered in experiments, the space-charge field depends on the writing beam ratio R, which is defined as the object beam intensity I_0 divided by the reference beam intensity I_r:

$$E_{sc} \propto \sqrt{R} : R < 1$$

$$E_{sc} \infty \sqrt{R} : R > 1 \qquad\qquad (11)$$

If the object beam intensity is larger than the reference beam intensity, then the space-charge field is proportional to the object beam amplitude. This is the usual case where hologram are recorded in the linear regime. If the reference beam intensity is lower than the object beam intensity, however, the space-charge field is proportional to the inverse of the object beam amplitude, and diffraction efficiency decreases with increasing object beam intensity.

This effect can be used advantageously for optical signal processing applications. For example, if the object consists of a periodic mask containing small defects such as specks of dust or missing section of lines, the Fourier transform of the object consists of a regualr array of strong spikes on which a weak, relatively wide defect signal is superimposed. In the case where the hologram is recorded with a weak reference beam matched in intensity to that of the defect signal, the diffracted beam will contain an enhanced signal component associated with the defect, at the expense of suppressing the periodic background pattern. As a result, the defect is made more prominently visible in the processed output image. Defects of submicron size can be detected with this method.

The experimental setup for obtaining defect enhancement is shown in Fig. 6. An argon-ion laser beam at 514,5 nm is collimated and split to provide two writing beams and a probe beam. The photorefractive element is a 8 mm cube of BSO to which a field of 4 kW is applied in the transverse direction. The combination of a half-wave plate, a polarizing cube beam-splitter and a second half-wave plate allowed the writing beam ratio to be adjusted while keeping the polarizations of the two writing beams constant. A lens placed midway between the object and the crystal produced the Fourier transform of the object at the crystal. The quality of the lens in this con-

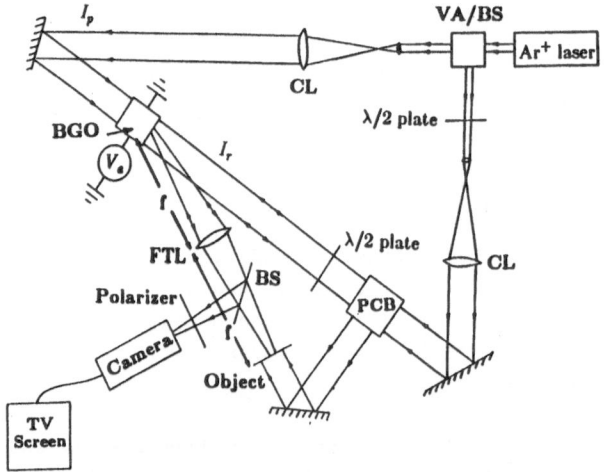

Fig. 6. Experimental setup for defect enhancement.

Fig. 7. Periodic mask with defects.

figuration does not have to be very good, because phase aberrations introduced by it are removed during the inverse transform operation.

Examples of the results obtainable with this system are shown in
Figs. 7 and 8. A mask consisting of a periodic array of 36 squares, each
250 μm in size, is contaminated with seven defects. The locations of the
defects are shown in Fig. 8, which is a direct image of the output of the
processor. Even the smallest defects (10 x 100 μm) are now clearly visible.
In a similar experiment the periodic mask is replaced by a readout head for
a magnetic disk, containing a 0.15 μm wide crack. In the direct image the
crack is barely visible in microscope image using a 200x magnification, as
shown in Fig. 9. With a slightly modified version of the experimental apparatus of Fig. 6, even these tiny cracks are made visible, as indicated in
Fig. 10. A digital image processing facility coupled to the output of the
optical preprocessor may be used to further inspect the defects. The
defect detection task is then substantially simplified, because the defects
have been enhanced and located in space, thus eliminating expensive search
procedures.

V.2. Tunable Image Velocity Filter

The photorefractive response of several important materials, including BGO and BSO, is significantly enhanced by illuminating the crystal with

Fig. 8. Enhancement of defects in Fig. 7.

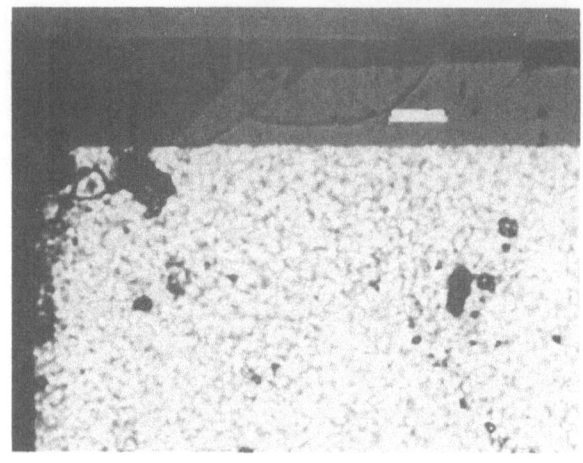

Fig. 9. Microscope image of a crack in a diskhead.

moving intensity gratings. In particular, it has been shown by several
authors that the magnitude of the photorefractive response increases substantially when the velocity of the moving grating reaches certain resonant values[10]. This effect allows a new class of real-time image filters
to be developed. As an example, this effect may be used to synthesize an
image velocity filter in which those components of a 2-dimensional image
moving in a certain direction and in a given range of speed are preferentially amplified with respect to the remainder of the image.

This technique may be understood by considering Fig. 11. A plane wave
of amplitude $\vec{E}_1$ illuminates an object normal to its surface. The object is
imaged onto a recording medium. Simultaneously a second reference beam of
amplitude $\vec{E}_2$ and propagation vector $\vec{k}_2$ illuminates the medium to produce
an interference pattern. If the object is translated parallel to its surface
with velocity $\vec{v}$, the grating is no longer stationary, but components in the
image with angular frequency $\vec{k}$ give rise to a grating component of spatial
frequency $|\vec{k}+\vec{k}_G|$ moving in the positive $(\vec{k}+\vec{k}_G)$ direction with velocity

Fig. 10. Enhanced image of a crack in a diskhead.

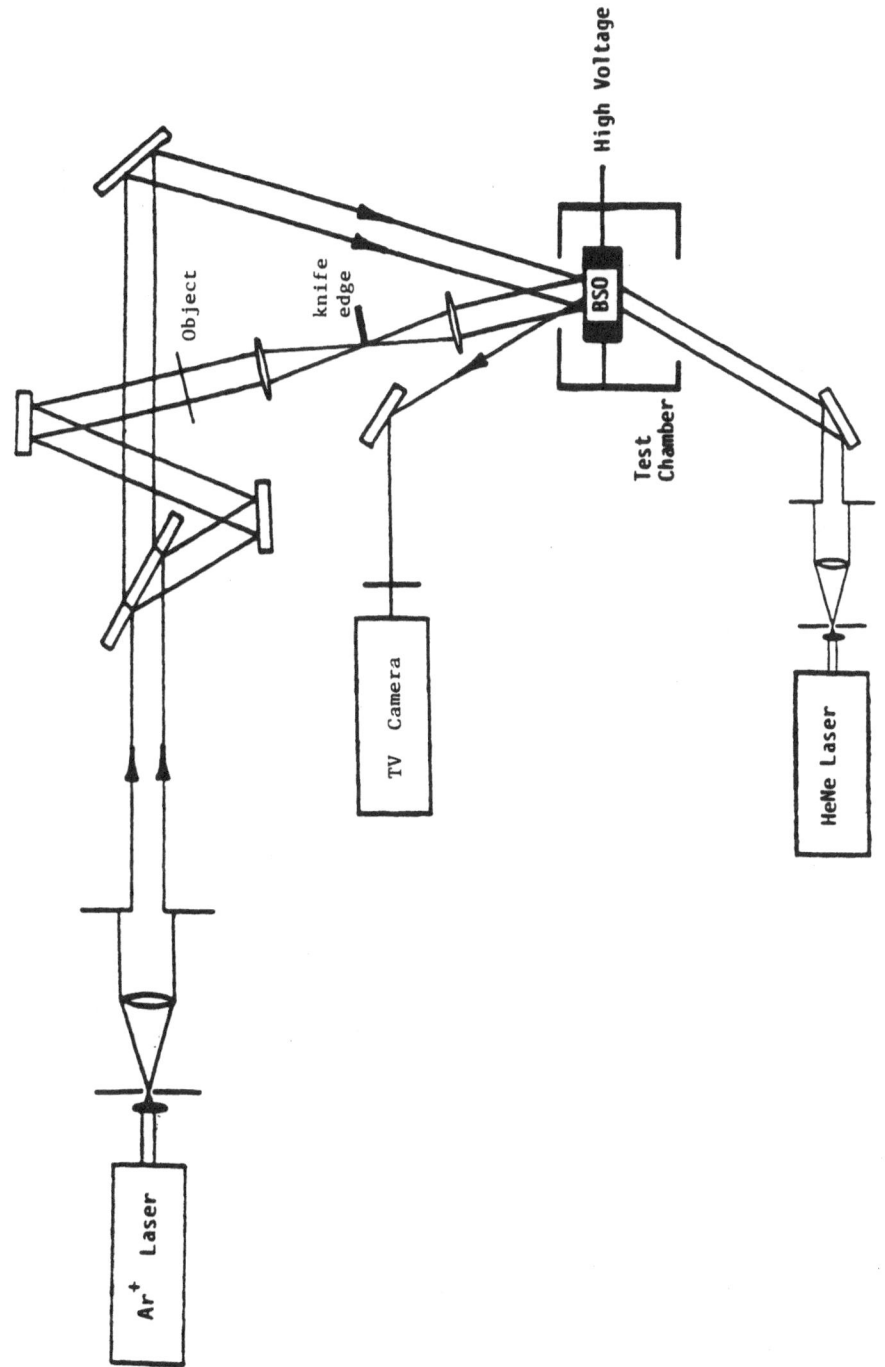

Fig. 11. Optical configuration for selective velocity enhancement of moving image components.

$$v(\vec{k}) \equiv \frac{\vec{k} \cdot \vec{v}}{|\vec{k} + \vec{k}_G|} \tag{12}$$

This moving grating component allows us to enhance specific moving components of the image.

Enhancement of the photorefractive response due to moving gratings occurs when the velocity of the grating meets the resonance condition:

$$v = \frac{sI_0}{Rk_G e_0} \equiv v_1 \tag{13}$$

where s denotes the photosensitivity of the medium, I_0 is the average incident intensity, $R \equiv N_A/N_D$ is the ratio of the charge acceptor and donor site densities in the photorefractive material, k_G the magnitude of the grating vector and $e_0 \equiv E_A/E_q$ denotes the normalized applied electric field. Here E_A is the magnitude of a DC applied electric field in the k_G direction and it is assumed to be sufficiently large so that drift dominates over diffusion in the material. E_q is the saturation value of the electric field within the photorefractive material.

Although enhancement of the fundamental component of the grating does not indeed occur for components of the velocity given by expression (13), a substantially improved enhancement occurs for higher order harmonics at the appropriate resonant velocities:

$$v_n \cong \frac{v_1}{n^2} \tag{14}$$

for the nth grating harmonic. By combining equations (12) and (14) we obtain local image amplification for those image components satisfying:

$$v(\vec{k}) \sim v_2$$

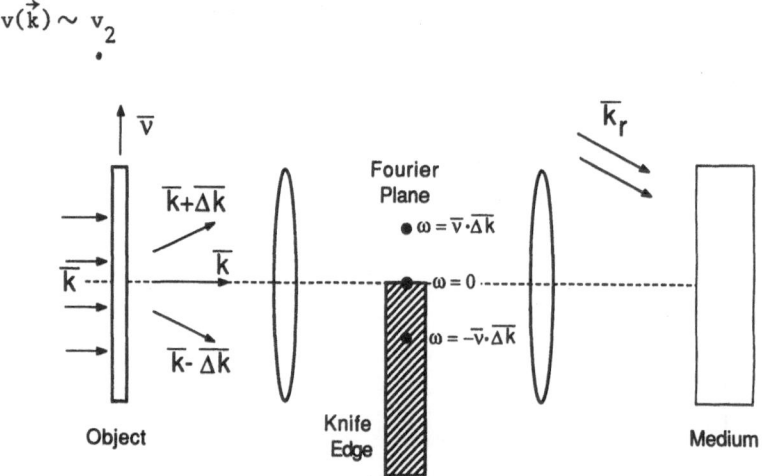

Fig. 12. Experimental apparatus for obtaining a holographic image velocity filter.

In addition, we see that v_2 is proportional to the incident intensity I_0, and the value of this enhancement velocity can be tuned simply by varying this intensity. In order to make the velocity filter directionally sensitive a schlieren stop is inserted between the two imaging lenses. Image velocity components in the direction perpendicular to the knife edge are either selectively enhanced or de-enhanced depending on the direction of motion.

The experimental apparatus used for this selective velocity filter is shown in Fig. 12. With this apparatus using a total writing intensity of 2.7 mW/cm^2 velocities of 0.25 mm/sec horizontally produced an average read-out image intensity 3 times greater than that obtained for $v = 0$, while similar speeds in the opposite direction yielded a drop-off in image intensity to 1/2 of that of the stationary case.

VI. ACKNOWLEDGEMENTS

It is a pleasure to acknowledge Drs. Ellen Ochoa and Fred Vachss and Craig Uhrich for their contributions to the image processing section of this work.

REFERENCES

1. J. Wilde, R. McRuer, L. Hesselink and J. Goodman, Proceedings of the SPIE, January 1987, Los Angeles
2. F. Vachss, Nonlinear holographic resonse in photorefractive materials, Ph. D. thesis Stanford University, April 1988
3. S. Redfield and L. Hesselink, Opt. Lett., vol. 13, no 10, 880 (1988)
4. L. Hesselink, S. Redfield, Opt. Lett., vol. 13, no 10, 877 (1988)
5. J. J. Hopfield, Proc. natl. Acad. Sci., USA 79, April 1982
6. R. Kostuk, J. W. Goodman and L. Hesselink, "Optical Interconnects", Springer Verlag (1989)
7. J. L. McClelland et al., Parallel distributed processing, MIT Press, Cambridge, MA vol. 1, Ch. 8 (1986)
8. C. Uhrich and L. Hesselink, Appl. Opt., vol. 27, no 21 (1988)
9. E. Ochoa, L. Hesselink and J. W. Goodman, Proc. SPIE, vol 613-633, p. 194 (1986)
10. F. Vachss and L. Hesselink, Appl. Opt., vol. 27, no 14 (1988)

MATERIALS AND DEVICES

ULTRAFAST ALL-OPTICAL SWITCHING IN OPTICAL FIBERS

S. Trillo, S. Wabnitz, and B. Daino

Fondazione Ugo Bordoni
Via Baldassarre Castiglione, 59, 00142 Rome, Italy

I. INTRODUCTION

Optical fibers have potential for generating and transmitting information at repetition rates of the order of a few THz[1], that is much above the capacity of signal processors based on electronic technology. This means that a substantial improvement in the speed of a communication or computing system can be achieved if basic logic operations and signal routing is performed in the optical stage, by means of optically controlled devices with sub-picosecond switching times. As a consequence, much research effort has been devoted to the design and implementation of optically activated switches in recent years. Initially, devices with an optical feedback, like the nonlinear Fabry-Perot, were proposed[2]. The fundamental limitation to the speed of an optically bistable element, where a nonlinear medium is placed in the cavity, are set by the trade-offs between cavity transit time and strength of the nonlinearity, which may lead to thermal heating problems.

An alternative approach to all-optical processing is provided by operating in the pipeline mode, whereby successive pulses of a traveling train may get switched simultaneously along the device, owing to their spatial separation. Therefore transit time no longer sets an upper bound to the switching speed, and the maximum processing repetition rate is fixed by the shortest input pulsewidth which can be fed into the device, without deteriorating its nonlinear switching characteristics. This entails that relatively long, fast-responding nonlinear media, such as optical fibers, can be employed. In this way, thermal heating can be eliminated since the non-resonant (electronic) nonlinear response of fibers is weak but virtualy absorption-free. Furthermore, the response time of the nonlinearity is of the order of a few optical cycles. Therefore, as we shall see in the following, the ultimate limitation to the input pulse width typically arises from the combined action of group velocity dispersion (GVD) and self and cross-phase modulation (SPM and CPM).

In this work we shall provide a general description of several schemes

based on optical fibers for all-optical switching, modulation and amplification of ultrashort light pulses. The present treatment is based on coupled
mode theory and permits to investigate the optimal characteristics of operation, as well to identify the dynamical properties of the field evolutions
in nonlinear dispersive coupling structures. Interesting dynamical phenomena result from the interplay between linear and nonlinear mode coupling in
several physically distinct devices: for example, spatial instabilities and
chaos. We shall also discuss experiments of self-switching of short pulses
(in the range 100 psec-100 fsec) in birefringent or periodic fibers and in
dual-core fibers. Finally, we shall examine numerical studies on the effects of GVD on nonlinear pulse propagation, showing the advantages and
limitations to the use of short pulses for all-optical switching in coupling
devices.

II. NONLINEAR SWITCHES AND COUPLERS

We consider guiding structures supporting two distinct copropagating
modes. This may include, for example, pulses propagating in two arms of a
fiber optic Mac-Zehnder interferometer, in a fiber directional coupler, or,
finally, in the two orthogonal polarizations of a fiber. In any case, one
may write the field as the superposition of two modes with slowly varying
envelopes u(z,t) and v(z,t), where z is the spatial coordinate in the propagation direction and t is time. In terms of normalized units, u and v
obey the coupled system of partial differential equations

$$iu_z \mp u_{tt} + \kappa v + (I_u + S|u|^2 + C|v|^2)u = 0$$

$$iv_z \mp v_{tt} + \kappa u + (I_v + S|v|^2 + C|u|^2)v = 0 \qquad (1)$$

where the subscripts indicate spatial and temporal derivatives, and the
minus (plus) sign holds in the case of normal (anomalous) dispersion.
Furthermore, κ is the linear coupling coefficient, $I_{u,v}$ are nonlinear
phase-shifts due to other copropagating waves, possibly at different wavelengths (these waves are supposed to be much more intense than u and v, so
that th ir own propagation may be considered undisturbed), whereas S and C
denote the SPM and CPM coefficients. In this section, we shall treat situations where the interacting pulses are sufficiently broad that dispersive
effects can be neglected, at least over the distances involved.

II.a. Nonlinear Phase-Shifters

Whenever linear coupling between the modes is absent, the continuous
wave (cw) solution to equations (1) is immediate: the power carried by each
mode is conserved. The only effect of nonlinear phase shift, that grows
linearly (provided that $S \neq C$) with distance z and with power difference
between the modes (or between I_u and I_v). Consider, for example, the nonlinear Mach-Zehnder interferometer, which is schematically displayed in
Fig. 1. A relatively weak (so that SPM can be neglected, while C = 0)
signal pulse is fed into the central input and couples to both arms in
equal measure. An intense controlling pulse may be injected into the upper
port and coupled to the corresponding guide. In the absence of control, a

218

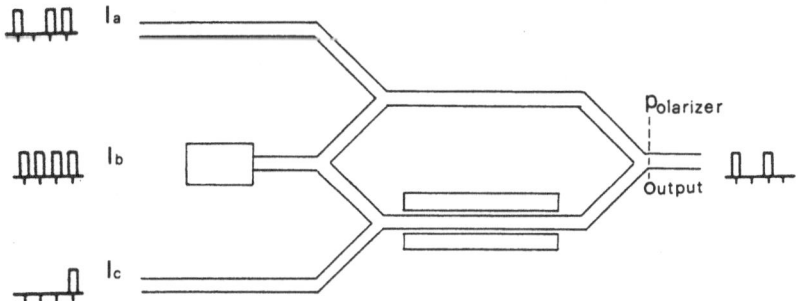

Fig. 1. Schematic of a nonlinear Mach-Zehnder interferometer. An electrooptic bias can be applied to one arm for tuning to zero the optical path length difference between the arms. In fiber optic implementations, the two arms may also be obtained by time-multiplexing the same polarization mode (see ref. 4).

phase-sensitive device (e.g., a polarizer) placed where the fields emerging from the two arms recombine, may be adjusted to obtain a zero output. Clearly, the phase shift I_u that is introduced by the presence of a copropagating pulse leads to a nonvanishing output. The third input port may be used for implementing the XOR function[3]. Experimental demonstrations have been reported by Haus and coworkers, using $LiNbO_3$ guides[4] or the polarization modes of a birefringent fiber[4]. Clearly, when the control pulse wavelength is different from that of the signal, a nonlinear interferometer is equivalent to a conventional Kerr shutter, where an intense beam changes the birefringence which is seen by a weak probe[5,6].

Also similar is the principle of operation of devices based on self-induced nonlinear birefringence of highly birefringent fibers. In this case one has $I_u = I_v = 0$, $S = 1$ and $C = 2/3$. The polarization state of a pulse emerging at the fiber output changes along the pulse profile, according to the instantaneous value of power. By placing a polarizer after the fiber output end, one obtains a power dependent transmission. This effect has been used in several fiber experiments for demonstrating pulse reshaping[7], intensity discrimination[8], and logic AND operation[9].

II.b. Nonlinear Mode Couplers

This class of switches, initially proposed by Jensen[10] and Maier[11], is characterized by the interaction between linear coupling, which dominates at low optical powers, and nonlinear phase shifts. Various implementations have been proposed and experimentally studied, in the context of both integrated optics and optical fibers. Restricting our attention here to fiber optical applications, we may mention the dual-core fiber coupler (see Fig. 2a)[12,13], the birefringent fiber coupler (Fig. 2b)[14,15], and the rocking rotator fiber filter (Fig. 2c)[16,17]. In the first case, the fiber is fabricated with two closely spaced cores which support two guided modes with identical polarization. The modes are linearly coupled through overlapping of the evanescent tails of the modal transverse fields. In linear conditions, the dual-core fiber constitutes a directional coupler: light couples back

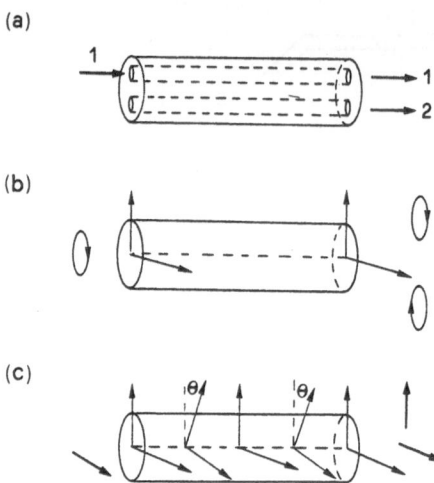

(a)

1

1

2

(b)

(c)

θ θ

Fig. 2. Three examples of nonlinear mode couplers: (a) the dual core
        fiber nonlinear directional coupler; (b) the birefringent fiber
        ($\kappa = (\beta_x - \beta_y)/2$); (c) the rocking rotator filter ($\kappa =$
        $= \theta(\beta_x - \beta_y)/4$).

and forth between the modes, with a beat length $L_b = 2\pi/\kappa$. SPM originates
from the intensity dependence of the refractive index of each core: $S = 1$,
while $C = 0$. In the birefringent fiber, linear coupling between two counter-
rotating circular polarization modes is due to fiber birefringence, whereas
both SPM and CPM result from the tensorial from of the nonlinear susceptibility (now $S = 2/3$, $C = 4/3$). At the fiber output, the modes can be separated
by means of a polarization splitter. Finally, in the rocking fiber linear
coupling is due to a periodic twist of the local birefringence axes, while
nonlinear birefringence appears at high powers. In this case, complete switching between the orthogonal linear polarization components may occur: again,
$S = 1$ and $C = 2/3$.

III. SPATIAL INSTABILITIES

     The mode coupling process inside nonlinear couplers is subject to
spatial instabilities. In linear conditions, the operation of a coupler is
completely characterized as soon as the two orthogonal eigenmodes and their
propagation constants are identified[18]. In the nonlinear regime, this simple
description of the spatial evolution of the modal amplitudes u and v in
terms of elementary functions remains valid only for devices that belong to
the class of nonlinear phase shifters. On the other hand, the solution for
u and v propagating in nonlinear couplers is expressible in terms of ellip-
tic functions. An important property is that an increase of the total input
power above a certain critical value, say P_c, leads to a symmetry-breaking
bifurcation[19,20]. As a result, above P_c two new (spatially stable) eigenmodes is spatially unstable (see Fig. 3). This means that small variations
in the excitation conditions may result in substantially different evolutions
and therefore large changes in the input transmission[11,14,20]. Fig. 3 displays the trajectors (for a given total input power below and above P_c, respectively, and different lanching conditions) that are followed along the z

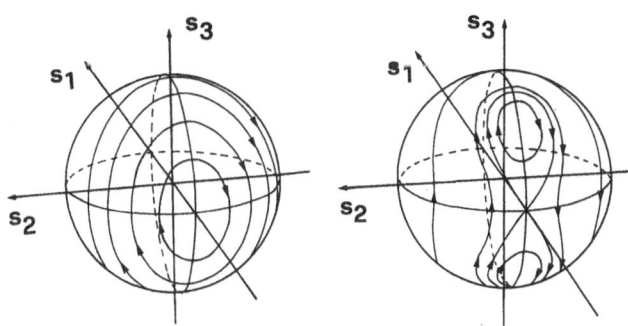

Fig. 3. Evolutions of the Stokes vector in a birefringent nonlinear
fiber coupler, when the input power $S_0 < P_c$ (left), and $S_0 > P_c$
(right).

coordinate by the tip of the real Stokes vector, whose components are

$$S_1 = iu^*v + c.c., \quad S_2 = u^*v + c.c., \quad S_3 = |u|^2 - |v|^2. \qquad (2)$$

As can be seen from the figure, when the input power $S_0 = |u|^2 + |v|^2 < P_c = 2\kappa/|S - C|$, the trajectories are slightly distorted but similar to the low
power ones. In particular, when just one mode is excited ($S_3(0) = 1$), complete
transfer still occur to the other mode after a certain coupling distance.
Conversely, when $S_0 > P_c$ a separatrix trajectory divides the region where
significant power exchange is permitted from two domains where the nonlinearly induced mismatch inhibits the coupling between the modes. Again,
in case of single mode excitation the transition between coupled and mis-
matched evolutions occurs precisely for $S_0 = P_s = 2P_c$. This effect is observable as a sharp "off-on" step in the power-dependent switching characteristics
of a coupler whose length is of the order of L_b or larger (see Fig. 4). The
display of Fig. 3 suggests also that tha transition could be activated at a
constant value of the input power, by means of a change in the relative
phase[21] or polarization[16,22] at the fiber input. An important application
of the instability is that it may provide the gain for the switching of an
intense beam between the two output ports, controlled by means of a weak
signal[21,23]. Fig. 4 also shows that the power-induced switching of the output state of a nonlinear coupler that is one coupling length long is not
that sharp, because nearby trajectories did not have enough distance to
separate due to the instability. However, the transmission exhibits a step-
like behavior that can be exploited for self-routing purposes.

We remark that the presence of both linear and nonlinear coupling is
necessary for the occurrence of symmetry breaking, but not of spatial or
polarization instability. In fact, spatial instabilities may exist even in
a linearly isotropic (but nonlinearly anisotropic) medium, such as a cubic
nonbirefringent crystal[24,25].

When couplers longer than L_b are involved, the effects of fabrication
imperfections may be relevant, to the point that the desired mode of operation is completely spoiled. The presence of uniform perturbations, such as
twists, may introduce a nonlinear nonreciprocal switching behavior[26,27].
Moreover, as discussed by Wabnitz et al.[28], for some input polarization states

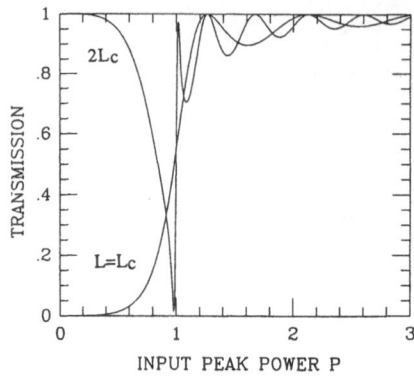

Fig. 4. Power dependent straight-through cw switching characteristics
in a nonlinear mode coupler of length $L_c = L_b/2$ and $2L_c$.

the spatial evolution of u and v may be highly sensitive to periodic imperfections ($\kappa = \kappa(z)$), so that for a wide range of total input powers an ir-
regular switching behavior results (due to hamiltonian chaos). Finally, the
effects of saturation, and of linear and nonlinear absorption, which is
associated to intensity dependent refractive index changes in a resonantly
enhanced nonlinear material, (e.g., a semiconductor doped glass fiber[29]) on
the switching characteristics and on the existence of spatial instabilities
have been the object of various theoretical investigations[30].

IV. EXPERIMENTS

In this section we shall discuss in some details two experiments involving reshaping and complete power-induced transfer inside intense picosecond pulses propagating in birefringent fibers. In the first case[15], a
train of Q-switched and mode-locked circularly polarized 80 psec pulses
(see Fig. 5) was injected into a birefringent fiber sample of length L_b
(see Fig. 2a). At the output, the intensity in the orthogonal polarization
component was detected. When the peak power of the envelope was raised
above P_s (equal to about 600 W), a sudden increase in transmission was
reported, in a good agreement with the theory (see Fig. 4). The achievable
switching was of about 40%, and not 100%, which is the prediction for an
ideally continuous wave (or for a square pulse). In fact, since the mode

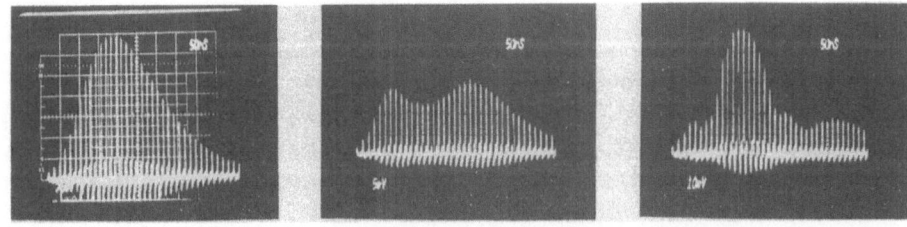

Fig. 5. Self-switching of Q-switched mode-locked circularly polarized
input pulses (left), emerging from a birefringent fiber switch
in the orthogonal polarization below and above the switching
power, respectively.

locked input pulses contain a continuous distribution of powers in their
envelope, the two components of the emerging pulses are obtained by calculating the respective cw transmission for each portion of the input pulse. The
result for a L_c = $L_b/2$ long coupler is shown in Fig. 6. When the response
time of the detector is slow with respect to the pulse rise time, the measured
transmission is averaged over the whole pulse emerging from the crossed analyzer . This degradation of the switching fraction, along with the severe
pulse reshaping, are detrimental in view of the possibility of operating
several switches in cascade. We shall see later on that a solution to this
problem may be provided by anomalous GVD.

A recent experiment reported self-switching of 6-2 psec pulses between
the two principal linear birefringence axes of a rocking rotator filter
(see also Fig. 2c)[31], whose length was L_c. At low powers, when the mean
optical frequency of the pulses was equal to the center of the filter bandpass (resonance condition), the fiber rotated by 90 degrees the polarization of an input pulse, that was initially aligned with one of the birefringence axes. By increasing the input peak power of the mode-locked pulses up
to one kW a substantial conversion to the original linear polarization was
obtained, which indicates that complete switching occurred for the central
portion of the pulses (see Figs. 6 and 7). The wavelength dependence of the
linear coupling deteriorated the switching fraction when the bandwidth of
the pulses approached the filter bandpass, and when the central frequency of
the pulses was detuned from resonance.

The self-switching experiment with the shortest pulses to date (100
fsec) has been performed by Silberberg et al.[32], using a short sample (0.5
cm) of dual-core fiber. The measured P_s was about 35 kW, and a considerable
broadening, due to the interaction of GVD and SPM, was also observed. When
such short pulses are involved, the cw analysis followed until now fails and
a solution of the partial differential equations (1) is necessary.

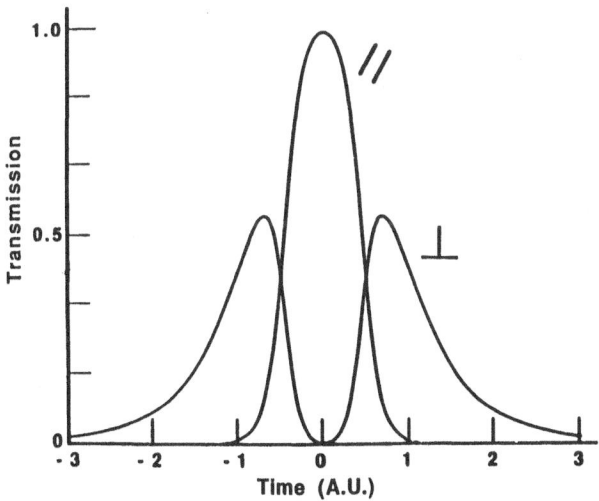

Fig. 6. Pulses emerging in the orthogonal polarization components of a
rocking fiber filter of length L_c, when the peak power is $1.25P_s$.

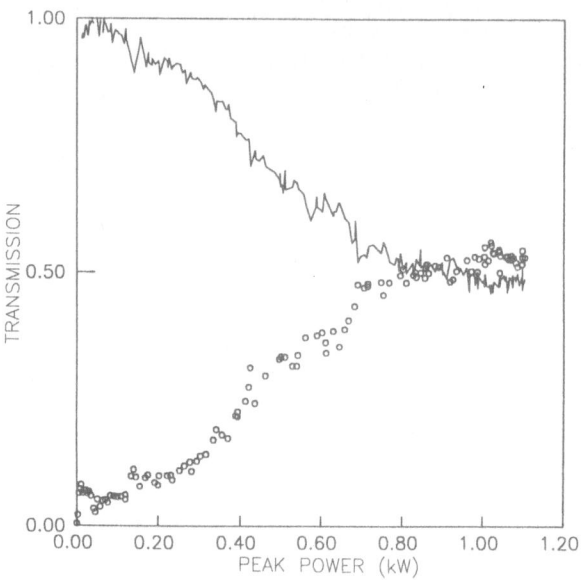

Fig. 7. On-resonance nonlinear transmission from a rocking filter,
averaged over the pulses.

V. ULTRASHORT PULSES

The effects of reducing the input pulsewidth dramatically depend on
the sign of GVD. When a short pulse travels in the normal dispersion regime
it experiences a linear frequency chirp due to SPM which GVD, may lead to
substantial spectral broadening. In time domain, distortion and temporal
broadening of the propagating pulse results. Numerical solutions of equations (1) in the normal dispersion regime show that, for a given coupler
length L_b, a critical value to the pulsewidth exists: for input pulses
shorter than this value, the power-dependent switching characteristics
(see Figs. 6 and 7) will severely deteriorate[33].

On the other hand, in the anomalous dispersion regime the nonlinear
chirp is such that it counteracts the dispersive broadening: when the peak
pulse power reaches a certain value, the undistorted propagation of a soliton
results. A typical feature of the soliton is that an unique phase shift,
proportional to the peak power, is experienced over the whole pulse profile.
T is fact may be exploited for achieving a great improvement in the switching efficiency of short pulses in nonlinear interferometers[34]. In nonlinear
mode couplers, spatial instabilities similar to the cw case occur for
soliton-like inputs[35]. Moreover, if the ratio of coupling to dispersion
distance is appropriate (i.e., the fundamental soliton power is approximately
equal to P_s), also the averaged switching characteristics recover the sharp-
ness typical of the cw case (see Fig. 8).

VI. CONCLUSIONS

The analysis of coupled mode representation of fiber nonlinear switches
revealed several intriguing physical phenomena, such as polarization insta-

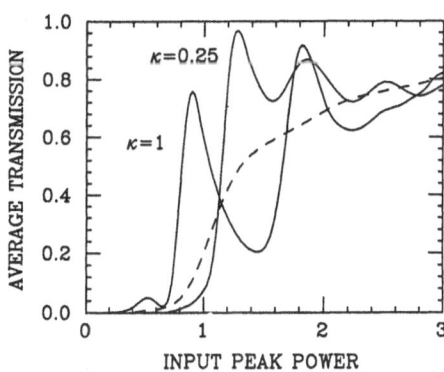

Fig. 8. Averaged nonlinear transmissions from a birefringent fiber coupler, when using long pulses (dashed line) or soliton-like input pulses (solid lines).

bility, chaos and soliton spatial instabilities. On the other hand, the investigation of the dynamics of nonlinear mode coupling provides an essential guideline in the development of new fiber devices and in the critical assessment of the ultimate performances of all-optical signal processors.

VII. ACKNOWLEDGEMENTS

This work was carried out in the framework of the agreement between the Fondazione Ugo Bordoni and the Istituto Superiore Poste e Telecomunicazioni.

REFERENCES

1. H. Itoh, K. Okamoto and K. Kubodera, Digest of IQEC'88, The Japan society of applied physics, Tokyo, p. 662 (1988)
2. See, e.g. "Optical Bistability III", edited by H. M. Gibbs, P. Mandel N. Peyghambarian and S. D. Smith, Springer, Berlin (1986)
3. A. Lattes, H. A. Haus, F. J. Leonberger and E. P. Ippen, IEE J. of Quantum Electron. QE-19, 1718 (1983)
4. M. Shirasachi, H. A. Haus, and D. Liu Wong, Digest of CLEO'87, Optical Society of America, Baltimore, p. 284 (1987)
    M. J. LaGasse, D. Liu Wong, J. G. Fujimoto and H. A. Haus, Digest of IQE'88, The Japan society of applied physics, Tokyo, p. 538 (1988)
5. R. H. Stolen and A. Ashkin, Appl. Phys. Lett. 22, 294 (1973)
    J. M. Dziedzic, R. H. Stolen and A. Ashkin, Appl. Opt. 20, 1403 (1981)
6. K. Kitayama, Y. Kimura and S. Seikai, Appl. Phys. Lett. 46, 623, (1985)
    N. H. Halas, D. Krokel and D. Grishkowsky, Appl. Phys. Lett. 50, 886 (1987)
    T. Morioka, M. Saruwatari and A. Takada, Electron. Lett. 23, 453 (1987)

7. B. Nikolaus, D. Grishkiwski and A. C. Balant, Opt. Lett. 8, 129 (1983)
8. R. H. Stolen, J. Botineau and A. Ashkin, Opt. Lett. 8, 512 (1982)
9. K. Kitayama, Y. Kimura and S. Seikai, Appl. Phys. Lett. 46, 317 (1985)
10. S. M. Jensen, IEEE J. of Quant. Elec. QE-18, 1580 (1982)
11. A. A. Maier, Sov. J. of Quant. Electron. 12, 1490 (1982)
    A. A. Maier, Sov. J. of Quant. Electron. 14, 101 (1984)
12. D. D. Gusovskii, E. M. Dianov, A. A. Maier, V. B. Neustruev, E. I. Shklovskii and I. A. Shcherbakov, Sov. J. of Quant. Electron. 15, 1523 (1985)
    D. D. Gusovskii, E. M. Dianov, A. A. Maier, V. B. Neustuev, V. V. Osiko, A. M. Prokhorov, K. Yu. Sitarki, and I. A. Scherbakov, Sov. J. of Quant. Electron. 17, 724 /1987)
    A. A. Maier, Yu. N. Serdyuchenko, K. Yu. Sitarkii, M. Ya Shchelev and I. A. Shcherbakov, Sov. J. of Quant. Electron. 17, 735 (1987)
13. S. R. Friberg, Y. Silberberg, M. K. Oliver, M. J. Andrejco, M. A. Saifi and P. W. Smith, Appl. Phys. Lett. 51, 1135 (1987)
14. B. Daino, G. Gregori and S. Wabnitz, Opt. Lett. 11, 42 (1986)
15. S. Trillo, S. Wabnitz, R. H. Stolen, G. Assanto, C. T. Seaton and G. I. Stegeman, Appl. Phys. Lett. 49, 1224 (1986)
16. A. Mecozzi, S. Trillo, S. Wabnitz and B. Daino, Opt. Lett. 12, 275 (1987)
17. S. Trillo, S. Wabnitz, B. Banyai, N. Finlayson, C. T. Seaton and G. I. Stegeman, Appl. Phys. Lett. 53, 837 (1988)
18. D. Marcuse, "Theory of dielectric optical waveguides", Academic press, New York (1974)
19. B. Daino, G. Gregori and S. Wabnitz, J. Appl. Phys. 58, 4512 (1985)
20. H. G. Winful, Opt. Lett. 11, 33 (1986)
21. S. Wabnitz, E. M. Wright, C. T. Season and G. I. Stegeman, Appl. Phys. Lett. 49, 11 (1986)
22. S. Trillo and S. Wabnitz, J. Opt. Soc. of Am. B 5, 483 (1988)
23. A. A. Maier, Sov. J. of Quant. Electron. 16, 892 (1986)
24. M. I. Dykman and G. G. Tarasov, Sov. Phys. Solid State 24, 892 (1986)
25. G. Gregori and S. Wabnitz, Phys. Rev. Lett. 56, 600 (1986)
26. F. Matera and S. Wabnitz, Opt. Lett. 11, 467 (1986)
27. S. Trillo and S. Wabnitz, Appl. Phys. Lett 49, 752 (1986)
28. S. Wabnitz, Phys. Rev. Lett. 49, 1415 (1987)
    E. Caglioti, S. Trillo and S. Wabnitz, Opt. Lett. 12, 1044 (1987)
29. D. Cotter, B. J. Ainslie, H. P. Girdlestone and C. N. Ironside, Digest of IQEC'88, The Japan society of applied physics, Tokyo, p. 536 (1988)
30. E. Caglioti, S. Trillo, S. Wabnitz, B. Daino and G. I. Stegeman, Appl. Phys. Lett. 51, 293 (1987)
    E. Caglioti, S. Trillo and G. I. Stegeman, J. Opt. Soc. of Am. B 5, 472 (1988)
31. R. H. Stolen, A. Ashkin, W. Pleibel and J. M. Dziedzic, Opt. Lett. 9, 300 (1984)
32. Y. Silberberg, A. M. Weiner, S. R. Friberg, B. G. Sfez and P. W. Smith, Digest of IGWO'88, Optical Society of America, Santa Fe, p. 336 (1988)

33. S. Trillo, S. Wabnitz, E. M. Wright and G. I. Stegeman, <u>Opt. Lett.</u>
    13, 672 (1988)

34. N. J. Doran and D. Wood, <u>J. Opt. Soc. of. Am. B</u> 4, 1843 (1987)

35. K. J. Blow, N. J. Doran and D. Wood, <u>Opt. Lett.</u> 12, 202 (1987)

NONLINEAR COUPLING TO ZnS, ZnO AND SDG PLANAR WAVEGUIDES:

THEORY AND EXPERIMENTAL STUDY

G. Assanto

CRES

Via Regione Siciliana 49, 90046 Monreale (PA), Italy

I. INTRODUCTION

Nonlinear integrated optics has been the object of intense studies in the past years due ist possible application to all-optical signals processing and its compatibility with optical fiber technology[1]. In addition, due to the extended interaction lengths and small cross sections which are achievable, nonlinear effects requiring large intensities can be studied at reasonable powers with high efficiencies. To date, several second and third order processes have been successfully implemented in guided wave structures. Effects such as second harmonic generation[2-3], sum and difference frequency generation[4], parametric amplification and oscillation[5-6] belong to the former category, while degenerate four-wave mixing[7-8], coherent Raman scattering[9], intensity-dependent refractive index[10] and nonlinear coupling[11] belong to the latter.

Third order processes, involving three waves interacting via a $\chi^{(3)}$ tensor, have attracted a great deal of interest because they can even occur in amorphous materials and, in some cases, provide an output signal at the same frequency as the inputs. Such a possibility provides a remarkable opportunity for signal manipulation. One of the simplest cases of this kind of interaction is an intensity-dependent refractive index in a distributed structure, by which an incident beam is coupled into guided modes of the structure. Prisms and diffraction gratings are usually employed to excite waves propagating in a planar optical waveguide and, in both cases, the coupling process has a distributed nature and leads to analogous effects. In this work we will focus our attention to nonlinear distributed coupling to slab waveguides, both for CW and pulsed excitation. Since the treatment is almost identical, emphasis will be placed on prism couplers with no loss of generality. Most of the experimental work presented hereby was, indeed, performed using prisms.

In a distributed input coupler, an efficient power transfer between incident and guided fields can be obtained only when the components of the corresponding wave-vectors match along the propagation direction. The optimum coupling angle is, therefore, the one which allows for such a synchronism. However, when the structure exhibits an intensity-dependent index of refraction - i.e., when at least one of the guiding media has a large $\chi^{(3)}$ coefficient - the guided wave-vector will change as the guided power grows under the coupler. This loss of synchronism does, in turn, reduce the coupling efficiency through a cumulative phase mismatch[12], shifts the optimum coupling angle and, in addition, introduces distortion in the pulse profiles in case of pulsed excitation[13]. According to size, sign and nature of nonlinearities, effects such as switching[14], limiting and bistability[15] can be predicted and have been observed[16-22].

Here we will summarize a theoretical treatment of nonlinear distributed coupling based on the travelling-wave-interaction formalism, and report on experimental results obtained in planar waveguides with one or two media exhibiting third-order nonlinearities. In particular, section II will be devoted to the case of CW excitation, and some relevant effects observed by prism-coupling to Zinc Sulphide waveguides will be presented, including switching and bistability. In section III we will treat the case of pulsed excitation, for both absorptive and Kerr-like nonlinearities; experimental results on Zinc Oxide and CdS_xSe_{1-x} semiconductor-doped-glass (SDG) waveguides will be reported and discussed.

II. NONLINEAR DISTRIBUTED PRISM COUPLING: CW EXCITATION

In the first part of this section we outline the travelling-wave formalism as a model for the distributed coupling process. The process is analysed with no time dependence, implying CW excitation of the guided modes. Emphasis will be placed on basic results relating to experiments in ZnS waveguides; more details can be found in Ref. (12). In the second part we will report on experimental results obtained by prism-coupling a CW Argon laser beam into ZnS waveguides on glass.

II.1. Basic theory

Let us consider a composite four-media structure with refractive indexes n_i satisfying the inequality:

$$n_p > n_f > n_s > n_c, \tag{1}$$

where subscripts p, f, s, and c indicate prism, film, substrate and cladding, respectively. We choose X along the propagation direction and X orthogonal to the guide plane, as sketched in Fig. 1. Assuming weak coupling, for a field incident onto the prism base at an angle θ:

$$E_{in}(r) = \frac{1}{2} a_{gi}(x) \, f_i(z) \, \exp\{j\,[\omega t - n_p k_0 (x\sin\theta - z\cos\theta)]\} + c.c., \tag{2}$$

a co-polarized TE guided field in the i-th medium can be written as:

$$E_{in}(r) = \frac{1}{2} a_{gi}(x) \, f_i(z) \, \exp\{j\,[\omega t + \beta_m(x)x]\} + c.c., \tag{3}$$

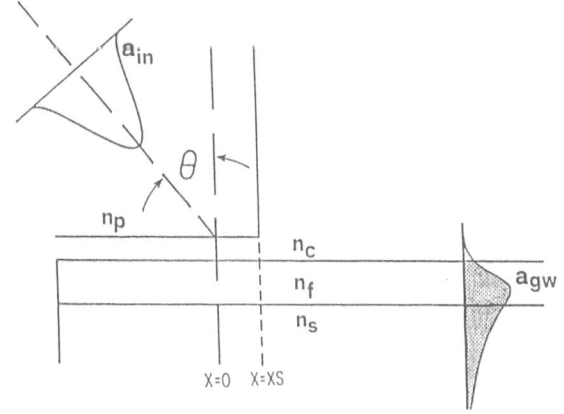

Fig. 1. Sketch of the guiding structure. X is along the propagation
direction and Z is orthogonal to the guide surface. XS
identifies the spot-position with respect to the prism edge.

where $k_0 = 2\pi/\lambda$ is the incident wave-vector, a_{in} and a_{gi} the field
amplitudes, f the transverse distribution and β_m the guided wave-vector
for the m-th mode. For TM modes, a polarization vector should be included
in (2) and (3)[12]. Due to the infinite extension of the structure along
the Y direction and assuming a line source, the amplitudes a(x) are normal-
ized so that $|a(x)|^2$ represents the power per unit distance along Y. In the
case of a Gaussian input beam, a_{in} and a_{gi} will be functions of both x
and y.

Letting t denote a transfer coefficient depending upon the actual
geometry of the structure and the field polarization, the coupled-wave
equation for the guided-wave amplitude is:

$$\frac{d}{dx} a_g(x) = t\, a_{in}(x)\, \exp\left[-j\phi(x)\right] -\left(\frac{\alpha}{2} + \frac{1}{\ell}\right) a_g(x), \tag{4}$$

with

$$\phi(x) = \frac{\Delta N}{2w_0} x - \phi^{NL}(x). \tag{5}$$

Here we have indicated by α a power-attenuation coefficient, ℓ a reradia-
tion distance for coupling back into the prism, w_0 the beam waist, $\Delta N = 2w_0(n_p k_0 \sin\theta - \beta_m)$ a detuning parameter and ϕ^{NL} a power-dependent phase
term. This last term brings in the nonlinear effects and causes the loss
of synchronism due to cumulative mismatch in (4). For a Kerr-like nonlinearity, with an intensity-dependent index of refraction:

$$n = n_0 + n_2 I, \tag{6}$$

the nonlinear phase term for a TE mode can be written[12]:

$$\phi^{NL}(x) \propto x\, I \int_{-\infty}^{+\infty} n(z)^2 n_2(z) |f_i(z)|^4 dz, \tag{7}$$

where the average over the transverse field distribution is necessary in
order to account for the actual waveguide geometry.

231

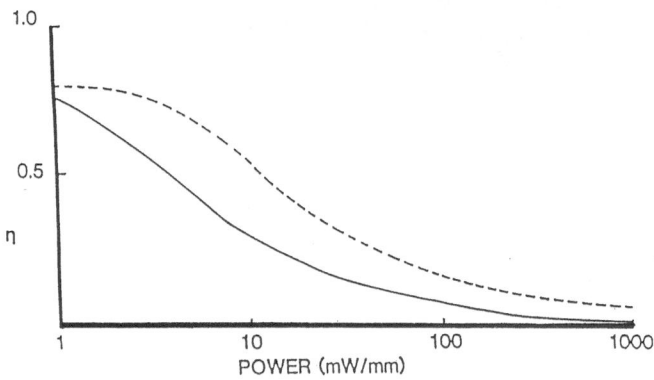

Fig. 2. Computed coupling-efficiency versus input-power for
nonlinear cladding with $n_2 = 10^{-9}$ m^2/W. The solid line is
obtained at the low-power optimum coupling angle, while the
dashed line corresponds to the optimum at each power.

The nonlinear coupling process can be described by referring to the
complex exponential in (4). When the guided power is low, i.e. at the
beginning of the coupling region x = 0 or for low incident powers, the
phase term has some constant value and the guided-wave amplitude a_g will
grow with x. However, when the guided power becomes non-negligible, the
phase term will change via the intensity-dependent index and out-of-phase
contributions will accumulate, with the net result being a decrease in
guided power. This can be easily understood in vectorial terms, comparing
the sum of several aligned vectors with the case of vectors forming an
angle with one another.

Equation (4) can be integrated numerically to yield coupling
efficiency versus power (for fixed angle of incidence) and coupling
profile versus detuning (for given power). Figs. 2 and 3 show some sample
calculations for the TE_0 mode: the efficiency was computed for $n_{2c} = +10^{-9}$
m^2/W, considering $\Delta N = 0$ (solid line) and the optimum ΔN (dashed line)[12];
the coupling profiles were calculated with reference to the ZnS case (see
section 2.2), assuming $n_{2f} = +10^{-12}$ m^2/W and no absorption[16].

II.2. Experimental study of ZnS waveguides

To investigate nonlinear prism-coupling with CW excitation we used
an Argon laser operating ad 514.5 nm, filtered, polarized, attenuated and
focussed at the base of a rutile prism on a ZnS slab waveguide. The waveguide, 1.3 μm thick, was fabricated by ion-assisted thermal evaporation
on a glass slide, in order to obtain a good quality low-absorbing film
reduce propagation losses. The waveguide was mounted on a goniometer and
computer controlled to rotations of 1 mdegree. Pyroelectric and/or
silicon detectors were used to measure the power out-coupled via another
prism, and to monitor the incident power.

Scanning at various powers around the low-intensity optimum coupling
angle, we could observe the predicted angular shift[12] and down-switching[14],

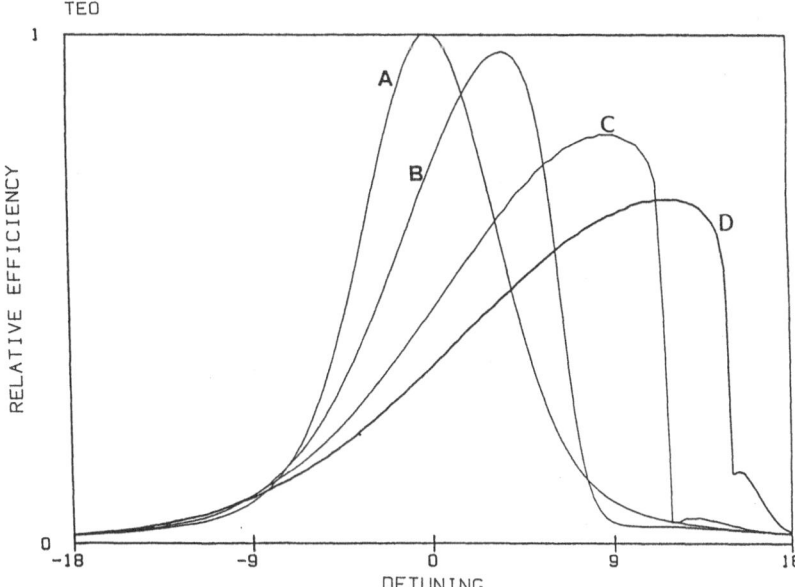

Fig. 3. Computed coupling profiles versus angle for TE mode in a guide with nonlinear film. $n_2 = +10^{-12}$ m^2/W; powers are 10 nW, 200 mW, 600 mW and 1 W for curves A, B, C and D, respectively. As expected, the coupling peak moves to higher angles.

as shown in Fig. 4[16]. The observed shift to higher angles is a confirmation of the positive sign of n_2, as expected from a thermal process. In addition, a sharp down-switching edge was obtained only at detunings larger than the angular width of the low-power coupling peak. This result can be explained on the basis of the previously outlined theory, keeping in mind that - at sufficiently high powers - the maximum efficiency is achieved at the angle for which wave-vector matching is recovered. As the angle is scanned towards the high-power optimum, more power gets coupled and hence absorbed, raising the film temperature. This in turn increases the effective index and moves the optimum angle even further. Once the optimum is reached a runaway switchdown can occur, because the in-coupled power is reduce, the temperature drops and the coupling angle moves back to smaller values (as measured from the linear optimum). This proceeds and eventually the coupling efficiency approaches the low-power value corresponding to that particular angle.

Experiments performed for both increasing and decreasing angles, at powers giving an angular shift of about five linewidths, showed that the switching occurs at smaller angles for decreasing than for increasing incidence angles, as in Fig. 5[16]. This phenomenon can be explained in terms of a power-dependent absorption considering that, since the guided power is much higher for increasing than for decreasing angles before switching, the nonlinearity n_2 is larger for the former than for the latter case.

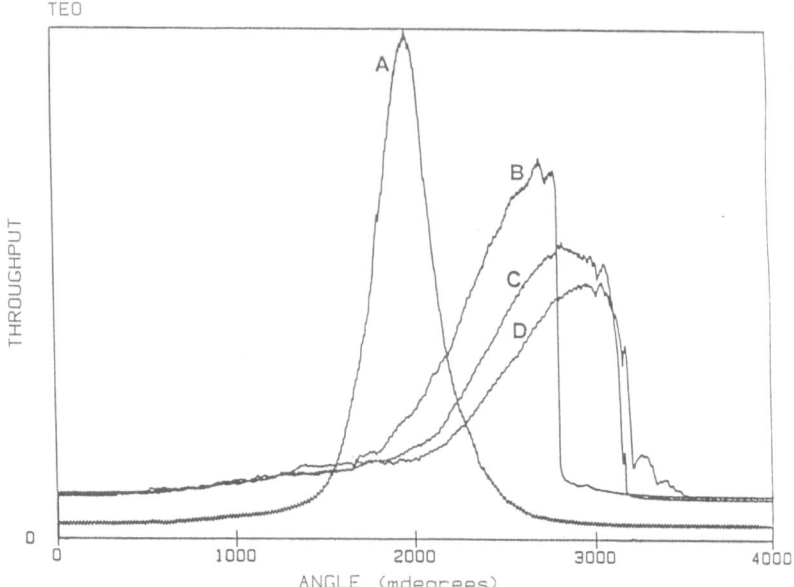

Fig. 4. Experimental angular profiles for the TE_0 mode in a ZnS
waveguide. Incident powers are 60 mW, 1 W, 1.7 W and 2.4 W
for curves A, B, C, and D, respectively.

This angular bistability can be modeled by taking $n_2 = n_2(I)$ in eq.
(6) and (7), and a sample calculation is presented in the insert of Fig.
5 for n_2 asymptotically varying from 1 to 3×10^{-12} m^2/W. From the angular
bistability a power bistability follows, and this was observed for an
incidence angle detuned 5 linewidths from the low-power optimum. Fig. 6
shows three scans of output versus input power and, aside from the noise,
demonstrates a large hysteresis with two well-defined stable states. The
stability was verified maintaining the system in the upper and lower
branches for times longer than one hour.

The collected data yield a value of $+10^{-12}$ m^2/W for n_2, and are in
agreement with the assumption of a thermal nonlinearity with power-
dependent absorption. Experiments performed at various lines of the Argon
laser did not show any resonant processes, and no qualitative difference
was found when exciting TM modes.

III. NONLINEAR DISTRIBUTED PRISM COUPLING: PULSED EXCITATION

Pulsed operation is of primary importance to demonstrate fast
processing capability of optical devices and to study the dynamics of
nonlinear materials. Following the same approach as the previous section,
we will discuss pulsed excitation of a prism coupler and then present
experimental results obtained with Zinc Oxide and semiconductor-doped-
glass (SDG) waveguides.

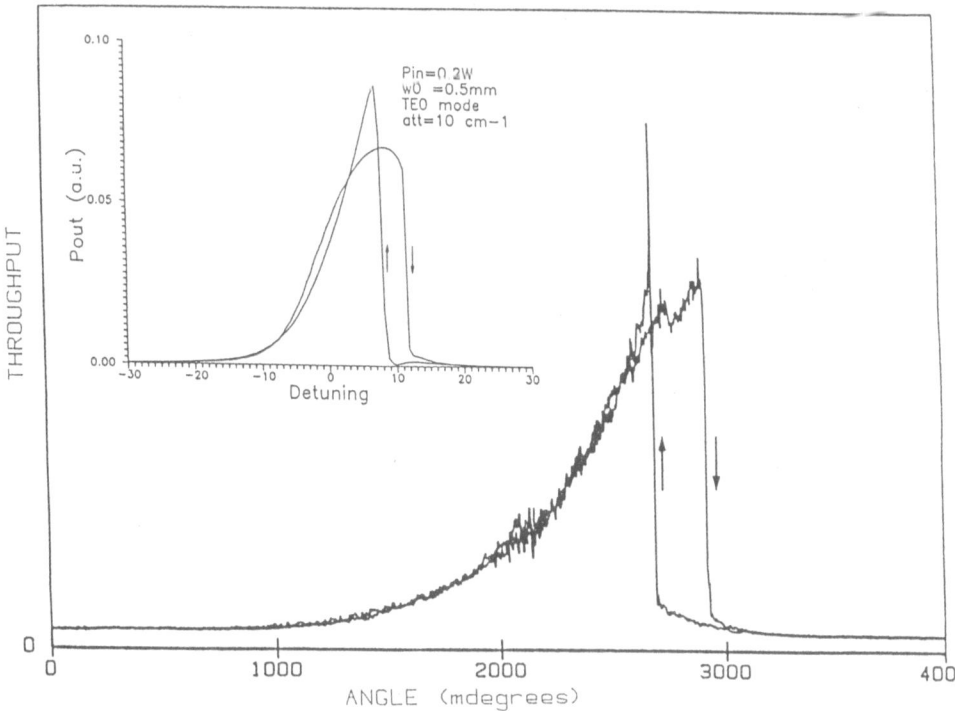

TEO   1 WATT

Fig. 5. Experimental angular profile for a fixed power of 1 W in the
TE$_0$ mode, scanning clockwise and counterclockwise around the
low-power optimum. Hysteresis is readily visible and can be
explained in terms of an intensity-dependent absorption. The
insert shows the result of a numerical simulation using this
assumption.

III.1. **Basic theory**

For the time-dependent case eqns. (2) to (5) hold, provided that all
the variables are to be considered functions of time. In particular, in
eq. (5) β(x) must be replaced by β(x, -∞). For semiconductors, we can
consider one and two-photon absorption with generation of carriers
N(x,z,t), and thermal effects with local temperature changes δT. In this
simplified treatment, we shall assume that no spatial diffusion takes
place over the duration Δt of a pulse. This hypothesis can be removed by
providing an internal feedback which can, in some cases, originate
optical bistability[15].

Using the subscript "e" for electronic, "th" for thermal, "1" and
"2" for one (q = 1) and two-photon (q = 2) processes respectively, and
indicating by τ typical response times and by σ the absorption cross-
section, we will discuss some simple cases in different time regimes. For
q = 1,2 and Δt >> τ_e :

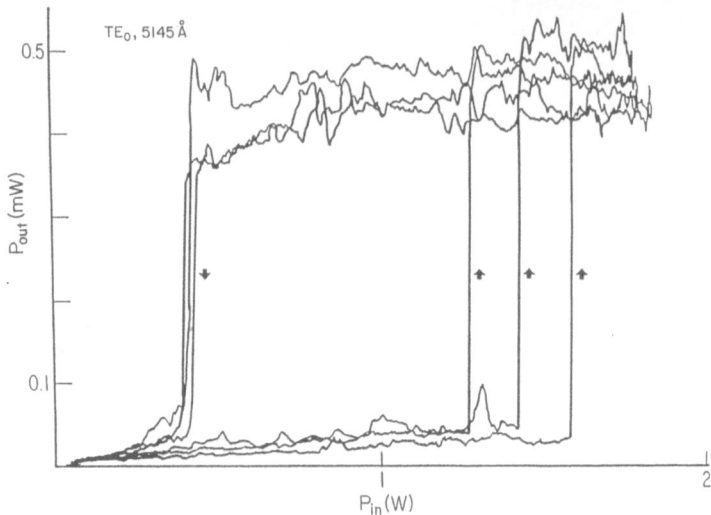

Fig. 6. Scans of output versus incident power at an angle detuned
five linewidths from the low-power optimum. The observed
bistability was verified for boths states and at various
scan-times.

$$\delta n_e(x,z,t) = -\frac{\partial n_e}{\partial N} \ \Sigma \ \frac{\sigma_q \tau_e}{qh\nu} \ I(x,z,t)^q \qquad (8)$$

and

$$n_{2q} = -\frac{\partial n_e}{\partial N} \ \frac{\sigma_q \tau_e}{qh\nu} \ . \qquad (9)$$

For a thermal nonlinearity with $\Delta t \gg \tau_e$:

$$n_2 = \frac{\partial n}{\partial T} \ \frac{\alpha_0 \ \tau_t}{\rho C_p} \qquad (10)$$

with ρ the density and C_p the heat capacity.

For a thermal nonlinearity with $\Delta t \ll \tau_{th}$:

$$\delta n_{th}(x,z,t) = \frac{\delta n}{\delta T} \ \frac{1}{\rho C_p} \ \int_{-\infty}^{t} \ dt' \ \alpha_0 \ I(x,z,t'), \qquad (11)$$

called an "integrating" nonlinearity[13]. Eq. (10) and eq. (11) together
give:

$$n_{2eff} \cong n_2 \ \frac{\Delta t}{\tau_{th}} \ . \qquad (12)$$

The effective thermal nonlinearity is, therefore, reduced by the ratio

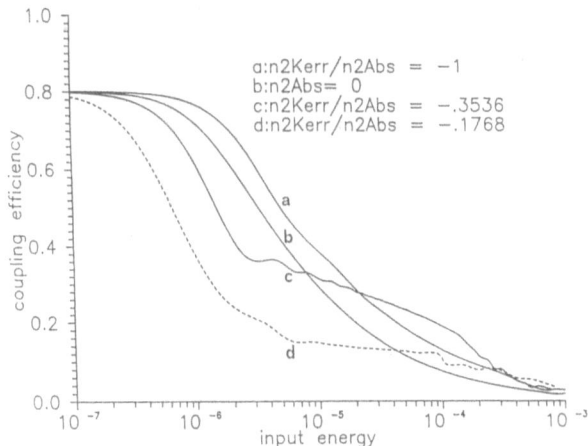

Fig. 7. Computer-generated coupling efficiency η versus input pulse
energy for a structure with both nonlinear film and substrate. n_{2Kerr} is kept at -10^{-14} m^2/W and the integrating
contribution is varied to show the possibility of interference effects. Notice the particular shapes of curves c)
and d).

of the pulse duration to the relaxation time.

Finally, for $\tau_{th} \gg \Delta t \gg \tau_e$:

$$\delta n(x,z,t) = n_2(q=1)I(x,z,t)+n_4(q=2)I(x,z,t)^2+\frac{n_{2th}}{\tau_{th}}\int_{-\infty}^{t} dt'I(x,x,t')$$

(13)

and, for $\tau_{th} \gg t$ and $\tau_e \gg \Delta t$:

$$\delta n(x,z,t) = \int_{-\infty}^{t} dt'(\frac{n_2(q=1)}{\tau_e} + \frac{n_{2th}}{\tau_{th}})I(x,z,t')+\frac{n_4(q=2)}{\tau_e} I(x,z,t)^2.$$

(14)

In all the other intermediate cases, a simultaneous solution of all the
time-dependent equations is necessary[13]. Notice also that the CW case,
treated in section II, can be included in this discussion considering a
Kerr nonlinearity with $\Delta t \cong \infty$ and/or $\tau \cong 0$.

To obtain the changes in the waveguide properties one needs to
average the nonlinear coefficients over the transverse field distribution
$f(z)$[23]. Letting A_i denote the weighted coefficients, the resulting nonlinear phase term is finally given by[13]:

$$\frac{\partial}{\partial x} \phi^{NL}(x,t) = k_0 \Delta(x,t) \left[A_1 + A_2 + A_t\right]$$

(15)

with $\Delta(x,t) = |a_{gw}(x,t)|^{2q}$ for $\Delta t \gg \tau$, and $\Delta(x,t) = \tau^{-1} \int_{-\infty}^{t} dt'|a_{gw}(x,t')|^{2q}$
for $\Delta t \ll \tau$. Saturation effects have been left out, but should be taken
into account as an upper limit to the achievable changes in ϕ^{NL}.

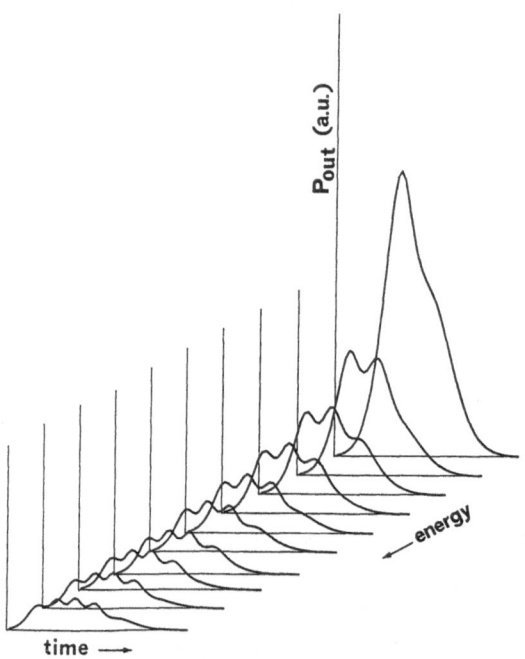

Fig. 8. Computer-generated temporal profiles versus input energy for
a line-source of trains of mode-locked pulses under a Gaussian
envelope and $n_{2f} = n_{2s} = n_{2th} = 2 \times 10^{-13}$ m^2/W. Coupling conditions are the optima at low power. Energies range from 5 to
95 µJ/mm with increments of 10 µJ/mm from top-right to bottom.
It is evident that the number of switching increases with
energy, while the efficiency decreases monotonically.

Computed coupling efficiency η versus input pulse energy is shown in
Fig. 7 for various ratios between Kerr ($\Delta t \gg \tau$) and integrating ($\Delta t \ll \tau$)
nonlinearities, assuming a negative sign for the former and considering
10 ns pulses Gaussian in time and along X. The prism edge x_s and reradia-
tion distance were set at their linear optimum values of $0.72w_0$ and $1.5w_0$,
respectively. Both the substrate and the film were considered nonlinear
(see section III.3). Temporal profiles versus input energies are presented
in Fig. 8 for the case of a solely thermal nonlinearity and a 220 ns Q-
-switch train of 80 ps mode-locked pulses (see section III.2 and III.3).
The graphs show severe distortion and multiple switching as the energy increases, with the initial switch occuring earlier and earlier in time. This
fact derives from the periodic character of the exponential term in (4) and
the cumulative effect of an integrating nonlinearity. Furthermore, the
number of peaks can be increased by the addition of a Kerr nonlinearity[13].

The study of pulse re-shaping due to nonlinear coupling can also be
done at a fixed imput energy, varying the detuning ΔN. This parameter
influences severely the pulse shape with relation to the optimum at that
particular energy. Sample calculations are shown in Fig. 9 for the previous
case, and major alternations in pulse profiles are readily seen. In some

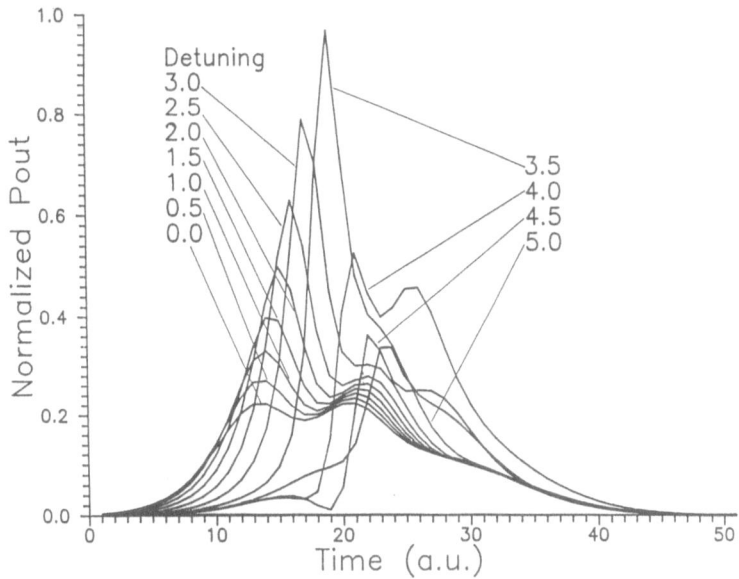

Fig. 9. Computer-generated temporal profiles for the above case, at
a fixed input energy of 19 μJ/mm and various positive
detunings.

cases pulse compression or 100% modulation can be obtained within a Q-switch
envelope[13].

III.2. Experimental study of ZnO waveguides

ZnO films were prepared to thickness of 0.6 μm by Magnetron RF-
sputtering on glass slides, and exhibited propagation losses as low as
1 dB/cm at 632.8 nm. The waveguides, placed on a precision rotary stage
with resolution of 1 mdegree, were excited by a Nd-YAG laser doubled to
532 nm, with Q-switch pulses 10 ns in duration or 500 Hz-trains of 80 ps
mode-locked pulses under a 220 ns envelope. For coupling, Strontium
Titanate prisms or ion-milled gratings were used, with linear efficiencies
of 22 and 4%, respectively[17].

The coupling efficiency exhibited a strong dependence on the input
energy/pulse, as shown in Fig. 10 for the 10 ns pulses. This behaviour,
with a 20-fold decrease, was substantially unmodified by the use of 220
ns pulses or different wavelengths from a dye laser, justifying the hypothesis of a non-resonant thermal process. A positive sign for the n_2
coefficient was established measuring the nonlinear angular shift. The
optimum coupling angle moved to larger values, as visible in Fig. 11,
were low and high-power profiles are shown[24]. Pulse energies did not
permit obtaining significant angular distortions other than a slight
broadening.

The experimental data[24] are all consistent with the presence of a
thermal nonlinearity (Eqs. (11) and (12)) with $n_2 \cong 10^{-14}$ m^2/W, although
a Kerr-like nonlinearity might contribute when using very short pulses

239

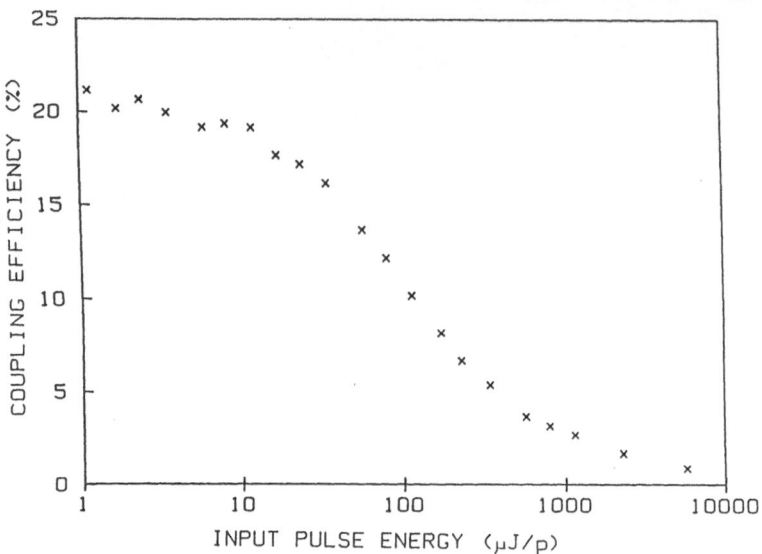

Fig. 10. Measured coupling efficiency versus input energy for a ZnO
waveguide excited by 10 ns pulses at 532 nm. A 20-fold
decrease is visible for energies up to 10 µJ/pulse.

($\neq$ps). Pulse profiles were, indeed reproduced integrating eq. (4) with
(15) and neglecting fast contributions from one- and two-photon absorption (A_1 and A_2). The measured relaxation time for the material was
1 µs, while typical switching times were about 1 ns. This demonstrates the
thermal origin of the nonlinearity and demonstrates the potential of
distributed couplers as fast devices.

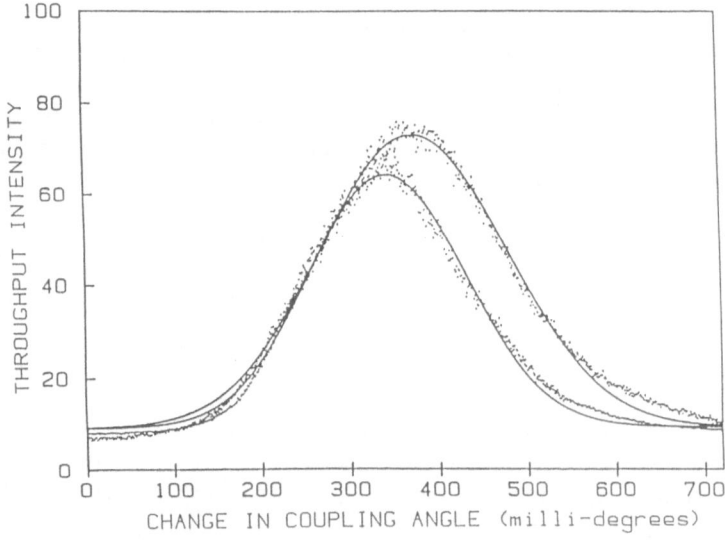

Fig. 11. Change in coupling angle for a ZnO waveguide at two different pulse energies. Notice the shift to higher angles,
corresponding to a positive n_2.

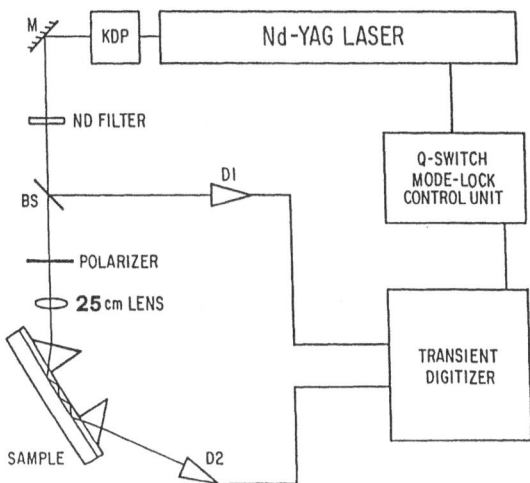

Fig. 12. Set-up used in the experiments with SDG waveguides. The
set-up is basically the same employed in the study of ZnO
waveguides.

III.3. <u>Experiments with SDG waveguides</u>

Sharp-cutoff color filters, containing microcrystallites of CdS_xSe_{1-x},
have been intensively studied with regard to their third-order nonlinear
properties since Jain and Lind's work in 1983[25]. These glasses combine a
relatively large $\chi^{(3)}$ coefficient with rapid response times[26-27], and are
commercially available and easily employed in waveguide fabrication[28].

Planar waveguides were realized with Corning $CdS_{0.9}Se_{0.1}$-doped
glasses by the K^+-Na^+ exchange technique[29]. The exchange was performed at
400 degrees Celsius for times longer than 20 hours. The latter was the
minimum required to obtain single mode guides. After characterization at
632.8 nm, the guides were placed on a precise goniometer and excited with
a Nd-YAG laser beam doubled to 532 nm. The laser output consisted of
trains of 80 ps mode-locked pulses under 220 ns Q-switch envelopes, at a
repetition rate of 500 Hz. For the experiments in the picosecond regime
32 ps pulses at 580 nm were used from a dye laser.

The experimental set up, quite similar to that employed in the study
of ZnO waveguides (sect. III.2), consisted of filtering and polarizing
optics, glass prisms for in- and out-coupling, detectors and a transient
digitizer (see Fig. 12). The coupling efficiency, for energies ranging
from 0 to 76 µJ per pulse, exhibited a substantial decrease from the low-
power optimum. Furthermore, as from Fig. 13a, it appears that two processes are present, with distinct energy constants. Examining the profile
of output power versus energy, shown in Fig. 13b, it is clear that, above
a threshold value, the net nonlinearity is altered by the influence of
Kerr-like effects added to thermal ones. This interpretation is in agreement with the outlined model, where nonlinearities with opposite signs
can combine with different time constants and contribute to ϕ^{NL}. Another
confirmation of the presence of electronic (band-filling) nonlinearities

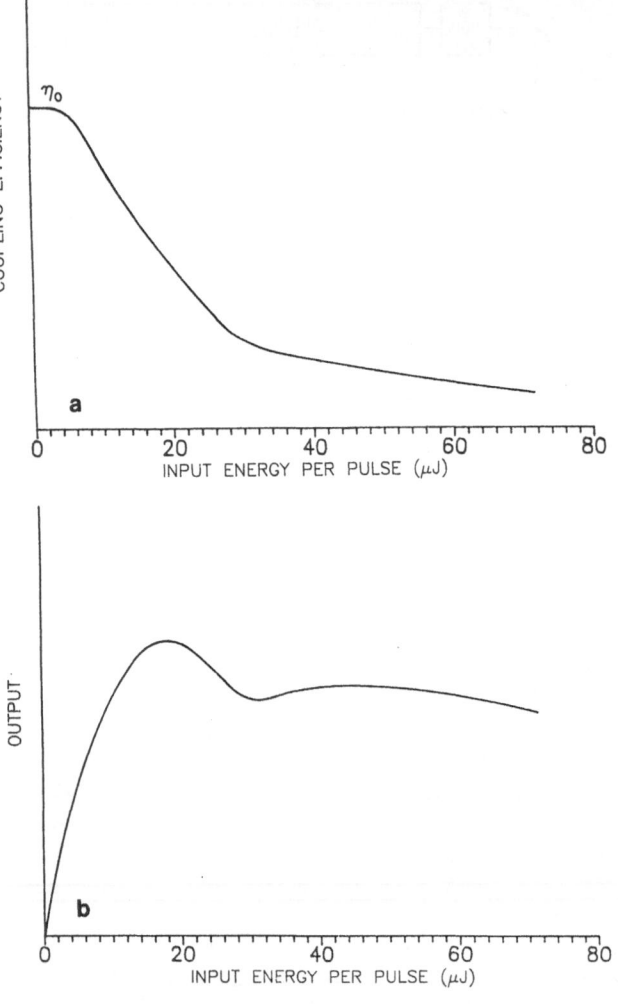

Fig. 13. (a) Measured coupling efficiency η and (b) transmitted
power for an SDG waveguide excited by a Nd-YAG laser at 532
nm. Notice the wavy behaviour in Fig. 14b, for qualitative
comparison with the graphs in Fig. 7c-d.

in the SDG waveguide was obtained coupling 32 ps pulses at 580 nm, in
order to reduce absorption and thermal effects. $CdS_{0.9}Se_{0.1}$ microcrystallites have an energy-gap of about 2.4 eV and are, therefore, low-absorbing to light at 580 nm. In Fig. 14 a plot of output power versus average
input power is presented, showing an evident decrease in coupling efficiency for peak powers of about 12 kW at a repetition rate of 500 Hz. This value
corresponds to the maximum in Fig. 13b, and this correspondence between the
two phenomena proves that the onset of fast nonlinearities is due to non-
-thermal effects. High peak powers are, as expected, required to emphasize
the role of a Kerr-like ($\tau \gg \Delta t$) mechanism competing with an integrating
($\tau \ll \Delta t$) one.

As predicted by the travelling-wave model, temporal distortion and
switching in time corresponded to the observed decrease in the efficiency.

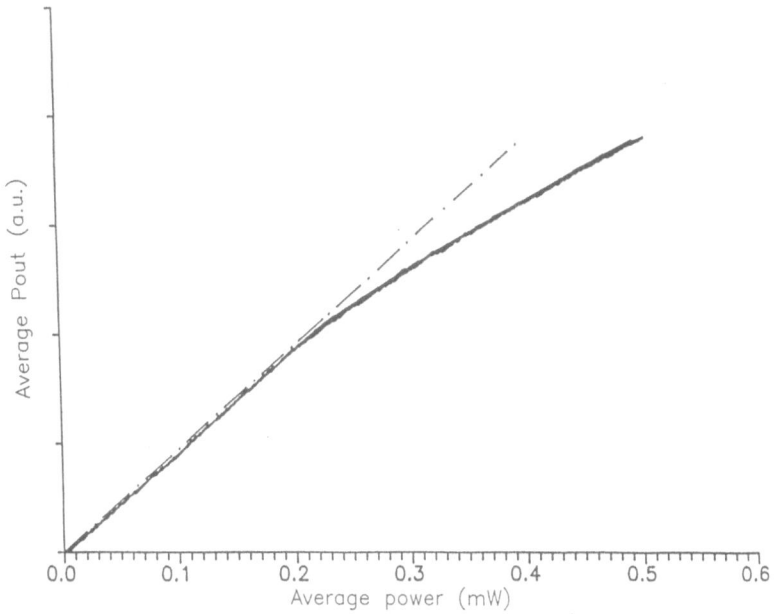

Fig. 14. Measured output versus incident average-power for an SDG
waveguide excited by 32 ps pulses at 580 nm with repeti-
tion-rate of 500 Hz.

According to the experiments, the agreement with the model is quite good,
and the value obtained for n_2 corresponds in order-of-magnitude to values
reported from bulk[25] and waveguide[6] measurements. This estimate is somewhat
altered by the presence of electronic effects, which tend to increase the
number of switching peaks[13]. Work is in progress to discriminate between the
two competing phenomena by data acquisition in different timescales.

IV. CONCLUSIONS

We have presented a number of experimental results and their comparison with a model based upon the travelling-wave interaction. It emerges
that distributed couplers, such as prisms and diffraction gratings,
provide an easy-to-use tool for the study of third-order effects in guiding structures. The results reported, both for CW and pulsed excitation
of waveguides with different nonlinear materials, demonstrate the possibility of achieving operations such as switching, bistability and limiting in integrated structures and, therefore, hold great promise in applications to all-optical signal processing and optical computing.

Despite the above considerations, further work is still required
towards material improvement and engineering. In particular, scattering
and absorption losses, saturation and (in SDGs) darkening at high fluences
are problems which need to be solved to permit device implementation.

ACKNOWLEDGEMENTS

The experimental work was carried on at the Optical Sciences Center-University of Arizona (Tucson, USA) in collaboration with Profs. G. Stegeman and C. Seaton. The author is indebted to them for valuable advice and encouragement. The author also gratefully acknowlegdges B. Svensson, R. Fortenberry and A. Gabel for the experimental work on ZnS, ZnO and SDG, respectively.

This work was supported by Italian C.N.R. (P.F. MADESS 1987), U.S. AFOSR and U.S. ARO.

REFERENCES

1. G. I. Stegeman and C. T. Seaton, J. Appl. Phys., 58(12), R57 (1985)
2. Y. Suematsu, Y. Sasaki, K. Furuya, K. Shibata, S. Ibukuro, IEEE J. Quant. Electron., 10, 222 (1974)
3. R. Normandin and G. I. Stegeman, Opt. Lett., 4, 58 (1979)
4. N. Uesegi, K. Daikoku, M. Fukuma, J. Appl. Phys., 49, 4945 (1978)
5. N. Uesegi, Appl. Phys. Lett., 36, 178 (1980)
6. W. Sohler and H. Suche, in: "Integrated Optics III", L. D. Hutcheson and D. G. Hall eds., Proc. SPIE 408, 163 (1983)
7. C. Karaguleff, G. I. Stegeman, R. Zanoni, and C. T. Seaton, Appl. Phys. Lett., 7, 621 (1985)
8. A. Gabel, K. W. DeLong, C. T. Seaton, and G. I. Stegeman, Appl. Phys. Lett., 51, 1682 (1987)
9. W. M. Hetherington III, N. E. Van Wyck, E. W. Koening, G. I. Stegeman, and R. M. Fortenberry, Opt. Lett., 9, 88 (1984)
10. G. M. Carter, Y. J. Chen, and S. K. Tripathy, Appl. Phys. Lett., 43, 891 (1983)
11. J. D. Valera, C. T. Seaton, G. I. Stegeman, R. L. Shoemaker, X. Mai, and C. Liao, Appl. Phys. Lett., 45 1013 (1984)
12. C. Liao, G. I. Stegeman, C. T. Seaton, R. L. Shoemaker, J. L. Valera, H. G. Winful, J. Opt. Soc. Am. 2, 590 (1985)
13. G. Assanto, R. M. Fortenberry, C. T. Seaton and G. I. Stegeman, J. Opt. Soc. Am., B 5(2), 432 (1988)
14. G. M. Carter and Y. J. Chen, Appl. Phys. Lett., 42, 643 (1983)
15. G. I. Stegeman, G. Assanto, R. Zanoni, C. T. Seaton, E. Garmire, A. A. Maradudin, R. Reinisch and G. Vitrant, Appl. Phys. Lett., 52, (11), 869 (1988)
16. G. Assanto, S. Svensson, D. Kuchibhatla, U. J. Gibson, C. T. Seaton, and G. I. Stegeman, Opt. Lett. 11, 644 (1986)
17. R. Fortenberry, R. Moshrefzadeh, G. Assanto, X. Mai, E. Wright, C. T. Seaton, and G. I. Stegeman, Appl. Phys. Lett., 49, 6987 (1986)
18. G. Assanto, A. Gabel, C. T. Seaton, G. I. Stegeman, C. N. Ironside, and T. J. Cullen. Electron. Lett., 23, 484 (1987)
19. W. Lukosz, P. Pirani and V. Briguet, in: "Optical Bistability III", Springer-Verlag, New York (1986)
20. F. Pardo, H. Chelli, A. Koster, N. Paraire, and S. Laval, IEEE J. Quantum Electron. Q2-23, 545 (1987)

21. H. Herominiek, S. Patela, Z. Jakubczyk, C. Delisle, and R. Tremblay,, J. Opt. Soc. Am. B 5(2), 496 (1988)

22. H. Sauer, N. Paraire, A. Koster, and S. Lavalle, J. Opt. Soc. Am. B 5 (2), 443 (1988)

23. G. I. Stegeman, C. T. Seaton, A. C. Walker, C. N. Ironside and T. J. Cullen. Opt. Commun., 61, 277 (1987)

24. R. M. Fortenberry, G. Assanto, R. Moshrefzadeh, C. T. Seaton and G. I. Stegeman, J. Opt. Soc. Am. B 5(2), 425 (1988)

25. R. K. Jain and R. C. Lind, J. Opt. Soc. Am., 73(5), 647 (1983)

26. S. S. Yao, C. Karaguleff, A. Gabel, R. M. Fortenberry, C. T. Seaton and G. I. Stegeman, Appl. Phys. Lett., 469(9), 801 (1985)

27. C. N. Ironside, T. J. Cullen, B. S. Bhumbra, J. Bell, W. C. Banyai, N. Finlayson, C. T. Seaton and G. I. Stegeman, J. Opt. Soc. Am., B 5, 492 (1988)

28. T. J. Cullen, C. N. Ironside, C. T. Seaton and G. I. Stegeman, Appl. Phys. Lett., 49, 1403 (1986)

29. G. L. Yip and J. Albert, Opt. Lett., 10(3), 151 (1985)

NONLINEARITY AT AN INTERFACE

F. Bloisi[a], L. Vicari[a], S. De Nicola[b], A. Finizio[b],
P. Mormile[b], G. Pierattini[b], A.E. Kaplan[c], S. Martellucci[d],
and J. Quartieri[d]

[a]Dip. di Fisica, N.S.M.F.A., Napoli, Italy
[b]Istituto di Cibernetica CNR, Napoli, Italy
[c]The John Hopkins University, Baltimore, USA
[d]Dip. di Ingegneria Meccanica, IIa Universitá degli Studi
  di Roma, Italy

Nonlinearity interfaces, i.e. those between an ordinary dielectric
and a dielectric material with an intensity-dependent index of refraction,
have recently played an important role as an element in optical bistable
devices[1,3]. The efforts of researchers are aimed at finding non linear materials with particular optical properties in order to fit very stringent
requirements. A very interesting nonlinear optical material is "latex"
which consists of an aqueous polystyrene microspheres suspension[4].

We report our experimental work aimed at obtaining bistable behaviour
in a device like that used by P. Smith and W. Tomlison[5]. We used a nonlinear interface, whose linear medium was a lithium fluoride (LiF) prism
and whose nonlinear medium was a polystyrene microsphere suspension (refractive index n=1.59) in ethanol. The value of the refractive index for
LiF at 5145 Å is n=1.3910 and for ethanol is n=1.3737. In order to analyze
the reflectivity versus the physical parameters of the medium we chose
several concentrations of polystyrene microspheres in ethanol. In order to

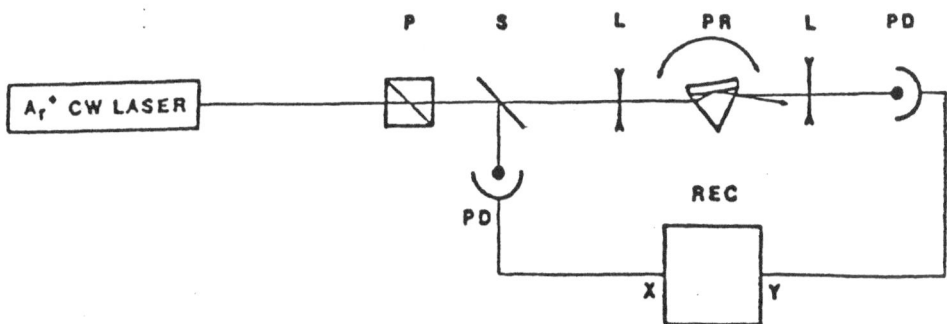

Fig. 1. Experimental arrangement for measuring output power versus
input power.

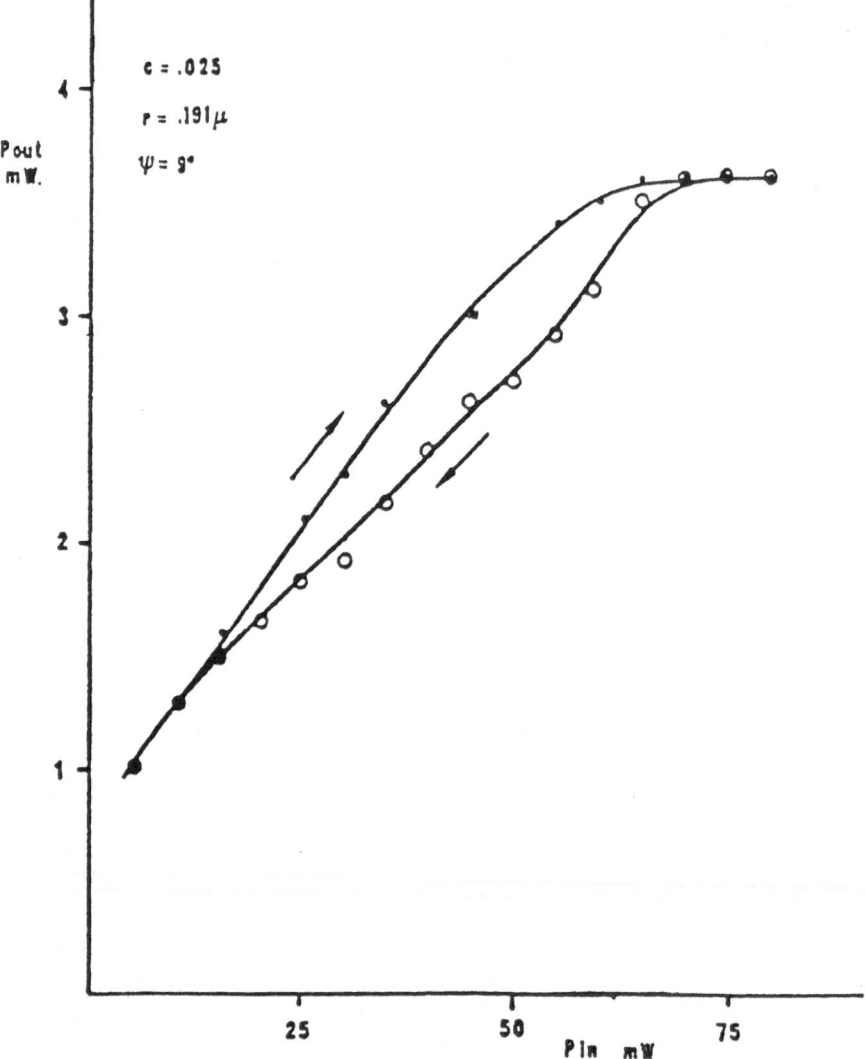

Fig. 2. Diagram of output power versus incident power: scanning of 20 min.

determine experimentally the response of the devices as a function of the nonlinearity of the medium, we used the setup shown in Fig. 1. Looking for bistability, we observed the output power and its relationship to the input power.

In the very first attempts we found no bistability, but by suitably choosing the latex concentration in the range .00025 to .5 and microspheres with radii from .0865 to .265µm, we obtained the hysteretic behaviour shown in Fig. 2. The hysteretic behaviour is enhanced when the microspheres radius is .1915µm and the concentration is .025. Under such conditions the nonlinear coefficient n is 2.42 x 10^{-2} cm^2/MW.

REFERENCES

1. A. E. Kaplan, <u>JEPT. Lett.,</u> 24, 3, (1976)

2. A. E. Kaplan, <u>Sov. Phys. JEPT,</u> 45(5), (1977)

3. G. Delfino and P. Mormile, <u>Optics Letters,</u> Vol. 10, 618, (1985)

4. P. W. Smith,, P. J. Maloney and A. Ashkin, <u>Optic Letters</u>, 7, 8, (1982)

5. P. S. Smith and W. J. Tomlison, <u>IEEE J. Quant. Electr.,</u> QE20, (1984)

SUGGESTION FOR FURTHER READING

POLARIZATION INSTABILITY IN CRYSTALS WITH NONLINEAR ANISOTROPY

AND NONLINEAR GYROTROPY

N.I. Zheludev

Laboratory of Nonlinear Optics - Faculty of Physics
Moscow State University - Moscow 119899, USSR

INTRODUCTION

These notes should be considered as a brief annotated bibliography on some new questions of polarization nonlinear optics of crystals with non-linear anisotropy and/or gyrotropy (Nonlinear Optical Activity and Nonlinear Eigenpolarization). The principal emphasis is on Nonlinear Optical Activity (NOA) and related subjects.

A strong light wave changes natural gyrotropy and optical anisotropy. The lecture treats the use of the NOA phenomenon in spectroscopy of crystals, especially transient spectroscopy near exciton resonances; NOA in light-by-light modulation; and the relationship of NOA to the violation of eigen-polarizations, and polarization instabilities.

I. EIGENPOLARIZATIONS IN NONLINEAR OPTICS

If the polarization state of an electromagnetic wave propagating in a crystal remains unchanged, we call this state an "eigenpolarization". Very important examples of eigenpolarizations in linear optics are "extraordi-nary" and "ordinary" rays in uniaxial crystals. An intense electromagnetic wave can experience self-induced anisotropy or/and self-induced gyrotropy. As a rule, these effects lead to the violation of some eigenpolarizations. Recently, the existence of stable and unstable eigenpolarizations has been recognised[1,2]. We should take into account that in the case of several interacting waves in a nonlinear medium, the concept of nonlinear eigen-polarizations is more complicated. In this case, one must consider an appro-priate set of dependent polarization states and intensities corresponding to the interaction configuration and the type of nonlinearity.

Example I.1

Any linearly polarized wave is an eigenpolarization in a nongyrotropic cubic crystal (m3m, m3) (linear optics). If the k-vector is directed along

Nonlinear Optics and Optical Computing
Edited by S. Martellucci and A. N. Chester
Plenum Press, New York, 1990

the <001> crystal axis, only <100>, <010>, <110> and <1$\bar{1}$0> polarization states are eigenpolarizations in the case of a third-order nonlinearity (nonlinear optics). It is significant that either crystal axes (<100>, <010>) or bisectors (<110>, <1$\bar{1}$0>) are stable or unstable eigenpolarizations. Polarization state stability depends on the sign of anisotropic part of the cubic nonlinear susceptibility[1,3].

Example I.2

As mentioned above, a linearly polarized light beam, which is launched in a crystal with its polarization along an "ordinary" or "extraordinary" direction maintains its polarization (linear optics). In an intense light beam the "fast" axis or birefringence becomes unstable, while the "slow" axis keeps its stability. The effect is important for optical fibers[4,5,6]. From our point of view the most promising objects for the study of polarization instabilities are crystals with a so-called "isotropic point" where the refractive index curves cross and the natural birefringence is small and easy to control[39] (see also paragraph VI).

Example I.3

In a nonlinear medium with non-instantaneous nonlinearity, the stability of eigenpolarizations of the two counterpropagating beams depends on the relation between the nonlinear response time and the round-trip-time across the crystal[7].

II. POLARIZATION SELF-ACTION

The problem of nonlinear eigenpolarizations is closely related to the polarization self-action phenomenon. Each violation of the eigenpolarization can be connected with some well known polarization effect. In isotropic media without optical activity the most important polarization effect due to self-induced birefringence has long been known as "ELLIPSE ROTATION", or the Maker-Terhune-Savage effect[8]. In a similar way, the difference between refractive indices for circularly polarized waves in a gyrotropic medium should vary in a strong electromagnetic wave. This can lead to the intensity dependent ROTATION OF PLANE OF POLARIZATION of a linearly polarized wave[9]. We call this effect "NONLINEAR OPTICAL ACTIVITY".

At present it is clear that two different causes of NOA are important: the spatial dispersion of medium nonlinearity (in gyrotropic liquids, for example), and the latent anisotropy of nonlinear absorption (in crystal of cubic symmetry, for example). It should be underlined, that both NOA mechanisms can be observed in crystals with or without natural optical activity, i.e., NOA can manifest itself as self-induced rotation[10]. Polarization interaction of several linearly polarized waves with various directions of propagation and/or frequencies can be considered as a generalized NOA phenomenon.

Since nonlinear ellipse rotation has been known for long time and is a well reviewed phenomenon (see ref. 11 and descending papers), the references 8 ÷ 11 mainly concentrate on linear optical activity.

254

III. PHENOMENOLOGICAL DESCRIPTION OF NONLINEAR OPTICAL ACTIVITY

Phenomenological description of "ellipse rotation" and "nonlinear optical activity" phenomena are based on the common solution of the wave and material equations.

When considering the amplitude effects of light self-action, the material equation is generally used in the form of an expansion of the electric induction in a power series of the electric field E, up to the third-order terms. There is no necessity for considering spatial dispersion, since in most cases the relative contribution of the amplitude effects related to the spatial dispersion of the medium (a/wavelength) is insignificant (a being the characteristic size in the medium-molecole diameter or the unit cell parameter of the crystal). However, when polarization dependent spatial dispersion is taken into account, the simplest form of the material equation is[9]:

$$D_i = H_{ij}^{(1)} E_j + H_{ijk}^{(2)} E_j E_k + H_{ijkl}^{(3)} E_j E_k E_l + \ldots$$

$$+ G_{ijk}^{(1)} \nabla_k E_j + G_{ij_{kl}}^{(2)} E_j \nabla_k E_l + G_{ijlmk}^{(3)} E_j E_l \nabla_k E_m + \ldots \quad (1)$$

Nevertheless, for light propagating along the optical axes, when the wave is taken to be transverse it is possible to get simple expressions for ellipticity B(z) and rotation angle r(z) of the major axis of the polarization ellipse:

$$r(z) = 1/2 \left[\text{Re } \{f\} \, z + \text{Re } \{Q(z)\} + 2 \, r(0) \right] \quad (2)$$

$$B(z) = \text{th} \left[\text{Im } \{f\} \, z + \text{Im } \{Q(z)\} + \text{Arth } (B(0)) \right] \quad (3)$$

$$Q(z) = k/(2n^2) \int_0^z \left[D_-^{nl}(z')/E_-(z') - D_+^{nl}(z')/E_+(z') \right] dz' \quad (4)$$

Here z is the sample thickness and f is the constant of linear gyrotropy. All "weak" polarization nonlinear self-action effects are contained in these equations[10]. An alternative approach to the problem is based on the powerful Stokes-vector formalism, where a complete analogy with the rigid-body dynamics was found[12,13,14].

Example III.1

NOA due to the spatial dispersion of nonlinearity in $\bar{4}3m$ symmetry group. Zincblende crystals are isotropic and they have no natural gyrotropy: $G^{(1)} = 0$. However, $G^{(3)} \neq 0$ and the contraction G(3)*EE can be regarded as an "induced" tensor G(1). For different initial polarizations, the self-induced rotation due to the nonlocal nature of the nonlinear response may differ in both magnitude and in sign; the polarization vector rotates asymptotically either toward the bisector between the fourth-order symmetry axes of the

255

crystal or towards the normal to this direction, depending on the sign of
G(3). Nonlinear optical rotation due to spatial dispersion changes nonlinear
eigenpolarization in cubic crystals.

Example III.2

NOA due to the latent nonlinear anisotropy in cubic crystals (m3m, m3).
Even if we neglect the spatial dispersion of nonlinearity, the nonlinear
response in cubic crystal depends on the polarization (k -vector $||<001>$).
The two-photon absorption coefficient is different for $E||<110>$. The polari-
zation is preserved only for waves polarized along the $<100>$, $<010>$, $<110>$,
or $<1\bar{1}0>$ axes. In all other cases, the polarization vector rotates asymp-
totycally towards the fourth-order axes or towards the bisectors between
them, depending on the sign of the anisotropic component of the cubic nonlinear susceptibility. The sign of the anisotropic part of the cubic nonlinear susceptibility may depend on wavelength and crystal temperature,
which can vary because of laser heating. Therefore, a flip-flop behaviour of
polarization due to spectral-polarization and thermalpolarization instabilities should be taken into consideration[31].

Example III.3

NOA mechanisms related to electronic nonlinearities do not necessarily
dominate. The main mechanism competing with fast NOA is thermal rotation due
to the heating up of a crystal by laser radiation.

IV. MICROSCOPIC MODELS OF NONLINEAR SUSCEPTIBILITIES, RESPONSIBLE FOR NOA

Ellipse self-rotation and different NOA mechanisms can be obtained
as particular cases of the simple Kuhn classical molecular oscillator
model[14,15]. An ensemble of molecules having neither a symmetry plane nor a
center of symmetry, and consisting of several charged coupled unharmonic
oscillators spaced by some distances (R) from one another and oscillating
in different directions, provides an easy to solve but fruitful model. We
can restrict the consideration by assuming cubic nonlinearity for the
oscillators forming the molecules. All nonlinear susceptibilities in the
material equation above can be effectively calculated here.

Example IV.1

An ensemble of randomly oriented Kuhn's molecules is a remarkably
useful model of an isotropic medium with spatial dispersion of nonlinearity.
The model permits us to obtain one important consequence: $G^{(3)}/H^{(3)} =$
$= G^{(1)}/H^{(1)} = R$. This means that a significant nonlinear rotation should
indicate strongly nonlinear crystals where nonlocality is important, i.e.,
in liquid crystals or near exciton and biexciton resonances in semiconductors.

Example IV.2

A set of similar independent nonlinear orthogonal oscillators at the

sites of a crystal lattice is a good model of a cubic crystal with anisotropy of the third-order nonlinear susceptibility. This model can be deduced
from the Kuhn model as well, but no orientational averaging is necessary,
and the interaction between the oscillators can be neglected.

Several authors have presented quantum-mechanical calculations of nonlinear susceptibilities of a medium with spatial dispersion of nonlinearity[16,17,18]. The most logical treatment is due to Svirko[19], who found, while
calculating the nonlinear response of the non-local system, that the magnetic-dipole moment M and the quadrupole moment Q played equal role in
forming of nonlinearity. He introduced an "OPERATOR OF SPATIAL DISPERSION OF
NONLINEARITY S", responsible for all first-, second- and third- order non-
local nonlinear effects:

$$S^{il} = i \left\{ \frac{d \ Q^{il}}{dt} - c \ e_{iml} \ M^m \right\}$$
(5)

V. SPECTROSCOPIC APPLICATIONS OF INTENSITY DEPENDENT
   POLARIZATION-PLANE ROTATION

Intensity-dependent change of polarization can be the subject of
spectroscopy: in particular, some interesting results on NOA spectroscopy
have been obtained during last few years. All chiral media, especially
liquid crystals or biological macromolecules, are the most promising
objects for this type of spectroscopy. NOA spectroscopy is uniquely suited
for investigation the spatial dispersion of the nonlinearity and latent
anisotropy in crystals. Such studies are of particular interest because of
the information they can provide concerning free excitons. The free exciton
lines in the spectra of the susceptibilities responsible for the nonlocal
character of the nonlinear response should be enhanced relative to the
spectra for ordinary nonlinearities. Since the exciton radius a(exc) and
the lattice parameter a(lat) are the only quantities with the dimensions of
the length, we can estimate:

$$G^{(3)}(exc)/G^{(3)}(nr) = \left[a(exc)/a(lat) \right] \left[H^{(3)}(exc)/H^{(3)}(nr) \right].$$
(6)

Example V.1

A "giant" one-photon exciton resonance in nonlinear optical activity
was detected experimentally in GaAs. The specific induced rotation was on
the order of 10000 deg/cm under 10 MW/cm^2 excitation[20].

When there is a high density of optically created free carriers the
electron-hole interaction is screened, the polarization exciton resonance
drops and self-rotation of the polarization plane disappears during the time
necessary for free carriers to recombine; thus the modulation of light by
light is possible[21]. The polarization plane of a probe beam will be rotated
or not at the output of the crystal when the optical generation of free
carriers by a pump beam in the probe-beam-crystal interaction region is
switched off or on, respectively. The exciton resonance self-induced optical
activity can probably be suppressed by a static electric field because of

257

exciton ionization, so this is a way to create an electro-optical polarization modulator, based on self-induced optical activity in GaAs or
related compounds.

Recently strong exciton resonance in the depolarization of picosecond
light pulses was detected in molecular beam epitaxy GaAs-AlGaAs multiple-
quantum-well (MQW) superlattices.

Example V.2

Two-photon resonances in different mechanisms of NOA in GaAs were
studied[22,23].

Example V.3

The NOA method provides a study of latent crystal anisotropy which is
a touchstone for different models of the ion interaction potential, and it
can be used for making more precise data on the interaction. Essential contributions of anharmonic short-range forces were found in NaCl, KCl, KBr
and LiF[24].

Example V.4

"Giant" thermal nonlinear optical activity of the order of 2000 deg/
(cm W) was found in gyrotropic ZnP_2 using a cw dye laser. It was found
that thermal NOA with cw lasers had a very curious property: the angle of
nonlinear rotation did not depend on the light beam intensity (for constant
beam power), but it depended on total power[25].

The physical description of TNOA (thermal nonlinear optical activity)
with pulsed and cw lasers differs in principle. Under short single light
pulse excitation, absorbed heat has insufficient time to leave the excited
zone during the light pulse. The temperature change depends on the density
of heat absorbed from the laser beam and thus on the laser beam intensity.
The nonlinear change of a specific constant of optical rotation is proportional to the temperature change and the temperature dependence of gyrotropy.

In stationary and quasi-steady state cases with cw lasers the situation is different: the temperature of the excited volume depends on the
balance between cooling of the sample and laser heating. If we change the
spot radius of a Gaussian laser beam without changing the total laser beam
power, we can find that intensity and cooling time respectively decreases
and increases, both as square of the radius of the laser beam. This means
that the temperature of the sample and the optical rotation in the center
of the laser spot do not depend on intensity with fixed power in the light
beam.

The intensity dependence means that the results of cw TNOA experiments
should be preferably presented in (deg/cm/W) units, but not in the (deg-

cm/W) units which are convenient for pulsed and fast "electronic" NOA data. At the same time, we should mention that power-dependent thermal NOA effects can be detected in gyrotropic crystals with relatively weak light intensity and thus without strong temperature gradients and related stresses in the sample.

Example V.5

"Giant" self-induced optical rotation due to latent anisotropy was found in colored cubic crystals with F-centers[26].

Example V.6

Intensity-dependent rotation of the plane of polarization of the light beam normaly reflected from a GaAs surface was found near exciton resonance with picosecond light pulses (nonlinear optical rotation "on reflection")[27].

In linear optics experimental measurements clearly show that spatial dispersion has no influence on the polarization of reflected light at normal incidence, supporting theoretical findings. This apparently means that no intensity-dependent change of polarization due to the spatial dispersion of nonlinearity can be detected under normal reflection. However, phenomenological considerations prove that the plane of polarization rotates on reflection (in first approximation in proportion to the intensity of light) because of anisotropy of nonlinear susceptibility of the third order, which is also responsible for the second mechanism of nonlinear optical activity "in transmission"[28]. For the simple case of weak absorption and nonlinearity, the angle of nonlinear rotation r and the degree of the induced ellipticity B of the reflected beam can be evaluated from the formula:

$$r^{ref} + iB = pqn/(n + 1)^4 \left[(n + 1)/(n - 1)\right] \text{SIN} (4\ r^{inc}) \tag{7}$$

where

$$p = 96\pi^2\ I^{las}\ \chi^{(3)}_{1221}/(c\ n_1\ n_2^2) \tag{8}$$

$$q = \Delta\chi^{(3)}/\chi^{(3)}_{1212} \tag{9}$$

$$\Delta\chi^{(3)} = (\chi^{(3)}_{1111} - 3\chi^{(3)}_{1221}) \tag{10}$$

The angles of NOA "in transmission" and the angle of NOA "on reflection" are in approximately the same ratio as the crystal thickness and the wavelength. This means that just as in natural optical activity "on reflection" one can hope for success only in crystals with strong natural rotation "in transmission". For this reason the most promising material for "reflection NOA" study now is GaAs, where "giant" excitonic nonlinear rotation of the order of 10000 deg/cm has found[20].

The NOA effect "in transmission" can be used for exciton spectroscopy in bulk semiconductors.

Example V.7

By choosing a configuration which is nondegenerated with respect to
frequency or wave vector, one can use NOA spectroscopy to study the time-
dependent nonlinear response, and to measure directly relaxation times for
optically induced anisotropy and gyrotropy. Various configuration can be
employed: for example, one linearly polarized ray at the "pump" frequency
alters the gyrotropy of the crystal, while the second weak ray at the same
or different frequency serves as a "probe"[29]. Polarization changes in a
self-diffracted beam, scattered from the grating recorded by two equally
polarized waves, are also connected with NOA effect and have been found in
GaAs[30].

In cubic crystals without natural optical activity a very effective,
background-free, and simple technique for NOA measurements can be employed.
The polarization effect is detected with quasi cw picosecond lasers and a
lock-in detector, with the sample rotated by a motor. The signal is detected
at a harmonic frequency of the rotation, corresponding to the orientation
dependence of the NOA effect (rotation-modulation polarization measurements,
RMPM)[29].

VI. POLARIZATION INSTABILITIES

Polarization self-action phenomena and nonlinear eigenpolarization
problems are closely related to polarization instabilities. Three different
types of polarization instabilities can be mentioned. The first one - polarization instability "in time" resulting from nonlinear interaction of a
stationary light wave with nonlinear medium manifests itself as polarization
self-pulsing or nearly chaotic time behaviour of polarization. Polarization
instability "in space", or spatial polarization instabilities of stationary
interacting waves, creates a "freezed" polarization distribution along the
rays with periodic, non-periodic and chaotic spatial polarization changes,
and this is the second type of polarization instability. To date, more than
a hundred papers have been publised in the field of polarization instabilities and polarization chaos "in time" and "in space", and a list of references can be found in Ref. (31).

A new third type of polarization instability in hybrid electro-optic
devices involves the formation of a complicated transverse polarization
distribution in the beam cross-section, with polarization domain structure,
self-pulsing and chaotic evolution of the "picture".

Two basic motions from classical mechanisms are used to describe polarization instabilities: the terms "hard" and "soft" polarization symmetry
breaking. A mechanical example of "hard symmetry breaking" is a deformation
of a bar compressed along its axis. When compression exceeds a threshold
value, the bar shape becomes unstable and a left or right bow-type configuration is formed. An optical example of "hard polarization symmetry breaking" is a violation of linear polarization of a light beam that leads to

formation of a left or right elliptical polarization as a result of fluctuations. The effect has an intensity threshold (see examples V.2 and IV.2).
An example of "soft polarization symmetry breaking" (a thresholdless effect)
is a violation of linear polarization launched into a cubic crystal with the
electric field vector along the unstable eigenpolarization. Small fluctuations of polarization parameters may lead to the appearance of left or right
elliptical polarization out of the crystal.

Example VI.1. Polarization instability in atomic systems[32-37]

A very simple example of a nonresonant system which exhibits optical polarization chaos "in time" was proposed. The description of a cell containing
a gas of atoms with J = 1/2 to J = 1/2 transitions with single mirror feedback in a stationary intense quasi-resonant light beam can be reduced to a
discrete series, and polarization period-doubling, chaos with symmetry-
recovering crises and polarization-related optical multistability can be
demonstrated.

Example VI.2. Polarization instabilities in cubic crystals[14,38-41]

The existence of periodic synchronized spatial as well as chaotic
distribution of the polarization of light was found in the mutual interaction of counterpropagating laser beams in nonlinear optical media having
anisotropy in the third-order nonlinearity, which is responsible for the
second NOA mechanism[38]. Such spatial optical polarization turbulence was
named "frustrated optical instability" and was experimentally observed in a
Fabry-Perot GaAs cavity.

Example VI.3. Polarization instabilities in gyrotropy crystals[14,42,43]

Oscillations of polarization and very complicated quasi-chaotic
behaviour of light polarization were predicted in gyrotropic crystals with
strong thermal NOA. The effect was similar to the McCall instability, and
related to competition between thermal and electronic nonlinearities[25].

Example VI.4. Polarization instabilities in crystals with strong two-photon resonance absorption[44-46]

Taking into account a two-photon selection rule in CuCl, a symmetry -
breaking instability was predicted in the change of the polarization ratio
of the transmitted light. The effect can be important near the biexciton
resonance.

Example VI.5. Polarization instabilities in isotropic media[7,14,47-50]

One of the first experimental papers on polarization instabilities
treated polarization ellipse rotation instabilities in liquids[50]. Such
instabilities can be explained by the interaction of the forward light beam
with the backward beam reflected from the sample window[7,49]. Also see
example I.3.

Example VI.6. Polarization instability in birefringent fibers and crystals[5,6,51-61]

See example I.2. In twisted birefringent fibers "hard" polarization symmetry breaking becomes "softer" and the polarization state of the transmitted light is twisted in direction[56].

REFERENCES

1. M. I. Dykman and G. G. Tarasov, Soviet JETP. 72:2246 (1977) (in Russian)
2. A. E. Kaplan, Opt. Lett. 8:560 (1983)
3. A. I. Kovrighin, D. V. Yakovlev, B. V. Zhdanov and N.I. Zheludev, Opt. Comm. 35:92 (1980)
4. K. L. Sala, Phys. Rev. A, 29:1944 (1984)
5. B. Daino, G. Gregori and S. Wabnitz, Opt. Lett. 11:42 (1986)
6. H. G. Winful, Optics Lett., 11:33 (1986)
7. A. L. Gaeta, R. W. Boyd, J. R. Akherhalt and P. W. Milonni, Phys. Rev. Lett. 58:2432 (1987)
8. P. D. Maker, R. W. Terhune and C. M. Savage, Phys. Rev. Lett. 12:507 (1964)
9. S. A. Akhmanov and V. I. Zharikov, Soviet JETP Lett. 6:664 (1967) (in Russian)
10. A. D. Petrenko and N. I. Zheludev, Optica Acta 31:1174 (1984)
11. N. Blombergen, "Non-linear Optics", W. A. Benjamin, Inc., New York - Amsterdam (1967)
12. N. V. Tratnik and J. E. Sipe, Phys. Rev. A. 35:2965 (1987)
13. G. Gregori and S. Wabnitz, Phys. Rev. Lett. 56:600 (1986)
14. S. A. Akhmanov, N. I. Zheludev and Yu. P. Svirko, Polarization instability in strong nonlinear media, in: "Proceedings of the Soviet Academy of Sciences", 46:1070 (1982) (in Russian)
15. S. A. Akhmanov and N. I. Zheludev, Spectroscopy of nonlinear optical activity in crystals, in: "Nonlinear Phenomena in Solids - Modern Topics", M. Borissov, ed., World Scientific Publishing Company, Singapore (1985)
16. D. Mukherjee and M. Chowdhyry, Physica 58:109 (1972)
17. V. N. Belyi and A. N. Serdukov, Soviet Optics And Spectroscopy 40::325 (1976) (in Russian)
18. P. W. Atkins and L. D. Barron, Proc. Roy. Soc. A 304:303 (1968)
19. Yu. P. Svirko, Quantum mechanical theory of spatial dispersion phenomena in nonlinear optics, XIII International Conference on Quantum Electronics, September 1988, Minsk, USSR
20. N. I. Zheludev, V. Yu Karasev, Z. M. Kostov and M. S. Nunuparov, Soviet JETP Lett. 43:747 (1986)
21. Z. M. Kostov and N. I. Zheludev, Optical and Quantum Electronics 20:103 (1988) (in Russian)
22. S. M. A khmanov, N. I. Zheludev and R. S. Zadoyan, Soviet Physics JETP 984:579 (1986)
23. M. G. Dubenskaya, R. S. Zadoyan and N. I. Zheludev, J. Opt. Soc. Am. 2B:1174 (1985)

24. L. B. Meysner, R. S. Zadoyan and N. I. Zheludev, <u>Solid State Comm.</u> 55:713 (1985)

25. N. I. Zheludev, I. S. Ruddock and R. Illingworth, Thermal nonlinear optical activity (TNOA) in BSO and ZNP , Eight National Quantum Electronics Conference, Sept.21-25, 1987, St. Andrews, UK, PD2

26. M. Ya. Valakh, M. I. Dykman, M. P. Lisitsa, G. Yu. Rudko and G. G. Tarasov, <u>Solid State Comm.</u> 30:133 (1979)

27. Z. M. Kostov and N. I. Zheludev, <u>Opt. Lett.</u>, 13, 640 (1989)

28. N. I. Zheludev, A. D. Petrenko and G. I. Trush, <u>Kristallografia</u> 32:393 (1987)

29. S. P. Apanasevich, D. N. Dovchenko and N. I. Zheludev, <u>Optics and Spectroscopy</u> 62:481 (1987) (in Russian)

30. Z. M. Kostov and N. I. Zheludev, Excitonic nonlinear optical activity in GaAs, Eight National Quantum Electronics Conference, Sept. 21-25, 1987, St. Andrews, UK, p. 91

31. N. I. Zheludev, <u>Soviet Physics Uspekhi</u> 31:706 (1988) (in Russian)

32. M. Kitano, T. Yabuzaki and T. Ogawa, <u>Phys. Rev</u> 46:926 (1981)

33. F. Mitschke, R. Deserno, W. Lange and J. Mlynek, <u>Phys. Rev.</u> 33A:3219 (1986)

34. H. J. Carmichael, C. M. Savage and D. F. Walls, <u>Phys. Rev. Lett.</u> 50:163 (1983)

35. F. T. Arecchi, J. Kurman and A. Politi, <u>Optics. Comm.</u> 44:421 (1983)

36. I. P. Aresev, N. N. Rosanov, T. A. Murina and V. K. Subashiev, <u>Opt. Comm.</u> 47:414 (1983)

37. C. Parriger, P. Hannaford and W. J. Sandle, <u>Phys. Rev.</u> 34A:2058 (1986)

38. J. Yumoto and K. Otsuka, <u>Phys. Rev. Lett.</u>, 54:1806 (1985)

39. V. E. Zhakarov and A. V. Mikhailov, <u>Soviet JETP Lett.</u> 45:279 (1987)

40. G. Gregori and S. Wabnitz, <u>Phys. Rev. Lett.</u> 56:600 (1986)

41. M. I. Dykman, <u>Soviet Physics JETP</u> 91:1573 (1986) (in Russian)

42. V. A. Makarov, A. V. Matveeva and M. M. Stolnitz, Polarization multi-stability and optical chaos in a nonlinear Fabry-Perot resonator, Proceedings of the Soviet Academy of Science, Physics 50:799 (1986) (in Russian)

43. V. A. Avetisov and S. A. Anikin, <u>Soviet Academy of Sciences Reports</u> 284:580 (1985) (in Russian)

44. H. H. Kranz and H. Haug, <u>Phys. Rev.</u> 34A:2554 (1986)

45. M. Inoue, <u>Phys. Rev.</u> 33B:1317 (1986)

46. H. H. Ritze and A. Bandilla, <u>Phys. Lett.</u> 78A:447 (1980)

47. M. V. Tratnik and J. E. Sipe, <u>Phys. Rev. Lett.</u>, 58:1104 (1987)

48. A. E. Kaplan and C. T. Law, <u>IEEE Journal of Quantum Electronics</u> qe-21:1529 (1985)

49. M. V. Tratnik and J. E. Sipe, <u>Phys. Rev.</u> 35:2976 (1987)

50. Ng. Phu-Xuan, J. L. Ferrier, J. Gasengel and G. Rivoir, <u>Opt. Comm.</u> 46:329 (1983)

51. S. Trillo, S. Wabnits, R. H. Stolen, G. Assanto, C. T. Seaton and G. I. Steagman, <u>Appl. Phys. Lett.</u> 49:1224 (1986)

52. K. Otsuka, J. Yumoto and J. J. Song, <u>Optics. Lett.</u> 10:508 (1985)

53. H. G. Winful, <u>Appl. Phys. Lett.</u> 47:213 (1985)

54. M. Roman, S. Kielich and W. Gadomski, <u>Phys. Rev. A</u> 34:351 (1986)

55. J. K. Blow, N. J. Doran and D. Wood, <u>Opt. Lett.</u> 12:202 (1987)

56. F. Matera and S. Wabnitz, <u>Opt. Lett.</u> 11:467 (1986)

57. A. Korpel and A. W. Lohman, <u>Applied Optics</u> 25:2253 (1986)
58. R. Cush and C. J. G. Kirbly, <u>Optics Comm.</u> 60:399 (1986)
59. S. Wabnitz, <u>Phys. Rev. Lett.</u> 58:1415 (1987)
60. A. Mecozzi, S. Trillo, S. Wabnitz and B. Daino, <u>Optics Lett.</u> 12:275 (1987)
61. S. Trillo and S. Wabnitz, <u>Phys. Rev.</u> 36 (1987)

SELECTED REFERENCES ON

OPTICAL COMPUTING USING PHASE CONJUGATION

G. J. Dunning and C. R. Giuliano

Hughes Research Laboratories
3011 Malibu Canyon Road, Malibu, California, USA

I. INTRODUCTION

There are many potential advantages for applying phase conjugation to optical computing. Phase conjugation can be used to provide optical amplification, thresholding, optical feedback and exact retroreflection. These properties, either singly or in combination can be used in various architectures to implement a host of computing algorithms.

The previous chapter of this volume on "Optical computing using phase conjugation" is structured to first give a background presentation of phase conjugation. We present there the unique properties of phase conjugation and delve in to the physical mechanism used for its generation. Because of the expanse and diversity of this growing field, rather than present a very brief survey of all of the different applications of phase conjugation in optical computing, we have chosen, instead to discuss at length two specific applications. These examples are designed to illustrate how one can capitalize on the unique properties of phase conjugation. The second section discusses a way to implement an all-optical analog to a neural network and the third section examines a high resolution imaging threshold detector.

Here, for those persons interested in pursuing additional topics in this expanding field, a bibliography with representative research of spatial and temporal applications of phase conjugation is included.

REFERENCES

1. D. Anderson, Coherent optical eigenstate memory, <u>Opt. Lett.</u>, 11, (1986)
2. J. AuYeung, Phase conjugation from nonlinear photon echoes, <u>in</u>: "Optical Phase Conjugation", R. Fisher ed., Academic Press, (1983)
3. Y. Bai, W. Babbit, N. Carlson and T. Mossberg, Real-time optical

waveform convolver/cross correlator, Appl. Phys. Lett., 45, Oct. 1984

4. J. Buchert, R. Dorsinville, P. Delfyett, S. Krimchansky and R. Alfano, Determination of temporal correlation of ultrafast laser pulses using phase conjugation. Opt. Comm., 52, Jan. 1985

5. N. Carlson, L. Rothberg, A. Yodh, W. Babbit and Mossberg, Storage and time reversal of light pulses using photon echoes, Opt. Lett., 8, Sept. 1983

6. N. Carlson, Y. Bai, W. Babbit and T. Mossberg, Temporally programmed free-induction decay, Phys. Rev. A, 30, Sept. 1984

7. H. Caulfield, Associate mappings by optical olography, Opt. Comm., 55, August 1985

8. A. Chiou and P. Yeh, Parallel image subtraction using a phase-conjugate Michelson interferometer, Opt. Lett., Vol 11, p.306, May 1986

9. M. Cohen, Coupled mode theory for neural networks: the processing capabilities of nonlinear mode-mode interactions at cubic and higher order, Proceedings of Neural Net Conference, Snowbird, UT, 1986

10. M. Cohen, Design of a new medium for volume holographic information processing, Proceedings of Neural Net Conference, Snowbird, UT, 1986

11. M. Cohen, Self organization, associaton, and categorization in a phase conjugation resonator, Proceedings of SPIE Optical Computing, 625, January 1986

12. G. Dunning and R. Lind, Demonstration of image transmission through fibers by optical phase conjugation, Opt. Lett., 7, Nov. 1982

13. G. Dunning, E. Marom, Y. Owechko and B. Soffer, An all-optical associative memory with shift invariance and multiple image recall, Opt. Lett., 12, May 1987

14. Y. Fainman, C. Guest and S. Lee, Optical digital logic operations by two-beam coupling in photorefractive material, Appl. Opt., 25, May 1986

15. Y. Fainman and S. Lee, Applications of photorefractive crystals to optical signal processing, in: "Optical and Hybryd Computing", SPIE Vol. 634, 1986

16. N. Farhat, D. Psaltis, A. Prata, and E. Paek, Optical implementation of the Hopfield model, Appl. Opt., 24, 1469 (1985)

17. G. Gheen and L. Cheng, Image processing by four-wave mixing in photo-refractive GaAs, Appl. Phys. Lett., 51, Nov. 1987

18. O. Ykeda, T. Sato and H. Kojima, Construction of a wiener filter using a phase conjugate filter, J. Opt. Soc. Am. A, 3, May 1986

19. R. Jain and G. Dunning, Spatial and temporal properties of a continuous-wave phase conjugate resonator based on the photo-refractive crystal $BaTiO_3$, Opt. Lett., 7, Sept. 1982

20. A. Kamshilin and M. Petrov, Sov. Tech. Phys. Lett., 6, 144, 1980

21. M. Kim and C. Guest, Adaptive 2D holographic associative processor, Proceedings of SPIE Optical Computing, 625, January 1986

22. M. Klein, G. Dunning, G. Valley, R. Lind, and T. O'Meara, Imaging threshold detector using a phase-conjugate resonator in $BaTiO_3$, Opt. Lett., 11, Sept. 1986

23. S. Kwong, G. Rakuljic and A. Yariv, Real-time image subtraction and

exclusive or' operation using a self-pumped phase conjugate
mirror, <u>Appl. Phys. Lett.</u>, 48, Jan. 1986

24. J. Marburger, Optical pulse integration and chirp reversal in
    degenerate four-wave mixing, <u>Appl. Phys. Lett.</u>, 32, March 1978

25. A. Marrakchi, A. Tanguay, J. Yu and D. Psaltis, Physical charac-
    terization of the photorefractive incoherent to coherent optical
    converter, <u>Opt. Eng.</u>, 24, Jan. 1985

26. E. Ochoa, L. Esselink and J. Goodman, Real-time intensity inversion
    using two-wave mixing in photorefractive $Bi_{12}SiO_{20}$, <u>Appl. Opt.</u>,
    24, 1985

27. S. Odulov and M. Soskin, Correlation analysis of images under
    degenerate four-wave mixing in colliding beams, <u>Sov. Phys.
    Dokl.</u>, 25, May 1980

28. T. O'Meara and A. Yariv, Time-domain signal processing via four-
    wave mixing in nonlinear delay lines, <u>Opt. Eng.</u>, 21, March 1982

29. Y. Owechko, G. Dunning, E. Marom and B. Soffer, A holographic
    associative memory with nonlinearities in the correlation
    domain, <u>Appl. Opt.</u>, 26, March 1987

30. D. Pepper, J. Yeung, D. Fekete and A. Yariv, Spatial convolution
    and correlation of optical fields via degenerate four-wave
    mixing, <u>Opt. Lett.</u>, 3, 1978

31. M. Petrov, S. Miridonov, S. Stepanov and V. Kulikov, Light diffrac-
    tion and nonlinear image processing in electrooptic $Bi_{12}SiO_{20}$
    crystals, <u>Opt. Comm.</u>, 31, Dec. 1979

32. L. Pichon and J. Huignard, Dynamic joint-fourier-transform
    correlator by bragg diffraction in photorefractive $Bi_{12}SiO_{20}$
    crystals, <u>Opt. Comm.</u>, 36, Feb. 1981

33. D. Psaltis, J. Yu and J. Hong, Bias-free time integrating optical
    correlator using a photorefractive crystal, <u>Appl. Opt.</u>, 24, Nov.
    1985

34. D. Psaltis and N. Farhat, Optical information processing based on
    an associative-memory model of neural nets with thresholding and
    feedback, <u>Opt. Lett.</u>, 10, Feb. 1985

35. D. Psaltis, J. Hong, and S. Venkatest, Shift invariance in optical
    associative memories, Proceedings of SPIE Optical Computing,
    625, Jan. 1986

36. D. Psaltis. D. Brady and K. Wagner, Adaptive optical networks using
    photorefractive crystals, <u>Appl. Opt.</u>, 51, May 1988

37. D. Psaltis and D. Brady, A photorefractive integrated otpical
    vector matrix multiplier, SPIE Proceedings, 825, Aug. 1987

38. H. Rajbenbach, Y. Fainmam and S. Lee, Optical implementation of an
    iterative algorithm for matrix inversion, <u>Appl. Opt.</u>, 26, March
    1987.

39. A. Rebane and R. Kaarli, Picosecond pulse shaping by photochemical
    time-domain holography, <u>Chem. Phys. Lett.</u>, 101, Oct. 1983

40. K. Sayano, G. Rakuljic and A. Yariv, Thresholding semilinear phase
    conjugate mirror, <u>Opt. Lett.</u>, 13, Feb. 1988

41. Y. Shi, D. Psaltis, A. Marrakchi and A. Tanguay, Photorefractive in-
    coherent-to-coherent optical converter, <u>Appl. Opt.</u>, 22, Dec. 1983

42. B. Soffer, G. Dunning, Y. Owechko and E. Marom, Associative
    holographic memory with feedback using phase-conjugate mirrors,
    <u>Opt. Lett.</u>, 11, Feb. 1986

43. Y. Tomita, R. Yahalom and A. Yariv, Real-time image subtraction with the use of wave polarization and phase conjugation, Appl. Phys. Lett., 52, Feb. 1988

44. K. Wagner and D. Psaltis, Multilayer optical learning networks, Appl. Opt., 26, Dec. 1987

45. H. White, N. Alridge and I. Lindsay, Digital and analogue holographic associative memories, Opt. Eng., 27, Jan. 1988

46. J. White and A. Yariv, Real-time image processing via four-wave mixing in a photorefractive medium, Appl. Phys. Lett., 37, July 1980

47. A. Yariv, Y. Tomita and Kazuo Kyuma, Theoretical model for modal dispersal of polarization information and its recovery by phase conjugation, Opt. Lett, 11, Dec. 1986

48. A. Yariv and S. Kwong, Associative memories based on message-bearing optical modes in phase-conjugate resonators, Opt. Lett., 11, March 1986

49. A. Yariv, S. Kwong, and K. Kyuma, Demonstration of an all-optical associative holographic memory, Appl. Phys. Lett., 48, April 1986

50. P. Yeh and A. Chiou, Optical matrix-vector multiplication through four-wave mixing in photorefractive media, Opt. Lett., Vol. 12, p. 138, February 1987

51. J. Yu, J. Hong and D. Psaltis, Photorefractive time integrating correlator and adaptive processor, OSA Topical Meeting on Photorefractives and Applications, L.A. Ca 1987

52. A. Zuikov, V. Samartsev and R. Usmanov, Correlation of the shape of light echo signals with the shape of the excitation pulses, JETP Lett., 32, Aug. 1980.

INDEX

Abbe's experiments, 154
Aberrated acoustic wave, 176
Absorbing etalons, 126
Absorption
  bleaching, 111
  coefficient, 10, 52, 90
  edge, 106
    shift, 52, 59
  losses, 243
  saturation, 84, 88, 89, 90, 93,
    120
  spectrum, 106, 116
Absorptive bistability, 37, 59
Absorptive grating, 123
Absorptive OB, 5
Acceptor density, 212
Acoustic frequency, 175
Active devices, 114
Active imaging, 173
Active material, 126
Adaptive optical networks, 267
Adaptive optics, 173
Address pulse, 8
Air Force resolution chart, 184
Airy function, 5, 12
All-optical associative holo-
    graphic memory, 268
All-optical associative memory,
    180, 182
All-optical computing schemes, 51
All-optical logic devices, 21
All-optical processing, 217
All-optical signal processors, 225
All-optical threshold detector,
    193
AlPc-F, 109, 111,
  thin films, 110
$Al_xGa_{1-x}As$ barriers, 120
$Al_{0.32}Ga_{0.68}$, 105

Ambipolar
  diffusion, 121
  coefficient, 124
Amplification, 180
Amplitude-phase encoding scheme,
    204
AND, 112
Angular bistability, 234
Angular detuning, 31
Angular multiplexing, 181, 182
Angular shift, 232
Anharmonic oscillators, 139
Anisotropic carrier diffusion, 129
Anomalous GVD, 223
Antiguiding parameter, 48
Anti-reflection coatings, 120
Aqueous polystyrene microspheres
    suspension, 247
Argon laser, 58, 193, 230, 234
Associate mapping, 266
Associative holographic memory,
    267
Associative memory, 180, 181, 266,
    268
Atomic resonances, 66
Atomic systems, 261
Atomic transition, 68
Auger
  effect, 30
  recombination processes, 40
Autoassociative memory, 204
Autoionizing levels, 63
Avalanche effect, 30
Average linear index, 127
Averaged absorption coefficient,
    123

B bands, 110
Backward field, 10

Backward stimulated Raman
  scattering, 175
Band edge absorption, 52
Band filling, 123
Band gap
  energy, 30
  resonant effects, 119
  resonant four wave mixing, 129
  resonant optical nonlinearities,
    126
Bandwidth, 134
Band-to-band
  recombination, 30
  transitions, 84, 89
Barium fluoride, 53, 54
Bar-state, 113
Basic logic operations, 217
$BaTiO_3$, 183, 266
  PCM, 193
Beat length, 112
Bessel function, 123, 143
BGO, 209
  crystal, 203
$Bi_{12}SiO_{20}$ crystals, 267
Biexciton, 84
  resonance, 261
Biharmonic laser pumping, 131
Binary collisions, 63
Binary digit sequences, 15
Binding energy, 99
Biological macromolecules, 257
Biological nervous systems, 159
Biological systems, 159
Birefringence, 254
Birefringent fiber, 219, 262
  coupler, 219
Bistability, 42, 131
Bistable c^3 laser, 46
Bistable devices, 24, 99
Bistable etalons, 127
Bistable interaction, 132
Bistable operation, 24
Bistable oscillators, 133
Bistable properties, 27
Bistable resonance, 137
Bistable semiconductors lasers, 37
Bloch equations, 87
Blue shift, 90, 106, 113
Bohr radius, 99, 106
Borosilicate glass, 106
Bound-exciton absorption, 84
Bow-type configuration, 260

Boxcar, 65
Bragg condition, 186, 198, 202
Brillouin
  gain, 176
  zone, 90
BSO, 199, 208, 209
Build-up
  dynamics, 104
  time, 21
Bulk
  case, 127
  GaAs, 93, 94, 100, 121

Cantor set, 16
Carrier
  confinement, 112
  creation, 31
  density, 112, 119, 121, 124
  dynamics, 119
  grating decay time, 126
  lifetime, 31, 84, 119, 120
  momentum relaxation, 91
  recombination, 121
    time, 124
  carrier collisions, 91, 93
    scattering, 94
Catastrophic self-focussing, 17
Cavity
  dumper, 120
  finesse, 12
  mistuning, 12
  mode frequency, 40
  OB, 5
  optical bistability, 5
  resonance frequency, 38
  response function, 15
    time, 12
  transit time, 217
CCD television camera, 161
CdHgTe, 51
CdS, 51, 106
$CdS_{0.9}Se_{0.1}$ microcrystallites,
    242
CdSe, 106
  quantum dots, 99, 106
CdS_xSe_{1-x}, 241
  semiconductor-doped-glass, 230
Channel structures, 112
Chaos, 15, 261,
Chaotic features, 76
Charge
  carrier injection, 103

exchange, 71
  collisions, 63
  separation, 126
Chemical potential, 115
Chiral media, 257
Classical electron radius, 134
Classical limit, 138
Cleaved coupled cavity laser, 37
Coherence time, 86
Coherent Raman
  scattering, 229
  acoustic wave, 175
Coherent response, 88
Collection time, 63, 72, 76
Collective effects, 72
Collective electron-ion interactions, 70
Collinear process, 175
Collision rate, 77, 92
Collisionless fluid approximation, 75
Colloids, 106
Colored cubic crystals, 259
Communication
  link, 199
  network, 197
Complex frequency synthesis chains, 142
Computational power, 180
Computing machines, 197
Conduction
  band, 30
  electrons, 133
Conductive properties, 109
Confinement
  effect, 100
  factor, 38
Confocal parameter, 192
Conical interconnect architecture, 203
Conjugate mirror, 174
Conjugate wave, 174
  generation, 175, 177
Connecting network, 151
Continuity
  conditions, 22
  equation, 73
Continuous wave solution, 218
Contrast switching, 25
Cooling, 93
Corning $CdS_{0.9}Se_{0.1}$-doped glasses, 241

Correlation noise, 183
Correlator, 161
CO_2 laser, 133, 141
Coulomb
  enhancement, 114
    factor, 116
  force, 152
  interaction, 91, 94
  potential, 114
  screening, 120
Coumarin 153, 64
  dye solution, 64
Counter-propagation, 18
Coupled mode theory, 218
Coupled-cavity semiconductor lasers, 37
Couplers, 218
Coupling conditions, 29
Coupling devices, 218
Coupling efficiency, 25, 31, 232, 242
Coupling parameter, 42
Coupling process, 229
CPM coefficients, 218
CPU, 11
Cracks magnetic storage, 207
CRAY, 11, 152, 154
Critical laser intensity, 145
Critical slowing down, 53
Critical switching, 56
  power, 52, 56
Cross well diffusion, 121, 126
Crossbar network, 156
Crosstalk, 151, 198, 203
  suppression, 185
Cross-phase modulation, 217
Cross-state, 113
Crystal
  anisotropy, 258
  axis, 254
  growth techniques, 100
  spatial frequency, 206
Crystallite size, 106
C_3
  InGaAsP laser, 44, 45
  laser, 38
CuBr, 106
CuCl, 106, 261
CuS, 51
Cubic crystals, 261
Cubic nonbirefringent crystal, 221

Cubic nonlinear susceptibility, 254
CW
  dye laser, 103, 258
  infrared laser, 142
  laser, 51
  regime, 133, 146
  transmission, 223
Cycle time, 8
Cyclo-Raman
  excitations, 137
  optical resonances, 142
  resonance, 132, 133, 135
  scattering, 142, 144
Cyclotron
  electron motion, 134
  excitation, 131, 132, 133
  frequency, 131, 134
  momentum, 136
  motion energy, 135
  orbits, 142
  radius vector, 135
  resonance, 131
Cylindrical einzel lens, 65

Damping parameter, 134
Darkening, 243
Decay dynamics, 104
De-excitation process, 29
Defect
  detection, 209
  enhancement, 207, 208
  signal, 199
Deformable mirror, 176
Degeneracy, 120
Degenerate four-wave mixing, 100, 175, 176, 229, 267
Dephasing time, 86, 94, 114
Desorption, 53
Destructive interaction, 144
Detector arrays, 180
DFWM 175, 176
  reflectivity, 187
Dielectric constant, 106
Difference frequency generation, 229
Differential transmission, 122
  spectroscopy, 108
    spectrum, 103, 104, 110
Diffraction, 17, 189
  efficiency, 123, 199, 203, 207
  grating, 121, 229

coupler, 22
-only calculations, 12
order, 24
Diffractive coupling, 17
Diffuser plate, 183
Diffusion
  length, 10, 16
  times, 126
  -and-diffraction calculations, 13
Diffusive-coupling model, 17
Digital electronic computers, 151
Digital image processing, 209
Digital information processing, 83
Digital optical computer, 151
Digital optical computing, 51
Digital processing, 83
Diode lasers, 83, 203
Dipole radiation, 145
Direct data linking, 201
Direct optical transitions, 89
Directional coupler, 22, 99, 112
Direct-band structure, 29
Direct-gap semiconductors, 92
Discrimination capability, 167
Dispersion regime, 224
Dispersive bistability, 37, 59
Dispersive broadening, 224
Dispersive coupling structures, 218
Dispersive effects, 29, 218
Dispersive limit, 6
Dispersive OB, 5
Dispersive optical bistability, 52
Distributed Bragg reflectors, 127
Distributed couplers, 240
DOC, 151
Donor density, 212
Doping
  modulation, 104
  superlattices, 103
Doppler
  effect, 135, 146
  nonlinear mechanism, 135
  phase modulation, 145
  -shifted scattered wave, 176
Down-switching, 232
Driving field, 138
DTS, 108, 110
Dual grating picture, 178

Dual-core
  fiber, 223, 218
    coupler, 219
Duffing oscillator, 139
Dwell-time, 8
Dynamic Burstein shift, 90
Dynamic equations, 143
Dynamic holographic interconnec-
    tion, 199
Dynamic holography, 160
Dynamic interconnect problem, 197
Dynamic nonlinearities, 28, 84, 88
Dynamic optical interconnects, 197
Dynamic optical nonlinearities, 84
Dynamic properties, 29
Dynamical effects, 33
Dynamical systems, 15
Dyson's series, 85

Edge
  -emitted luminescence, 95
  -filter function, 16
Effective cyclotron frequency, 140
Effective mass, 99
Effective mass approximation, 106
Effective mass of electron, 131
E-h
  pairs, 92
  plasma, 90
Eigenpolarizations, 253
Eigenstates, 83
Electric dipole approximation, 85
Electrical feedback, 51
Electron
  energy, 131
  excitation, 142
  -hole Coulomb correlation, 116
    attraction, 91
    excitations, 114
    interaction, 257
    pair, 28, 84, 100
    plasma, 10, 90, 93, 114, 120
    recombination, 29, 111
    transition, 106
  occupation factor, 90
  -phonon interactions, 93
  rest mass, 131
  thermal motion, 77
  diffusion, 17
Electronic dynamics, 110
Electronic effects, 30, 32, 242
Electronic gain, 39

Electronic implementation, 160
Electronic nonlinearity, 17
Electronic nonlinearity, 52
Electronic temperature, 90
Electro-optic coefficient, 183,
    207
Electro-optic effects, 51
Electro-optical polarization
    modulator, 258
Electro-striction, 175
Ellipse
  rotation, 254
  self-rotation, 256
EM wave, 131
Energy
  band gap, 28
  eigenstates, 83, 86
  eigenvalue, 14
  exchange, 83
Enhanced nonlinear optical phe-
    nomena, 119
Enhanced nonlinearities, 127
Epitaxial growth techniques, 83
Equation of motion, 75
Erase beam, 187
  optics, 190
Erasure sentivitiy, 203
Etalon, 126
  devices, 111
Etalons fabrication, 54
Etched coupled cavity laser, 37
Ethanol, 247
Exact retroreflection, 265
Excimer laser, 66
Excitation
  density, 10
  -dependent mass-effect, 146
  energy, 143
Exciton, 84, 99, 106, 119
  absorption, 102
  Bohr radius, 100, 101, 106
  doublet, 120
  ionization, 258
  oscillator strength, 100
  radius, 257
  resonance self-induced optical
    activity, 257
  saturation, 100
  spectroscopy, 259
  wavelength, 120
Excitonic absorption, 100, 119
  resonance, 120

Excitonic effects, 91, 100
Excitonic lines, 91
Excitonic nonlinear refraction, 119, 129
Excitonic nonlinearities, 119
Excitonic optical nonlinearities, 119
Excitonic region, 96
Excitonic saturation, 123
Exotic materials, 153

F-centers, 259
F-number, 189, 190
Fabry-Perot
  bistable etalons, 126
  cavities, 6, 10, 21, 23, 37
  etalon, 51
  GaAs cavity, 261
  interferometer, 23
Far-field distribution, 47
Fast Fourier transform, 157
Fast Hankel transform method, 11
Fast nonlinearities, 81
Fast optical switching, 21
Fast switchings, 33
Feedback, 51
  image, 163
Femtosecond
  light pulses, 92
  pulses, 91
  regime, 94
Fermi-Dirac distributions, 90
Fermi
  energy, 90
  functions, 114
Fermions, 152
FFT, 157
  algorithm, 11
FHT method, 11
Fiber
  directional coupler, 218
  nonlinear switches, 224
  optic Mac-Zehnder inter-
    ferometer, 218
Field-free solution, 75
Filter designs, 53
Finesse, 56, 127,
Fine-structure constant, 138
Finite-difference technique, 10
First-order nonlinear force, 139
Flip-flop behaviour, 256

Flouro-aluminum phtalocyanine, 100, 109
Fly's eye lens arrays, 53
Folded cavity, 6
Forward field, 10
Fourier
  components, 86
  transform, 85
    hologram, 163, 181, 184
    limit, 92
Four-photon interactions, 132
Four-wave mixing, 119, 120, 122, 176, 178, 267
  phase conjugate mirror, 200
FP, 23
Free carrier, 119
  absorption, 30, 39
  generation, 28
  gratings, 120
  thermalization, 30
Free electron, 131
Free exciton lines, 257
Free-induction decay, 266
Frequency
  chirp, 224
  domain description, 86
  drifting, 180
  mixing, 84
    processes, 89
  multipliers, 142
  standards, 142
Fresnel number, 189
Frustrated optical instability, 261
Full adder circuits, 53

GaAs, 51, 89, 115, 257, 259
  -AlGaAs hetero n-i-p-i super-
    lattice, 105
    MQW, 99, 100, 105, 112 113, 119, 129
    multiple-quantum well-super-
      lattices, 258
  -based epitaxial structures, 97
  etalons, 126
  MQWs, 93, 94, 95, 120, 127, 129
  n-i-p-i, 103
    structures, 99
  NLDC, 113
  quantum-well material, 115
Gain
  coefficient, 40
  in quantum wells, 95

saturation, 37
spectrum, 116
Gallium arsenide, 151
Gaussian
  beams, 192
  profile, 75
Ge, 84
Generation processes, 28
Glass matrix, 106
Glasses, 106
Graded index designs, 54
Grating
  coupler, 22, 24
  coupling, 22
  decay rates, 124
    time, 126
  harmonic, 212
  modulation depth, 25, 31
  period, 29, 121
  spacings, 124
  vector, 212
Green's functions, 91
Group
  index, 38
  velocity, 38
    dispersion, 217
Guided wave structures, 229
GVD, 217, 223
Gyrotropic liquids, 254
Gyrotropy of the crystal, 260

Half-beat length, 112
Half-wave plate, 208
Hamiltonian chaos, 222
Hamiltonian dynamics, 15
Hard polarization symmetry break-
      ing, 260
Heat
  capacity, 30
  exchange, 29
  boss, 56
  sinking, 52, 53
Heaviside function, 85
HeNe, 203
  laser, 143
Hetero-associations, 183
Hetero n-i-p-i superlattices, 104
High bandwidth interconnects, 201
High brightness lasers, 173
High finess limit, 127
High-order cyclotron subharmonics,
      132

High-order nonlinear effects, 134
High-order optical subharmonics,
      139
High-order subharmonics, 133, 141
High power oscillators, 180
High Q-factor cavity, 21
High speed
  optical signal processing, 99
  switching, 108
Hold level, 8
Hole-burning, 92
  feature, 94
Hole
  excitons, 123
  occupation factors, 90
Hologram, 160, 161, 180
Holographic associative processor,
      266
Holographic crossbar, 198, 204
Holographic information proces-
      sing, 266
Holographic interconnect, 199
  scheme, 197
Holographic interconnections, 160
Holographic lenslets, 53
Holographic recording, 198
Holographic storage medium, 199
Hopfield model, 205, 266
Hybrid electro-optic devices, 260
Hybrid systems, 22
Hydrogen-atom-like quasi
      particles, 99
Hypercube network, 151
Hysteresis, 42, 131, 143
  effect, 138
  loop, 27, 33, 45, 48
Hysteretic branch, 145
Hysteretic cyclotron resonance,
      131, 134, 136
Hysteretic excitation, 132
Hysteretic jumps, 138
Hysteretic resonance, 131, 135

Image
  amplification, 212
  intensifier, 164
  processing, 129, 198, 266, 267,
      268
  transmission, 266
Imaging threshold detector, 186
Impurity, 14
InAs, 51

Indirect-band structure, 29
Induced ellipticity, 259
InGaAsP, 42
Inhomogeneous current injection, 37
Injection, 40
  rate, 104
  -locked diode lasers, 198, 199
Inorganic semiconductors, 99, 111
Input-output characteristics, 164
InSb, 10, 11, 51, 84
Instabilities in liquids, 261
Instability strip, 16
Integrated electro-optic modulators, 22
Integrating nonlinearity, 236
Intensity-dependent refractive index, 229
Intensity-dependent rotation, 259
Intensity discrimination, 219
Interaction length, 192
Interband
  interaction, 99
  polarization, 114
  recombination, 94
  transitions, 89
Intercavity coupling, 38, 49
Interconnect configuration, 201
Interconnect pattern, 197
Interconnection
  matrices, 198, 206
  network, 151
Interconnects, 149
Interelectronic collisions, 90
Interference filter, 10
Interference filter etalons, 52
Interference pattern, 210
Intraband exchange effects, 114
Intracavity
  lens, 189
  optical distortion, 180
Intrawell diffusion, 125, 126,
Intrinsic distributed feedback systems, 22
Invariance capability, 169
In-well ambipolar diffusion coefficients, 124
Ion
  -milled gratings, 239
  spectrometer, 65, 70
  thermal motion, 77
Iris, 190

Isolas, 132, 142, 144
Isotope separation, 63
Isotropic point, 254
Iteration, 183

Joint density of states, 90

K-space, 91
K-vector selection rule, 89
KBr, 258
KCl, 258
Kerr
  cavity, 7, 9
  effect, 5
  -like effects, 241
    nonlinearity, 230, 231, 234
  limit, 6
  medium, 17
  nonlinearity, 10, 237
  shutter, 219
Knife edge, 212
Kogelnik analysis, 207
K^+-Na^+ exchange technique, 241
Kramers-Kronig transformation, 100
Krypton laser, 55
Kuhn
  classical molecular oscillator model, 256
  model, 257
  molecules, 256

Laser
  annealing, 63
  excited electron-hole pairs, 106
  beating, 256, 258
  induced ionization, 63, 67
  isotope separation, 63
  -matter interactions, 78
Latex, 247
Lattice
  heating, 29, 31, 126
  parameter, 257
  temperature, 29, 90, 93
LCLV, 160
Lead fluorides, 53, 54
Least-square-fit, 108
L/I curve, 44
LiF, 8, 84, 247, 258
Lifshitz-Slyozov law, 106
Light
  echo signals, 268
  field polarization, 90

modulators, 160
pulse duration, 31
-matter coupling, 91
function, 5
Limiting resolution, 190
LiNbO$_3$, 200
guides, 219
Linear approximation model, 181
Linear coupling, 218
coefficient, 218
Linear gyrotropy, 255
Linear susceptibility, 90, 256
Liquid crystal, 257
light valve, 160
Lithium fluoride, 247
LO-phonon
broadening, 100
collisions, 120
Logic AND operation, 219
Logic functions, 112
Logic interactions, 152
Logic operations, 111
Long-range interactions, 63
Lorentz
equation, 133
nonlinear mechanism, 135, 139
radiation force, 133, 135, 136,
146
linefunctions, 90
lineshapes, 90
Low-level parallelism, 154
Low-order subharmonics, 141
Low-speed optical computing, 103
Luminescence, 95

Magnetic-dipole moment, 257
Magnetic disk, 209
Magnetic field, 133, 144
Magnetron motion, 136
Maker-Terhune-Savage effect, 254
Manley-Rowe relationships, 145
Many body
collisions, 63
effects, 91
theory, 114
Many photon interactions, 132
Massive parallelism, 21
Material equation, 255
Material nonlinearities, 101
Maxwell-Bloch equations, 5
MBD, 52
filters, 52

layers, 54
technique, 52
ZnSe etalon, 55
spacers, 53
MBE, 100
McCall instability, 261
MCP, 65
Mean-field model, 17
Memory
elements, 51
state, 16
Metallic-coated dielectric grat-
ing, 26
Metalorganic chemical vapor
deposition, 100
Metal-strip-loaded GaAs struc-
tures, 112
Microcrystallites, 106, 241,
Microprocessor chips, 199
Microwave, 131
frequency standards, 142
oscillator, 140
Micro-holograms, 157
MOCVD, 100, 103
Mode hopping, 180
Mode-locking, 119
Modulation
depth, 27, 124, 206
instability, 17
Modulators, 83
Molecular beam
deposition, 52
epitaxy, 100, 120
Momentum relaxation, 91
Monolithic system, 24
MQW, 93,94, 95, 100, 119
strip-loaded channel wave-
guide, 112
structures, 126
systems, 100
Multichannel plate, 65
Multiphoton
effects, 134
interaction, 133
ionization, 63
processes, 131
Multiple beam diffraction, 207
Multiple exposure hologram, 181
Multiple quantum well, 83, 89,
100
structures, 100, 119
Multiple switching, 238

Multiplexed holograms, 197
Multiplexing, 197
  capability, 198
  method, 203
Multiresonant dynamic nonlinear-
    ities, 84
Multistable interaction, 132
Multi-wave mixing, 145
  effects, 142

N-i-p-i crystals, 103
  structure, 103
N-photon process, 132
Na, 64
  D line, 65
  vapor, 64
  vapour, 3
Na_2 ions, 68
NaCl, 258
Narrow-gap semiconductors, 133
Nd-YAG laser, 110, 239, 241
Near-field distribution, 47
Neural architectures, 159
Neural computer, 159
Neural network, 149, 159, 180,
    266
Neuron gain, 165
Neurons, 159
Nitrogen-pumped dye laser, 107
NLDC, 112
NN propagation configuration, 137,
    139
NOA, 253
  effect, 260
  in trasmission, 259
  method, 258
  on reflection, 259
  spectroscopy, 257, 260
  spectroscopy, 260
Noise effects, 18
Noncyclotron
  component, 134
    of momentum, 143
  motions, 134
Non-ensemble quantum theory, 146
Nongyrotropic cubic crystal, 253,
    254
Nonlinear Fabry-Perot, 217
Nonlinear absorption, 119, 222,
    254
  spectra, 100
Nonlinear anisotropy, 253

Nonlinear chirp, 224
Nonlinear coupling, 229
Nonlinear cyclotron resonance, 139
Nonlinear directional coupler, 112
Nonlinear eigenpolarization, 253
Nonlinear ellipse rotation, 254
Nonlinear etalons, 111
Nonlinear guided waves, 21
Nonlinear gyrotropy, 253
Nonlinear interaction, 83, 137
Nonlinear interface, 247
Nonlinear interferometers, 224
Nonlinear interferometry, 100
Nonlinear material, 51, 109
Nonlinear medium, 21
Nonlinear mode coupler, 219, 222
Nonlinear optical activity, 253,
    254
Nonlinear optical devices, 99,
    111
Nonlinear-optical effects, 131
Nonlinear optical materials, 99
Nonlinear optical response, 85
  functions, 97
Nonlinear optical rotation, 256
Nonlinear optics, 131, 145, 173
Nonlinear organics, 109
Nonlinear parametric systems, 139
Nonlinear phase-shifters, 218
Nonlinear photon echoes, 265
Nonlinear physics, 144
Nonlinear refraction, 51, 90, 119,
    120, 127, 129
  cross section, 123
Nonlinear refractive index, 84
Nonlinear response, 87
  functions, 85
Nonlinear susceptibilities, 87,
    256
Nonlinear switches, 218
Nonlinear transmission spectrum,
    103
Nonlinear wave phenomena, 4
Nonlinear waveguide, 99
  devices, 21, 27
Nonlinearity interfaces, 247
Non-parametric processes, 84
Non-resonant thermal process, 239
Non-thermalized distributions, 91
NOR, 112
NOR-gate loop, 54
NP propagation configuration, 137

Numerical instability, 75

OB, 3
  amplifier, 7
  arrays, 13
  devices, 3
  media, 18
  memory array, 15
  systems, 3, 9
  theory, 18
  threshold, 11
Occupation factors, 90
OFF state, 37, 186
ON state, 37, 186
One-electron states, 91
One-photon exciton resonance, 257
On-resonance nonlinear trans-
       mission, 224
OPC, 173
Operating power, 21
Operator of spatial dispersion of
       nonlinearity, 257
Optical amplification, 265
Optical architectures, 160
Optical associative loop, 162
Optical associative memory, 159,
       160, 204, 267
Optical biharmonic, 140
Optical biharmonic laser, 142
Optical bistability, 1, 3, 37, 51,
       55, 58 , 83
  in guided-wave structures, 22
Optical bistable devices, 28, 247
Optical communications, 99, 173,
Optical computing, 51, 149, 151,
       173, 265
Optical connections, 151
Optical converter, 267
Optical crossbar switch, 198, 201,
       204
Optical data processing, 173
Optical detectors, 197
Optical devices, 111
Optical digital logic opera-
       tions, 266
Optical effects, 83
Optical eigenstate memory, 265
Optical emission, 145
Optical excitation, 121, 131, 133
Optical feedback, 265
Optical fiber, 217, 254
  technology, 229

Optical gain, 37
Optical gates, 84
Optical image processing, 198, 207
Optical implementation, 160
Optical information
  processing, 267
  storage, 13
Optical instabilities, 235
Optical interconnects, 197
Optical interference filters, 51
Optical laser, 133
Optical logic gates, 83, 97, 99
Optical loop, 168
Optical memory, 3, 16,
Optical networks, 151, 155
Optical neural computer, 160
Optical neural networks, 180
Optical nonlinearity, 1, 51, 83,
       97, 104
Optical parallel logic, 152, 154
Optical parametric oscillator, 141
Optical phase conjugation, 173
Optical phonon, 94
  emission, 94
Optical pixellation, 13
Optical processors, 151
Optical resonance, 178
Optical signal processing, 208,
       266
Optical Stark effect, 112
Optical Stark-shift, 142, 143
Optical subharmonic resonance, 132
Optical susceptibility, 114
Optical switching, 34, 197, 201
  devices, 22, 119
Optical transmission, 103
  bandwidth, 201
Optical traveling waves, 142
Optical waveguide, 21, 22,
Optically activated switches, 217
Optically bistable element, 217
Optically bistable loops, 58
Optically controlled devices, 217
Optically coupled states, 90
Optimum coupling angle, 230
Optimum thickness, 127
Optoelectronic devices, 83
Optogalvanic technique, 63, 66
Optothermal devices, 56, 59
OR, 112
Ordinary Bessel function, 140
Ordinary mirror, 174

Organic polymers, 99
Organic semiconductors, 99, 109
Organic thin films, 99
Organics, 109
Oscillator strength, 101
Outer-product interconnection
        matrix, 204, 206
Oxide-stripe structure, 47

Pade' approximation, 114
Parallel computing, 151
Parallel digital loop circuits, 53
Parallel processing, 180
Parallel processors, 151
Parametric amplification, 84, 229
Parametric conversion, 173
Parametric effects, 84
Parametric interactions, 28
Parametric oscillations, 141
Parametric processes, 83, 87
Passive optical bistability, 52
PC, 109
   propagation configuration, 137,
      142
PCM, 181
PCR, 179
Peirod masks, 207
Penning trap, 133, 136
   potential, 133
Perfect shuffle network, 157
Periodic fibers, 218
Phase
   conjugate, 121
      Michelson interferometer, 266
      mirror, 129, 163, 181
      reflectivity, 184
      resonator, 179, 266
   conjugation, 83, 84, 119, 129,
      173, 265
   factor, 205
   multistability, 133, 14
   -sensitive device, 219
   space, 120
   transition theory, 8
Phonon emission, 28, 29,
Photocarrier, 103
   density, 29
Photochemical time-domain holo-
      graphy, 267
Photoconductive application, 109
Photodarkening effect, 107
Photodeposition, 63

Photodetectors, 83
Photoelectrochemical applica-
        tions, 109
Photoexcitation, 84, 90
Photo-excited carriers, 93, 94
Photo-excited multiple quantum
        well, 96
Photoluminescence spectra, 91
Photon
   decay rate, 38, 49
   echoes, 266
   energy, 120
Photonic switching, 100
Photons, 153
Photorefractive crossbar switching,
      204
Photorefractive crystal, 183,
      160, 197
Photorefractive effects, 197
Photorefractive gratings, 186
Photorefractive phase conjugate
      resonator, 186
Photorefractive response, 210, 212
Photorefractive transmission holo-
      grams, 198
Photosensitivity, 212
Photovoltaic applications, 109
Phtalocyanines, 109
π-conjugated electrons, 109
π-electron system, 109
π-π^* excitations, 110
Picosecond light pulses, 90
Pipeline mode, 217
Pixel packing density, 16
Pixellation, 13
Planar guides, 112
Planar optical waveguide, 229
Planar writing, 202
Plane-wave
   approximation, 38
   excitation, 12
Plasma
   density, 91
   dynamics, 94, 95
   effects, 71
   interactions, 72
   phenomena, 63
Point defect, 14
Poisson equation, 75
Polarization, 85
   effects, 198
   exciton resonance, 257

instability, 253, 260, 261, 262
  in space, 260
  in time, 260
  period-doubling, 261
  self-action, 254
  signal, 92
  switching, 201
  vector, 137, 231
  -related optical multistability, 261
Polarizing cube beamsplitter, 208
Polycrystalline phtalocyanine films, 110
Polycrystalline ZnSe, 59
Polystyrene microsphere suspension, 247
Population
  density, 121
  gratings, 178
  kinetics, 88
  lifetime, 86
  relaxation time, 108
Potential energy, 14
Power-broadened Lorentzian line-shape, 89
Power-dependent absorption, 233
Power-dependent switching, 221, 224,
Power-dependent thermal NOA effects, 259
Power dependent transmission, 219
Prism coupling, 22, 230,
Probe-beam-crystal interaction, 257
Probe wave, 177
Processing
  capability, 234
  gain, 183
  repetition rate, 217
Propagation
  configuration, 137
  equations, 10
  mode, 29
Pulse
  distortion, 224
  duration, 86
  reshaping, 219, 238
Pulsed excitation, 234
Pulsed regime, 87
Pump-and-probe technique, 103
Pump-probe spectroscopy, 100
Pump waves, 177

Q band, 110
QD, 107
  optical nonlinearities, 108
  -size dependence, 107
Quadrupole
  moment, 257
  potential, 136
  system, 136
Quantization regime, 106
Quantum
  balance, 145
  -confined CdSe microcrystal-lites, 106
  confinement, 106
    effects, 100
  dots, 106
  limit, 133, 138
  -mechanical calculations, 257
  -mechanical perturbation theory, 83
  optical Bloch equation, 114
  well cases, 127
  wells, 81, 94, 121
  wires, 106
Quasi-bistable effects, 30
Quasi-chemical potentials, 114
Quasi-CW regime, 133
Quasi-particle self-energy, 91
Quasi-particles, 99, 106
Quasi-2d GaAs quantum-well, 115
Quasi-two dimensional quantum-well structures, 114
QW, 113
  layers, 113
Q-switched Nd:Yag laser, 31
Q-switching mode locking, 222

Rabi frequency, 87
Radiation forces, 143
Rate equations, 38
Read-out process, 177
Read-write cycle, 200
Real-time
  holography, 175
  image filters, 210
  recording, 197
Recall process, 166
Recombination, 40, 108
  coefficient, 29
  processes, 28
Recording sensitivity, 200
Recording/reconstruction cycle, 198

Red-shift, 91
Reflection
  coefficient, 24, 27
  hologram, 200
  states, 51
Refractive index, 119
Regenerative feedback, 180
Relativistic change of mass, 131
Relativistic cyclotron frequency,
    143
Relativistic mass-effect, 131, 139
Relativistic nonlinear effect, 137
Relativistic nonlinear mechanism,
    135
Relativistic nonlinear optics, 131
Relativistic nonlinearities, 133
Relativistic shift of mass, 134
Relativistically-hysteretic cyclo-
    tron resonance, 137
Relaxation
  rate, 126
  time, 87, 92, 237
  -modified Lorentz equation, 133
Resolution, 180
  condition, 189
Resonant detuning parameter, 137
Resonant driving wave, 143
Resonant excitation, 29
Resonant light-matter interac-
    tions, 87
Resonant nonlinearities, 51
Response function, 7
RF-sputtering, 239
Ridge waveguides, 112
RMPM, 260
Rocking fiber filter, 223
Rocking filter, 224
Rocking rotator fiber filter, 219
Room temperature operation, 28
Rotation
  invariance, 170
  -modulation polarization measure-
    ments, 260
Round trip cavity phase, 127
Runaway switchdown, 233

Sapphire, 120
  device, 31
Saturable absorber, 37
Saturable absorption, 5
Saturable gain, 180

Saturated absorption coeffi-
    cient, 90
Saturation, 222, 243
  carrier concentration, 101
  effect, 111
  intensity, 41
  model, 108
SBN, 198, 201,
SBS, 175
Scaling invariance, 169
Scattering, 87, 108, 243
  absorption, 39
  processes, 94
  rates, 91
Schlieren
  stop, 212
  system, 199
Schrodinger equation, 14
Screened Coulomb potential, 114
SDG, 229
  waveguide, 230, 241
Second harmonic generation, 84,
    173, 229
Second-order nonlinear force, 142
SEED, 99
Selection rules, 91
Selective etching, 120
Self-action effects, 255
Self-crossing, 145
Self electro-optic effect, 99
Self-focussing nonlinearity, 17
Self-generated electric field, 63,
    70, 75
Self-generated field, 77
Self-induced anisotropy, 253
Self-induced birefringence, 254
Self-induced gyrotropy, 253
Self-induced optical activity, 258
Self-induced optical rotation, 259
Self-induced rotation, 254
Self-phase modulation, 217
Self-pulsing, 260
Self-saturation, 107
Self-switching, 218, 223
Self-synchronization, 141
Semiconductor, 28, 51, 235
  bistable etalons, 51
  diode lasers, 111
  doped glass fiber, 222
  -doped-glass waveguides, 234
  etalon, 111
  lasers, 119

layer, 29
  microcrystallites, 106
  micro-structures, 106
  waveguides, 112
Sharp-cutoff color filters, 241
Shift invariance, 169
Short-range interactions, 63
Side mode suppression ratio, 38
Sigmoid function, 164
Signal
  processing, 99
  processors, 217
Silicon, 29, 151,
  device, 31
  waveguide, 29
Single-beam problems, 12
Single cyclotron electron, 146
Single domain crystal, 187
Single electron, 131
Single-fluid
  approximation, 73
  expansion, 76
  solution, 75
Single-mode
  GaAs/AlGaAs strip-loaded wave-
      guides, 112
  guides, 241
Single-particle like equation, 75
Single slightly-relativistic
      cyclotron electron, 131
Slowing-down phenomenon, 8
Soft polarization symmetry break-
      ing, 261
Solitary wave, 18
  patterns, 17
Soliton, 224
  -like structures, 17
  solutions, 17
Soret band, 110
Space-charge
  effects, 64, 78
  field, 207
  potential, 103, 104
Spatial chaos, 15
Spatial confinement, 119
Spatial diffusion, 235
Spatial filtering, 154
Spatial instabilities, 218, 224
Spatial light modulators, 180
Spatial resolution, 200
Spatial solitons, 17
Spectral bistability, 45

Spectral broadening, 224
Spectral hysteresis, 45
Spectral polarization, 256
Spectroscopic applications, 83,
      257
Speed of light, 131
Split-step FFT method, 11
SPM, 223
  coefficients, 218
Spotsize dependence, 12
SRS, 175
Stability
  analysis, 143
  condition, 15
Standing wave, 136
  configuration, 139
  pattern, 139, 143
Stark shift, 146
Stationary nonlinear operation, 23
Steady-state
  excitation, 144
  OB, 10
  solution, 88, 137
Step function, 16
Stimulated Brillouin scattering,
      175
Stimulated cyclo-Raman scattering,
      132, 142, 146
Stimulated emission rate, 38
Stimulated recombination, 40, 94
Stimulated Raman scattering, 145,
      173
Stokes vector, 221
  formalism, 255
Storage
  capacity, 182
  crystal, 197
  density, 160
  devices, 199
  matrix, 206
Strip-loaded structures, 112
Strontium Barium Niobate, 198
Strontium Titanate, 239
Submillimeter range, 131
Subnanosecond switchings, 32
Substrate thermal conductivity, 56
Sub-picosecond
  switching times, 217
  time resolution, 119
Super computers, 151
Superconducting magnet, 136
Superlattices, 103

Super-saturated solid solution, 106
Surface recombination, 126
Susceptibility, 86, 115
Switching, 151
  criteria, 127
  devices, 103, 126
  down, 12
  dynamics, 8
  efficiency, 224
  irradiance, 56
  machine, 155
  power, 52, 58, 126
  speed, 52, 58, 217
  time, 21, 32, 46, 53
  up, 12
  waves, 12
Symbolic substitution, 53, 154
Symmetry-breaking instability, 261
Symmetry group, 255
Synchro-cyclotron, 141
Synchrotron, 141
  electron, 137
  radiation, 133, 145
Synthetic microstructures, 83

TEM_{00} mode, 190
Temporal broadening, 224
Temporal effects, 119
TEO modes, 29, 32, 33, 232
Thermal effects, 21, 31, 235
Thermal equilibrium, 90
Thermal exchange, 30
Thermal expansion, 76
Thermal heating, 217
Thermal motion, 76
Thermal NOA, 258, 261
Thermal nonlinear optical activity, 258
Thermal nonlinearity, 17, 234, 236
Thermal phenomena, 30
Thermal rotation, 256
Thermal time constant, 4
Thermalization, 30
Thermalized distribution, 94
Thermalized regime, 90, 93
Thermalpolarization instabilities, 256
Thermo-optic devices, 52
Thermo-optic effects, 51
Thermoplastic film, 183
Thermoplastic plates, 163

Theta
  logic, 155
  modulation, 154
ThF4, 52
Thin films, 109
Third harmonic generation, 84, 110
Third order nonlinear susceptibility, 84, 257
Third order processes, 229
Third order susceptibility, 176
III-V compounds, 29
  semiconductors, 83, 97
Three photon
  ionization, 63
  interactions, 132, 143
  process, 142
Three-wave down conversion, 175
Threshold
  amplitude, 141
  carrier density, 41
  detector, 186
  intensity, 138
  ionization experiments, 78
  point, 138
Thresholding, 180, 265
  capability, 193
Time-division multiplexing, 206
Time-of-flight drift tube, 65
Time
  reversal, 174
  -space evolution, 73
  translation, 85
TNOA, 258
TOF, 65, 68
Trace analysis, 63
Transfer coefficient, 231
Transformed-limited pulse, 87
Transient digitizer, 241
Transient grating, 123
  technique, 119
Transition energy, 114
Transmission
  coefficient, 27
  holography, 204
  mask, 189
  states, 51
Transparencies, 160
Transphasor regime, 8
Transport communication, 152
Transverse Laplacian, 10
Transverse cooling, 58

Transverse coupling, 10
Transverse effects, 10, 12, 17
Trapping potential, 136
Travelling wave, 136
  interaction, 243
  formalism, 230
  model, 242
Tunable absorption coefficient,
      103
Tunable dye laser, 120
Tunable femtosecond light pulses,
      91
Tunable image velocity filter, 209
Tunable photoluminescence spectrum,
      103
Turn-on delay, 37
Twin-strip
  laser, 48
  semiconductor lasers, 47
Twisted birefringent fibers, 262
2-band systems, 89
Two-beam switching, 113
Two-dimensional gray-scale image,
      183
Two-fluid system, 75
Two-level systems, 86, 87
2LS
  approach, 89
  picture, 91
Two-photon
  absorption, 235
    coefficient, 256
  resonance, 258
    absorption, 261
  resonant absorption, 68
  selection rule, 261
II-VI semiconductors, 51
Two-step process, 199
Two-transverse-dimension system,
      18
Two-wave mixing, 267
Two-wavelength dynamic optical
      interconnect, 201

UHV film fabrication, 53
UHV/MBD techniques, 52
Ultrafast all-optical switching,
      217
Ultrafast dynamics, 94
Ultrafast laser pulses, 266
Ultrafast recombination, 94
Ultrafast relaxation processes,

Ultrafast response time, 109
Ultrahigh rf-range, 141
Ultra-high vacuum system, 52
Ultrashort light pulses, 86, 87, 218
Uniaxial crystals, 253
Unified theory, 132
Unperturbed hamiltonian, 86
Unstable eigenpolarizations, 253

Vander Lugt correlators, 160
Variable carrier lifetime, 103
Vector model, 87
Velocity filter, 199
Vibration hysteresis, 139
Video camera, 189
Video image analyzer, 189
Volume
  effects, 72
  holographic, 207
  phase hologram, 207
  recording medium, 204

Wannier
  approximation, 99
  equation, 114
Wave front conjugation, 175
Waveform convolver/cross corre-
      lator, 266
Waveguide
  devices, 112
  thickness, 29
Wave-vector matching, 233
Weighting coefficient, 206
Wiener filter, 266

XOR function, 219

Zinc Oxide, 230, 234
Zinc Sulphide waveguides, 230
Zincblende crystals, 255
ZnO, 229
  films, 239
  waveguide, 239, 240
ZnP_2, 258
ZnS, 52, 229
ZnSe, 52, 53, 58
  Fabry-Perot filters, 52
  absorption coefficient, 55
  bistable etalons, 53
  etalons, 59
  filters, 53
ZnSe/ThF4 mirrors, 53